Topics in Applied Physics Volume 56

Topics in Applied Physics Founded by Helmut K. V. Lotsch

The Physics of
Hydrogenated
Amorphous Silicon II

Electronic and Vibrational Properties

Edited by J. D. Joannopoulos and G. Lucovsky

With Contributions by D. Allan D. K. Biegelsen
J. D. Joannopoulos L. Ley G. Lucovsky N. F. Mott
W. B. Pollard R. A. Street T. Tiedje

With 203 Figures

Springer-Verlag Berlin Heidelberg GmbH 1984

Professor *John D. Joannopoulos,* PhD

Massachusetts Institute of Technology, Department of Physics
Cambridge, MA 02139 USA

Gerald Lucovsky, PhD

Department of Physics, North Carolina State University,
Raleigh, NC 27607, USA

ISBN 978-3-662-31164-6   ISBN 978-3-540-38847-0 (eBook)
DOI 10.1007/978-3-540-38847-0

Library of Congress Cataloging in Publication Data. Main entry under title: The Physics of hydrogenated amorphous silicon. (Topics in applied physics; v. 55–56) Contents: 1. Structure, preparation, and devices – 2. Electronic and vibrational properties. 1. Silicon. I. Joannopoulos, J.D. (John D.), 1947–. II. Lucovsky, G. III. Series. QC611.8.S5P49 1983   537.6′22   83-16732

Originally published by Springer-Verlag Berlin Heidelberg New York in 1984.
Softcover reprint of the hardcover 1st edition 1984

Typesetting: Schwetzinger Verlagsdruckerei, Schwetzingen
Offset printing and bookbinding: Brühlsche Universitätsdruckerei, Giessen
2153/3130-543210

Preface

During the past several years there has grown an enormous experimental and theoretical activity associated with amorphous silicon and its alloys. This is based on the exciting possibilities emerging from the doping of hydrogenated amorphous silicon. Experimental and theoretical efforts have been directed at obtaining an understanding of the underlying physics of a variety of interesting and unusual phenomena associated with this material. In addition, major effort has also been expended towards the exploration of the technological consequences of these phenomena. This two part series presents a broad, as well as in-depth, overview of the entire field of amorphous silicon and its alloys. At the present, sufficient progress and understanding exist that such volumes should be useful and timely. Briefly, the preceeding Volume I concentrates on structure, preparation techniques, and device applications. The present Volume II concentrates on theoretical and experimental investigations of a variety of electronic and vibrational phenomena. Each chapter is written as a critical review with a conscious effort to separate fact and interpretation. In addition, the contributions represent constructive reviews that help define future directions of research whenever possible. The contributions are written at the level of a graduate student which should be helpful to both students and scientists who are not experts in this area. Finally, in an attempt to add an archival flavor to the reviews, many representative results and comprehensive lists of citations are presented.

Cambridge, Raleigh
October 1983

J. D. Joannopoulos · G. Lucovsky

Contents

Contributors

Allan, Douglas
  Massachusetts Institute of Technology, Department of Physics
  Cambridge, MA 02139, USA

Biegelsen, David K.
  Xerox Palo Alto Research Center, 3333 Coyote Hill Road
  Palo Alto, CA 94304, USA

Joannopoulos, John D.
  Massachusetts Institute of Technology, Department of Physics
  Cambridge, MA 02139, USA

Ley, Lothar
  Max-Planck-Institut für Festkörperforschung, Heisenbergstraße 1
  D-7000 Stuttgart 80, Fed. Rep. of Germany

Lucovsky, Gerald
  Department of Physics, North Carolina State University
  Raleigh, NC 27607, USA

Mott, Nevill F., Sir
  Cavendish Laboratory, University of Cambridge
  Cambridge CB 30 HE, England

Pollard, William B.
  Department of Physics, Morehouse College
  Atlanta, GA 30314, USA

Street, Robert A.
  Xerox Palo Alto Research Center, 3333 Coyote Hill Road
  Palo Alto, CA 94304, USA

Tiedje, Thomas
  Exxon Research and Engineering Company – Corporate Research
  Laboratory, P.O. 45
  Linden, NJ 07036, USA

1. Introduction

John D. Joannopoulos and Gerald Lucovsky

This volume is the second member of a two part series that deals with the preparation, characterization, physics and technological importance of hydrogenated amorphous silicon. The first volume [1.1] emphasizes the preparation, physical and chemical characterization, and device technology associated with amorphous silicon and closely related alloy materials. The volume is self-contained in the sense that it focuses on material properties, their dependence on deposition techniques and the manner in which they find their way into the device application area. This second volume is qualitatively different in as much as it presents discussions of the basic physical properties and phenomena that are associated with the deposited films which form the basis of the device physics that underlies the technological applications. In this context Volume II provides a wealth of basic information that is useful for both the device technologist and the scientist interested in fundamental phenomena of disordered solids.

Chapter 2 by D. Allan and J. D. Joannopoulos presents a theoretical basis for understanding the electronic structure of the material. Chapter 3 by L. Ley emphasizes the optical and photoemission properties that are associated with transitions involving the electronic states and as such draws from the base established in Chapter 2. Chapter 4 by Sir N. F. Mott is a general discussion of localization phenomena and serves as a basis for understanding the defect states and the transport properties. Chapters 5 by R. A. Street and D. K. Biegelsen and 6 by T. Tiedje emphasize, respectively, the defect states and the carrier transport phenomena. Finally, Chapter 7 by G. Lucovsky and W. B. Pollard deals with the vibrational properties of the material with particular emphasis on local impurity and alloy atom structure.

Volume I and II then serve complementary roles in treating the recent scientific and technological progress that has been made in hydrogenated amorphous silicon and other closely related alloys. Since the physical properties and some of the unique phenomena derive from the local atomic structure at the alloy atom, impurity atom, dopant atom, and native defect sites, and since these in turn may be different for different preparation techniques, it is impossible to draw a clean line between the science and technology of this class of materials. In this context Volumes I and II should prove to be helpful reading for both the conscientious scientist and the serious technologist. The material is also appropriate for graduate students and post doctoral fellows who are embarking on research in this area of disordered materials.

Abbreviations Frequently Used in the Text

a-Si	amorphous silicon
c-Si	crystalline silicon
a-Si : H	hydrogenated amorphous silicon
a-Si : F	fluorinated amorphous silicon
CVD	chemical vapor deposition
PES	photoemission spectroscopy
XPS	x-ray photoemission spectroscopy
UPS	ultraviolet photoemission spectroscopy
EDC	energy distribution curve
DOS	density of states
JDOS	joint density of states
LDOS	local density of states
VDOS	vibrational density of states
CB	conduction band
VB	valence band
CBM	conduction band minimum
VBM	valence band maximum
LRO	long range order
SRO	short range order
CRN	continuous random network
RDF	radial distribution function
BL	Bethe lattice
CBLM	cluster Bethe lattice method
ETB	empirical tight binding
LCAO	linear combination of atomic orbitals
OLCAO	orthogonalized linear combination of atomic orbitals
CPA	coherent potential approximation
GVB	generalized valence bond
ESR	electron spin resonance
LESR	light induced electron spin resonance
NMR	nuclear magnetic resonance
ODMR	optically detected magnetic resonance
PL	photoluminescence
DLTS	deep level transient spectroscopy
EXAFS	extended x-ray absorption fine structure

Reference

1.1 J. D. Joannopoulos, G. Lucovsky (eds.): *The Physics of Hydrogenated Amorphous Silicon I*, Topics in Appl. Phys., Vol. 55 (Springer, Berlin, Heidelberg, New York, Tokyo 1984)

2. Theory of Electronic Structure

Douglas C. Allan and John D. Joannopoulos

With 28 Figures

In spite of an excellent theoretical understanding of electronic states in crystalline materials (crystalline Si in particular), amorphous systems pose special and difficult problems for theoretical work. The theory of electronic states in amorphous materials is in a seminal stage. This chapter will focus on newly developing methods and models which have been applied to amorphous systems. In presenting results of theoretical work, we restrict ourselves mainly to recent studies of hydrogenated amorphous silicon (a-Si : H).

2.1 Background

Hydrogenated amorphous silicon presents a system in which a number of fundamental physical questions can be addressed. How is the doping [2.1] mechanism to be understood? How is doping reconciled with *Mott's* [2.2] plausible argument regarding the inability to dope amorphous solids? What is the nature of the energy gap? What is the nature of the bonding conformations that may exist? What is the nature of the geometric structure? Studies of electronic structure constitute a microscopic probe which provides insight and an understanding of the basic physics underlying the answers to these questions.

The presence of long-range order (LRO) or translational symmetry in crystals greatly simplifies theories of electronic states. With loss of LRO in amorphous materials, Bloch's theorem no longer strictly applies (k is not a good quantum number) so that a classification of states by a band structure $E(k)$ is not useful. One can, nevertheless, measure the same electronic response functions, e.g., the reflectivity, dielectric function and photoelectric response. Moreover, in many cases certain features of the response functions are the same for corresponding crystalline and amorphous materials. Thus, one aim of theory should be to compute the various response functions based on some picture of the electronic states. The expression for photoelectric response $R(E, E_f)$ is essentially

$$R(E, E_f) \propto \sum_i M_{if} \delta(E - E_i - E_f) , \qquad (2.1)$$

where E_i is the energy of a one-electron state in the system and M_{if} is the probability of making a transition from E_i to E_f under the influence of photon absorption. A useful special case arises when the probabilities M_{if} are nearly independent of energy. In that case structure in the response function is attributed to the electronic density of states $N(E)$,

$$N(E) \propto \sum_i \delta(E - E_i) . \tag{2.2}$$

$N(E)$ is the number of electronic states in the system at energy E per unit energy. These expressions are equally well defined in both crystalline and amorphous materials; the structure and bonding of the material will determine the spectrum of values E_i. In photoemission experiments, matrix elements usually vary slowly enough with energy so that, with some care, such spectra can be directly interpreted in terms of $N(E)$.

In addition to an overall density of states $N(E)$, it is exceedingly helpful to utilize a local density of states $N_a(E)$ defined by

$$N_a(E) \propto \sum_i |\langle \phi_a | \psi_i \rangle|^2 \delta(E - E_i) . \tag{2.3}$$

Here ϕ_a is an orbital localized about some given atom (or bond) a and $\{\psi_i\}$ are eigenstates of the system. $N_a(E)$ is conveniently obtainable theoretically and contains much information. We digress here to review the significance of $N_a(E)$ and its interpretation.

From definition (2.3), $N_a(E)$ is the relative probability that an electron has energy E and is found on (localized) orbital ϕ_a. Thus, it is effectively the density of states of an electron when it is near atomic site a. Consider the

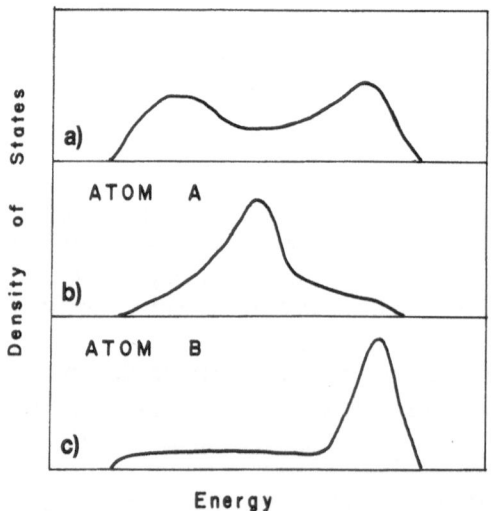

Fig. 2.1a–c. Densities of states for an example system discussed in the text. (a) Overall $N(E)$ for the entire system. (b) Local $N_A(E)$ on atoms in geometric environment "A". (c) Local $N_B(E)$ on atoms in geometric environment "B"

following simple example of how the local density of states can be used to obtain a variety of information. In Fig. 2.1 we present an illustrative sketch of the density of states for a hypothetical amorphous system consisting of like atoms. The top panel shows the total $N(E)$ for the system. The next two panels show local densities of states $N_A(E)$ and $N_B(E)$ computed for two arbitrary atoms which we label A and B. Note that an electron near site A would most likely be found near the middle of the energy band, while an electron near site B would most likely be found near the top of the band. By comparing with the overall $N(E)$, we can deduce that the system is not predominantly made up of atoms in environment A. We can furthermore suggest that B sites are found more commonly than A sites in the hypothetical system. Moreover, we can suggest that the B sites are responsible for the hump at the top of the band. In addition, there clearly must be other kinds of bonding sites in the system which contribute to the states near the bottom of the band. This we know because $N(E)$ is the sum over all orbitals a of $N_a(E)$, and the top curve (E) is not obtainable from purely $N_A(E)$ and $N_B(E)$. By theoretically computing $N_a(E)$ for atoms α in a variety of proposed geometries and comparing with the observed $N(E)$, we can rule out or suggest an abundance of certain geometries.

Particular kinds of sites (e.g., point defects) in real systems will have distinctive features in $N_a(E)$. The above discussion illustrates that by comparing characteristic features in an experimental spectrum with theoretical local densities of states, one may obtain important atomic bonding and structural information. This proves especially important in nonperiodic systems for two reasons. (i) Nonperiodic systems are usually harder to structurally characterize, so that a sensitive probe of microscopic structure is needed. (ii) $N_a(E)$ remains a well-defined observable, as mentioned above.

Complete loss of LRO in an amorphous system is not as drastic a perturbation as it may seem. It turns out that the short-range order (SRO, the bonding out to second or third-neighbor distances) generally dominates the electronic properties. Until the past decade or so, however, LRO has tended to dominate the theories. There are in print several reviews of recent theoretical efforts to recognize and utilize SRO in the theory of electronic states. These range from introductory reviews [2.3, 4] to early models [2.5–7] to more recent, detailed models and calculations ([Ref. 2.8, Chap. 2] and [2.9–17]). Reviews [2.8, 17, 18], and proceedings of recent conferences [2.19–26] emphasize the present experimental and theoretical[1] effort to characterize and understand the consequences of the SRO of a-Si:H.

The bond lengths and tetrahedral bond angles of amorphous Si (a-Si) are nearly identical to crystalline Si (c-Si). One finds fourfold coordination and a small ($\pm 10°$) variation in bond angles in the amorphous samples. (Electron-diffraction experiments find strong first and second-neighbor peaks in the

---

1 A Green's function approach can be found in [2.27]

radial distribution function, but the third-neighbor peak is apparently washed out by bond angle and dihedral angle variations [2.28].) Furthermore, probes of electronic states in a-Si ([Ref. 2.8, Chap. 2] and [2.17, 18]) indicate qualitatively the existence of filled valence and empty conduction bands separated by a gap of about the same magnitude as in c-Si. These general features are an important part of our theoretical understanding of electronic states in a-Si:H. Theoretical work will seek, in principle, detailed relationships between the geometrical structure of a-Si:H and its electronic structure.

2.2 Brief Review of Experimental Facts Regarding Electronic States

In this section we review a selected set of experimental observations which provide information on electronic states in a-Si:H. We group the observations into those probing primarily the bands of the material, such as photoemission, and those probing primarily the band edges and energy gap, such as spin resonance, transport and optical experiments.

2.2.1 Electronic States Within the Energy Bands

To investigate electronic states in the valence band of hydrogenated amorphous silicon, we begin by investigating pure amorphous and crystalline silicon. The experimental x-ray photoemission spectra (XPS) for c-Si and c-Ge are presented as the top curves of Fig. 2.2a, b [2.29]. Note in each case the characteristic three-peak spectrum. The theoretical density of states $N(E)$ for Si and Ge, computed by the empirical pseudopotential method [2.15], is shown as the third curve of Fig. 2.2a, b. To compare with experiment, the theoretical curves broadened by the experimental energy resolution are given as the middle curves of Fig. 2.2a, b. There is excellent agreement with the experimental peak locations and widths. Note that the densities of states for c-Si and c-Ge are nearly identical. The differences in the peak heights of the experimental spectra for Si and Ge are a consequence of the variation with energy of scattering matrix elements. In general, when comparing theoretical densities of states with experimental photoemission results, one expects the peak locations to be well predicted, while peak heights may vary.

From the theoretical calculations we know the orbital character of the three observed humps [2.15]. The lowest hump consists predominantly of *s*-like bonding states, the middle hump is a mixture of *s* and *p* and the hump near the top of the valence band consists predominantly of *p*-like bonding states. Knowledge of the orbital character is helpful in understanding what happens to these states when the lattice is disordered, as in the amorphous materials.

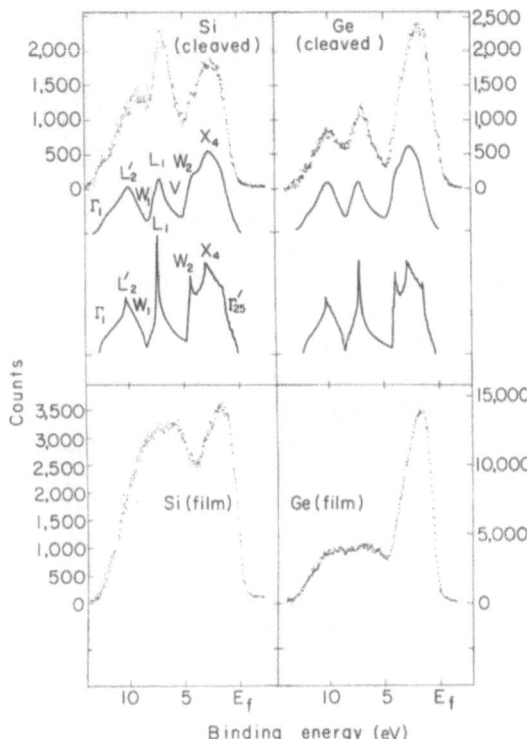

Fig. 2.2 a–d. Experimental XPS results and related theoretically calculated densities of states. (a) Experimental curve (···) for crystalline Si [2.29]; density of states for crystalline Si obtained by the empirical pseudopotential method (*bottom curve*) [2.15]; theoretical curve broadened by the experimental resolution (*middle curve*) [2.15]. (b) XPS for amorphous Si [2.29]. (c) Same as (a), but for Ge. (d) Same as (b), but for Ge. The relative sizes of the humps in the Si and Ge XPS differ because of the differences in scattering cross section of the Si(3s), Si(3p), and Ge(4s), Ge(4p) electrons, respectively

We now compare the crystalline XPS spectra with those for pure (nonhydrogenated) a-Si and a-Ge (Fig. 2.2c, d). Two salient changes appear. (i) The p-like hump moves slightly upward in energy as the valence band edge steepens. (ii) The two lower humps of the crystalline spectrum merge into a single featureless hump. The same features are noted in both Si and Ge. The energy dependence of the scattering matrix elements, as in the crystalline case, is evident. A full theoretical discussion and explanation of these trends is given in [2.15] and references therein. The steepening of the valence band edge is found to be a consequence of varying bond angles in the amorphous network (to be demonstrated in Sect. 2.5). The significant loss of structure below about −5 eV has an interesting origin, which we will now explain.

The states which derive from atomic s-states in a bonded, covalent system are sensitive to the bonding topology (i.e., ring statistics) of the system. This holds because the interaction topology among the s-states is the same as the bonding topology of the underlying system. This does not hold for the p-like states, which explains why the p-like states near the top of the valence band are much less sensitive to the change in topology between the crystalline and amorphous samples. In fact, in the Weaire-Thorpe tight-binding model the

p-derived density of states is merely a δ-function and completely insensitive to underlying topology. The influence of rings of bonds, furthermore, is to place boundary conditions on the allowed wave functions. This quantizes the energies of states which can exist on a given structure and tends to put peaks in the density of states. It is the presence of sixfold rings of bonds in c-Si, for example, and not necessarily the presence of long-range order, which produces the two-hump structure in the lower valence band (we will show this in Sect. 2.5). When a variety of ring sizes exist, as in a-Si, then the peaks occur over a range of energies and can result in the smooth and nearly featureless spectrum observed.

When hydrogen is bonded into films of a-Si, several new features arise. Recent ultraviolet photoemission spectroscopy (UPS) experiments [2.30–32] on a-Si:H have revealed spectra which fall roughly into two categories. Figure 2.3a shows the first kind of spectrum, associated with films of poorer quality (higher density of gap states and lower film density), in which peaks A and B have been linked with SiH_2, SiH_3 and $(SiH_2)_n$ bonding [2.30, 31]. Figure 2.3b, c both refer to the second kind of spectrum, associated with high quality ("device grade") films. Curve b was taken after annealing the sample which gave spectrum a, while curve c was taken for a sample deposited onto a higher temperature substrate. The peak C has been linked with monohydride (SiH) bonding [2.30] by comparison with spectra from the monohydrated Si(111) : SiH surface and by comparison with theoretical calculations. Suggested explanations of the full spectrum will be presented in detail in Sect. 2.5 where theoretical results are presented. There we show that atomic structure information contained in UPS and XPS spectra may help explain how doping is possible.

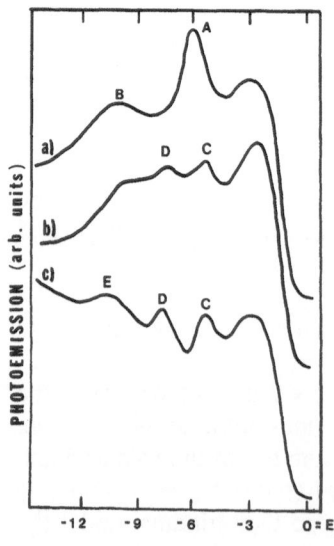

Fig. 2.3. He II (40.8 eV) UPS spectra for hydrogenated amorphous Si [2.30, 31]. (*curve a*) Glow-discharge sample deposited at a substrate temperature of about 250 °C. (*curve b*) Same sample annealed at 350 °C. (*curve c*) Sample sputtered with hydrogen at a substrate temperature of about 350 °C

2.2.2 Electronic States Near and in the Band Gap

An understanding of band-tail and gap states in a-Si : H is crucial from a technological perspective in order to understand and control transport and optical properties. The following is a brief selection of observations to be addressed by the theory sections of this chapter.

It is generally agreed that a-Si produced by sputtering or evaporation, without H, contains a large number ($\sim 10^{20}$ cm^{-3}) of dangling bonds as seen in electron spin resonance (ESR) experiments. Gap states associated with these dangling bonds may explain the insensitivity of a-Si to dopants. When H is added to the network, a number of important changes occur. The ESR signal is reduced (to $\leq 6 \times 10^{15}$ cm^{-3}), apparently as a consequence of H bonding to the dangling bonds and removing the singly occupied states from the gap. It has been typically observed ([2.33–37] and [Ref. 2.8, Chap. 7]) that at least an order of magnitude more H than the number of observed spins must be added to quench the spin signal. This is explained by an argument proposed by *Phillips* [2.38]. He investigated the strain associated with a fourfold coordinated continuous random network (CRN) which he found is relieved by the occurrence of broken bonds in arrays. The electronic structure of arrays of broken bonds in a-Si can be understood in terms of the similar c-Si surface problem, where atomic reconstruction can pair electrons and eliminate most of the free spins. Several models of atomic and spin reconstruction of Si(111) have recently been investigated [2.39–45]. In particular, *Pandey* has proposed a new π-bonded chain model [2.39, 40] in which the surface reconstruction places dangling bonds at nearest neighbors. This model of reconstruction, which is favored by self-consistent pseudopotential total energy comparisons [2.41, 42], allows the dangling bond states to interact strongly. Electrons are thus paired in bonding states and the spin signal is removed. When H chemisorbs on Si(111), it typically reverses the structural relaxation and forms bonds with all the unreconstructed broken Si bonds. Thus, the H added to a-Si may be chemisorbing on surfaces of voids where atomic reconstruction formerly allowed essentially only one spin per void [2.13]. As H is added, the optical gap increases with H content from about 1.5 eV in a-Si to around 2.0 eV at 30 at.% H [2.17, 18]. The photoemission experiments show that the gap is apparently widened by removal of states from the top of the valence band [2.30, 31]. This observation suggests a mechanism for localized states to exist at the valence band edge as a consequence of fluctuations in H concentration. This will be discussed in Sect. 2.5.

When a-Si : H is doped, EXAFS (x-ray absorption fine structure) studies have shown that, for example, only 20% of As atoms bonded in the structure are fourfold coordinated with Si, and that there is an apparent lack of any order beyond the first-neighbor shell about the As atom ([Ref. 2.8, Chap. 8] and [2.46]). In addition to EXAFS studies, more recent P NMR experiments on doped films reveal some interesting results [2.47]. In P doped films, about 20% of bonded P atoms are fourfold coordinated while the rest are threefold coordinated. Most importantly, an analysis of P–H and H–H dipolar couplings indicates that the fourfold coordinated P atoms are located near clusters of bonded H atoms. This evidently occurs at internal hydrogenated surfaces providing strong evidence connecting doping with a clustered hydrogen microstructure. It is also possible that dopants introduce more defects and gap states into the material as suggested by decreases in luminescence efficiency and photovoltaic performance [2.48, 49]. The relations among microstructure, gap states and doping need further investigation.

We now consider a number of features observed in the electron spin resonance (ESR) and light-induced ESR (LESR) spectra in a-Si:H ([Ref. 2.8, Chap. 7] and [2.50–63]). There is a characteristic resonance line with $g = 2.0055$ (the value for a free electron is 2.0023) corresponding to dangling bond states whose concentration can vary from $\leq 10^{20}$ cm^{-3} down to the level of detectability $\leq 10^{15}$ cm^{-3} depending on preparation conditions. It is noteworthy that the line shape is very nearly independent of preparation conditons. This is evidently caused by a population of strongly localized paramagnetic centers, consistently interpreted as dangling bonds [2.50]. This signal decreases and eventually disappears with n or p-type doping as the dangling bond states become filled or empty, respectively. The ESR signal seen in p-type films has a single broad resonance at $g = 2.013$ which is attributed to self-trapped hole states [2.51]. A narrower resonance with $g = 2.004$ is observed in n-type films; this is attributed to conduction band tail electrons [2.51]. The LESR signals from these three types of films are as follows. Undoped films show both the hole and the electron resonances; the n-type samples show only the dangling bond resonance, while the p-type samples show both the dangling bond and $g = 2.013$ resonances [2.51]. In addition, higher levels of doping decrease the LESR signal. This reduction presumably occurs because of increased nonradiative recombination at defects introduced by doping [2.52]. These results are conventionally explained in terms of the following band-tail model [2.50]. In undoped films at equilibrium, the paramagnetic centers are half-filled dangling bond states. Under illumination, transitions can occur both from the dangling bond state near midgap to a state in the conduction band tail (trapping an electron), or from the valence band into the dangling bond state (forming a doubly occupied dangling bond and trapping a hole in the valence band tail). The observed LESR resonances seen in undoped films purportedly arise from electrons and holes in localized tail states. Under doping conditions, consider an n-type material in which the Fermi energy lies in the conduction band tail. Then the majority of the dangling bond states are doubly occupied in equilibrium and provide no ESR signal. Furthermore, the electron tail states are now occupied and provide the observed $g = 2.004$ equilibrium ESR signal. Since electrons are the majority carriers, photoinduced holes immediately combine with the excess electrons. When the doubly occupied dangling bonds trap holes, they become paramagnetic centers. Hence, there is only the dangling bond resonance observed in LESR. *Adler,* however, has argued against the band-tail interpretation [2.53]. He notes, among other things, that in both n and p-type doping, the $g = 2.004$ and $g = 2.013$ signals, respectively, go through a maximum and then decrease with increasing dopant concentration [2.54, 55], which is hard to interpret in terms of uniform band tails. A defect-related band of states near the band edge seems more likely. In addition, at a given concentration of P or B dopants, the Fermi energy moves toward the majority carrier band edge with increasing temperature [2.55]. *Adler* [2.53] points out that since E_F must move away from large densities of states with increasing temperature, these results for both n and p-type doping are inconsistent with the band-tail interpretation. If an exponential band tail were the principle feature in $N(E)$ near the band edges, for example, one would expect the opposite temperature dependence of E_F.

In another type of ESR experiment, *Cohen* et al. [2.64] observed the changes in the dark absorption ESR signal as a result of modulating the width of the depletion region in n-type a-Si:H diode junctions. They identify a band of defect states just above midgap (0.85 eV below the conduction band) as the energy level of the second electron bound to the dangling bond defect. This same band of states has been seen in other probes of gap states to be discussed below. They furthermore conclude that there exists a positive correlation energy U of at least 0.2 eV between the singly and doubly occupied dangling bond centers [2.64].

A fascinating metastable reversible photoelectronic effect was discovered by *Staebler* and *Wronski* [2.65, 66]. Optical exposure for a period of several hours decreases both the photoconductivity and the dark conductivity, the latter by nearly four orders of magnitude, while annealing above 150 °C reverses the process. Since an increased density of $g = 2.0055$ spins [2.67] is associated with the photoinduced metastable state, the effect has been conventionally interpreted as a photocreation of dangling bonds. There are several problems with this approach [2.53], not the least of which is that the optical energy available to the lattice is less than the band gap (< 2.0 eV), which is less than the strength of a Si–Si bond (about 2.35 eV). A better understanding of the electronic states and creation energies of defects is needed to proceed toward explaining these results.

The reduction in gap state density accompanying hydrogenation allows the use of certain experimental techniques which purport to measure directly $N(E)$ in the gap (these techniques cannot be used with pure a-Si because the high gap state density inhibits any measureable response). Typical field effect measurements ([2.68–70] and [Ref. 2.8, Chap. 9]), for example, find peaks in $N(E)$ at 0.4 eV and 1.2 eV below the conduction band edge, with the lower energy peak larger by an order of magnitude. The uniqueness of the data reduction, however, in particular the presence or absence of sharp structure in $N(E)$, has been challenged [2.71]. In any case, deep-level transient spectroscopy (DLTS)[2] measurements [2.73–75] indicate a much lower

---

2 For details see [2.72]

overall $N(E)$ in the gap, suggesting that the field effect measurements may be hampered by surface states. Recent DLTS work [2.73] finds a deep minimum in $N(E)$ between 0.3 eV and 0.6 eV from the conduction band edge and a broad shoulder of states extending from the valence band edge up to midgap, with resolvable differences among samples produced under different conditions. Photoemission results [2.31, 76, 77] as well as a correlation of conductivity, photoconductivity and photoluminescence [2.78] are also consistent with a large population of hole-trap states in the lower half of the gap. Discovering the source of these hole-trap states is especially important because their elimination would greatly improve device performance.

Photoluminescence (PL) experiments provide an additional sensitive probe of states in the gap (for review, see *Street* [2.79]). These experiments generally find a dominant peak near 1.4 eV [2.17, 18, 79–80] which has been interpreted as band-edge transitions with an approximately 0.5 eV Stokes shift [2.81–85]. Absorption near 1.4 eV [2.86, 87] has been cited as evidence against a Stokes shift. Observed correlation, however, between the 1.4 eV absorption shoulder and the spin density, as well as the strong dependence of the size of the shoulder on deposition conditions, may indicate that the absorption is really defect-related and not an intrinsic film property [2.88, 89]. The comparison of PL data with the absorption results, in fact, seems to show that strong 1.4 eV PL is inversely related to absorption near 1.4 eV, so that the underlying mechanisms may be independent [2.88]. Photoluminescence excitation (PLE) experiments have, in fact, found rather convincing evidence of the independence of the near 1.4 eV PL and absorption [2.88]. In films of the type which show the absorption shoulder, no inflection or kink is seen in the PLE spectrum near an excitation energy of 1.4 eV. Thus, processes which contribute to absorption near 1.4 eV do not contribute to the excitation of the 1.4 eV PL band [2.90]. Optical absorption measurements which extend into the gap region [2.89] attribute the absorption shoulder to dangling bond defect levels located about 1.3 eV below the conduction band edge. They also observe that doping creates defect states at about the same energy, but that compensation removes these states [2.89]. It is also suggested that these are the same states seen in the lower part of the gap in DLTS and field effect experiments [2.89]. It could be consistent to place the singly occupied dangling bond level at about 1.3 eV below the conduction band and the doubly occupied level at about 0.45 eV above it, with a correlation $U \sim 0.45$ eV. A correct picture of electronic levels in this system in the presence of fairly strong correlations, however, may require abandoning the one-electron picture altogether [2.14].

Additional information has recently emerged from time resolved PLE experiments [2.91]. These experiments investigate the time evolution of the 1.4 eV PL band as a function of excitation frequency, including frequencies in the band-tail region. The results are inconsistent with an early time spectral shift based on correlation between recombination energy and lifetime. Tunneling recombination is a possible explanation, as long as one site is neutral and one is charged (implied by the absence of a Coulomb spectral shift) [2.91].

In addition to the intrinsic 1.4 eV peak, an extrinsic PL peak at 0.9 eV is observed under conditions of doping or of increased defect density [2.58, 63, 85, 87, 92]. This has been interpreted as a transition between an electron in a doubly occupied dangling bond state (about 0.5 eV below the conduction band edge with distortion energy about 0.2 eV) and a valence band-tail hole [2.92]. It is significant that the 0.9 eV PL band can be excited by photon energies lying above the absorption shoulder, which do not excite 1.4 eV PL. In addition, both the absorption shoulder and the 0.9 eV PL are related to a high spin density. It has therefore been suggested [2.90, 92] that at least part of the absorption shoulder arises from absorption at the defect responsible for 0.9 eV PL. To investigate these questions theoretically one must be able to compute not only the electronic states and total energies of defects, but also the relaxations that accompany the capture of electrons and holes. The inclusion of electronic correlations as well as structural relaxations is becoming evidently important in interpreting the experimental results. Related calculations will be presented in Sect. 2.5.

2.3 Elementary Concepts and Models Based on Chemical Bonding

In this section we review concepts and models which can provide a qualitative understanding of electronic states. A chemical bonding picture of interactions among atoms indicates that only the arrangement of nearby atoms (the short-range order) will dominate the formation of electronic states in the material. We discuss a structural model for "ideal" amorphous silicon and present a simple scheme for estimating electronic state energies and total energies of defects in the ideal structure.

2.3.1 Dominant Importance of Short-Range Order

SRO is the local arrangement of atoms about a given reference atom. Specification of SRO must include (i) the number and type of immediate neighbors, (ii) their separation from the reference atom and (iii) their angular distribution [Ref. 2.8, Chap. 8]. We have mentioned that the SRO of a-Si and c-Si is very similar and that as a consequence, the gross features of their XPS spectra are similar (Fig. 2.2). It is known ([2.7, 93] and [Ref. 2.21, p. 119]) that properties of solids on an atomic scale, whether crystalline or amorphous, should be dominated by local chemical valence considerations as in the chemistry of molecules. Thus, the valence electrons of Si atoms in c-Si or a-Si are driven energetically to form sp^3 hybrid orbitals and bond covalently in a locally tetrahedral structure. Theoretical techniques and models borrowed from quantum chemistry are, in fact, becoming important in the study of amorphous materials. One may characterize the types of disorder present in models of amorphous semiconductors into the classes quantitative and topological, with various subclassification schemes ([Ref. 2.8, Chap. 2] and [Ref. 2.21, p. 119]). Quantitative disorder mainly refers to potential fluctuations which may be caused by bond length and bond angle variations, or by the presence of alloy or impurity atoms. Topological disorder refers to variation in bonding topology or ring statistics (e.g., there may be 5 or 7-fold rings present in a-Si which are not found in c-Si). Clearly the separation is somewhat artificial, and the more realistic models and Hamiltonians will consider both types of disorder simultaneously.

The effects of purely topological disorder have been studied using the simple model Hamiltonian of *Weaire* and *Thorpe* [2.94–96] which is of the form

$$H = V_1 \sum_{i,j' \neq j} |\phi_{ij}\rangle\langle\phi_{ij'}| + V_2 \sum_{i' \neq i,j} |\phi_{ij}\rangle\langle\phi_{i'j}| . \tag{2.4}$$

The atoms of a tetrahedrally coordinated structure are labeled by i and the bonds by j, and the localized orbitals $|\phi_{ij}\rangle$ represent sp^3 hybrids on each atom. Matrix elements V_1 are intrasite ("banding") and V_2 are intersite ("bonding") interactions, and interactions beyond nearest neighbors are assumed zero. When V_1 and V_2 are taken as constant (zero quantitative disorder) over the entire infinite structure, the mathematically rigorous statement holds that ranges of the ratio V_1/V_2 exist at which the electronic structure separates into valence and conduction bands with an energy gap ([Ref. 2.8, Chap. 2] and [2.96]). Thus, even for arbitrary topological disorder and hence loss of LRO, an energy gap can exist in this simple model. This was one of the first rigorous demonstrations that SRO (i.e., the constancy of V_1 and V_2) can dominate the electronic structure and create a gap. These types of model calculations have been improved by utilizing more realistic tight-binding Hamiltonians which include more of the interactions among the local orbitals and by treating matrix element variations due to quantitative disorder [2.97–102]. Multiple-scattering theory for clusters of atoms [2.103] has been utilized to obtain an expression for the density of states $N(E)$ which has no dependence on LRO. Numerical calculations using a realistic self-consistent one-electron potential showed the development of a pseudogap (region of very low $N(E)$) for fairly small clusters (about 8) of atoms. Computational problems have limited the practicality of extending this approach [2.5]. Models, methods and theories which probe or predict microscopic geometries and interactions in a local basis are needed for understanding amorphous semiconductors.

2.3.2 Random Network Picture of Ideal Amorphous Silicon and Hydrogenated Amorphous Silicon

Specification of the SRO does not fully determine the structure of an amorphous semiconductor. One must also determine the network connectivity or topology (i.e., the distribution of ring sizes), which can have important consequences for electronic states ([2.15, 97–102, 104] and [Ref. 2.8, Chap. 8]). Ideal (homogeneous) models of the geometrical structure of a-Si are historically divided into two classes: microcrystalline and random network ([2.15], [Ref. 2.8, Chap. 8] and [Ref. 2.21, p. 1]). In the microcrystalline model, regions of crystalline-like order are connected by disordered boundaries, while in the random network model there are, in principle, no regions of crystalline order. The important role of the connective tissue in the microcrystalline model, especially for smaller (on the order of a few lattice constants) microcrystallites, can be a problem in this type of model [Ref. 2.8, Chap. 8]. For microcrystallites smaller than a few lattice constants in size, the structure is essentially a random network.

In real a-Si:H, the experimental evidence seems to indicate important structural inhomogeneities, even in the best "device grade" materials ([2.17, 18] and [Ref. 2.21, p. 1]). The evidence for columnar microstructure, for example, was presented in [Ref. 2.105, Chap. 2]. It is apparently possible to produce high quality a-Si:H films without the approximately 10 Å voids which are found in a-Si and in lower quality a-Si:H [2.17]. Nuclear magnetic resonance (NMR)[3] studies, however, find two superposed resonance lines of different temperature-independent width in a variety of films both with and without detectable microstructure [2.17, 107–112]. This indicates a two-phase compositional inhomogeneity in which H atoms are found in two kinds of environment: these are identified as (i) clustered monohydrides as would be found on the surfaces of voids or in connective tissue, or the multiply bonded H conformations SiH_3, $(SiH_2)_n$, and (ii) randomly distributed noninteracting monohydrides. It is especially interesting that the broad NMR line due to clustered hydrogens persists even when small-angle x-ray and neutron scattering experiments show no evidence of microstructure [2.107]. Thus, it is not obvious that the clustered monohydrides are associated with voids. It is worth noting that recent proton NMR experiments [2.113, 114] suggest a small ($\sim 1\%$ of the hydrogen) concentration of molecular hydrogen in a-Si:H films. This would be consistent with models for the growth of the films and would possibly indicate a structure of voids or vacancies throughout the film [2.114]. The characterization of structure is progressing, although it is far from complete. Theory must consider the possible influence of voids which may permit H adsorption, and of H-rich regions of connective tissue which should increase network flexibility and also have an important influ-

---

3 The principles of NMR are treated in [2.106]

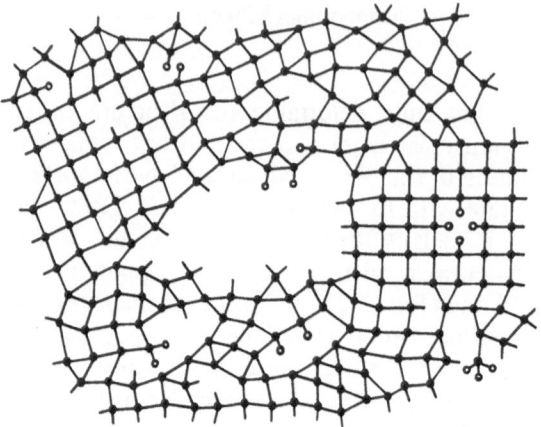

Fig. 2.4. Two-dimensional sketch of a random network incorporating two microcrystalline regions, a large void, several dangling bonds and several Si–H bonds in various configurations

ence on transport properties. The basic picture of ideal a-Si:H as a kind of continuous random network (CRN), essentially homogeneous, is an appropriate starting point for modeling the structure. One may then add special local defects such as dangling bonds or large bond angle variations, and one may add voids, heterogeneous regions, or regions of microcrystalline order. A schematic illustration of two microcrystalline regions, a void, a number of dangling bonds and bonded H atoms in a two-dimensional random network is given in Fig. 2.4.

2.3.3 Simple Models of Electronic States and Total Energies of Defects

The chemical bonding picture invoked in Sect. 2.3.1 is again used to argue for the existence of well-defined defects in the structure of an amorphous solid [2.14]. Since the short-range environment of an atom should be determined by its valence, one expects Si atoms typically to bond tetrahedrally (sp^3). Strains developed during film preparation, however, can force certain atoms to develop undesirable bond angles or to be undercoordinated or to bond in other higher energy local configurations. Since a-Si:H is never in thermal equilibrium during preparation, many defects can be frozen in with no opportunity for relaxation. These defects should have well-defined energy states typically occurring in the energy gap, and it is a goal of theoretical work to characterize the gap states of a-Si:H in terms of specific network defects.

A zeroth-order starting point which permits comparison of the total creation energies and electronic states of a variety of defects is the chemical bonding picture of *Adler* [2.14, 53, 115–117] and *Elliott* [2.118]. In this approach, the creation energies of defects are estimated in terms of their valence bonding energies which are obtained or guessed from relevent bond strengths of crystals and molecules involving the atoms studied. An attempt is also made to include electron correlation energies which is considered to be an important part of the model. The defect is generally assumed to relax fully into the hybridized bonding configuration expected from local chemical considerations, which should be a source of some error when compared with the actual constrained defects expected in a-Si : H. The attempt to estimate electron correlations is important because the presence of strongly correlated defects renders the effective one-electron picture of energy states invalid [2.53]. For example, when energetically significant structural relaxations can occur, quasi-particle excitation energies must be calculated explicitly and compared only with experiments which specifically probe that excitation. This model has been used to suggest that a twofold coordinated Si atom (denoted T_2^0) is one of the lowest energy defects in a-Si. It furthermore predicts a negative effective correlation energy for the dangling bond (T_3^0) defect [2.117]. These predictions, however, can be challenged by the results presented in Sect. 2.5.3. These models have also been used to suggest that a variety of neighboring defects which transfer charge and then interact electrostatically will be present in the material. These defects, which may further relax after capture of electrons or holes, will tend to pair spins. This type of defect could possibly explain the large infusion of H needed to quench the ESR signal and could explain other evidence of spinless defects in the material [2.53]. Owing to the uncertainty of conclusions based on these types of calculations, the usefulness of the chemical bonding approach appears to lie in its simplicity and applicability to a wide variety of defects, the details of which can later be checked by more sophisticated and typically more cumbersome theoretical methods.

There have been other special defects proposed which would possibly pair spins. An interesting example is the Si–H–Si three-center bond (TCB) proposed by *Ovshinsky* and *Adler* [2.119], and by *Fisch* and *Licciardello* [2.120]. By using diborane (B_2H_6) bonding as a model, it was suggested that a single H atom could bond to two Si atoms in a situation where the bond was slightly strained before the H atom was available. It is also assumed that sterically there is insufficient freedom to accommodate two hydrogens and form a pair of monohydrides. It is then argued that structural relaxations accompanying the capture of an electron [2.120] will lead to a negative U_{eff}. This suggested defect with accompanying negative U has been challenged for several reasons. Firstly, an unpinned Fermi level indicates that there cannot be a significant number of negative U defects of any kind. Secondly, molecular orbital mean field (X_α) calculations [2.121] indicate that insertion of H atoms into Si–Si bonds should occur predominantly for pairs of H atoms. The configuration of two H atoms slightly off axis between two Si atoms has been found to stabilize the bond over a range of Si–Si bond lengths. This type of bond, based on dissociated molecular H_2, is found to be much more stable than a single-H TCB. Thus, the concentration of single-H TCB's is expected to be negligibly small [2.121].

2.4 Realistic Theoretical Methods

Realistic theoretical methods yield results directly comparable with experiment. We seek a quantitative theory or model which, by comparison with experiment, can disallow certain explanations and limit the remaining ambiguity. Hence, a realistic method must make detailed predictions about observable quantities. Present investigations of a-Si:H are hampered by the lack of detailed and well understood theoretical calculations as well as the lack of structure in experimental spectra, especially when compared with related investigations in crystalline systems. The past decade, however, has seen tremendous progress. Theoretical methods are being developed which rely on local geometrical models of the structure and which stress SRO. A detailed and fairly comprehensive review of recently developed theoretical methods is given in [2.9–12]. Here we briefly review methods which have been applied to a-Si and a-Si:H.

2.4.1 Model One-Electron Hamiltonians and Basis Sets

a) Empirical Tight-Binding Methods

Empirical tight-binding calculations have enjoyed a "renaissance" [2.9] recently since the method has been placed on a more secure theoretical footing [2.9, 122–124] and since many quantitatively successful calculations have been performed [2.10, 125]. The basic approach of empirical tight-binding (ETB) is the following [2.126, 127]. One takes as a basis set a finite set of localized atomic-like orbitals $\{\phi_a\}$, expanding the one-electron wave functions as

$$\psi_i = \sum_a c_{ia} \phi_a .$$
(2.5)

The Schrödinger equation is then written

$$\sum_b (H_{ab} - E_i S_{ab})c_{ib} = 0$$
(2.6)

and the eigenvalues E_i are solutions of the secular equation

$$\det (H - ES) = 0 .$$
(2.7)

H is an effective one-electron Hamiltonian and S is the overlap matrix of the atomic-like basis functions. In practice, in the empirical approach the ϕ_a are never explicitly known or evaluated. Instead, one regards the H_{ab} and S_{ab} as adjustable parameters to be fitted to known experimental data or accurate theoretical band structure calculations [2.126]. Further assumptions are usually made to reduce the number of free parameters, such as (i) use of only the lowest valence states on each atom such as $3s$ and $3p$ on Si, (ii) neglect of interactions between localized orbitals separated by more than some given distance (e.g., the nearest-neighbor approximation keeps only interactions between neighboring orbitals), (iii) neglect of multicenter integrals (the "two-center approximation" [2.126]) and (iv) neglect of overlaps S_{ab} (assumption of orthogonal orbitals). These parameters, which are often fitted to accurately known bulk crystalline data, are then transferred without adjustment to the amorphous system. If orbitals ϕ_a can be chosen sufficiently localized and the parameter-reducing approximations made are not too severe, this scheme should be successful because the parameters are largely determined by SRO, which is the same in both systems. The method is intuitively appealing and is easily implemented in numerous model calculations described in Sect. 2.4.2. The theoretical justification for the existence of such local and approximately transferable orbitals (which are not exactly atomic orbitals since, for example, orthogonality is often assumed) was given in [2.10] and references therein.

   Krieger and *Laufer* have noted an important caution for the use of the ETB method for vacancy and surface type calculations [2.128, 129]. Care should be taken in modeling the removal of an atom simply by deleting from the basis set orbitals localized on that atom without any readjustment of the ETB parameters of interacting orbitals near the removed atom. In particular, such a model leads to ambiguous results for the location of states in the energy gap [2.128, 129].

   Within the context of the ETB method for computing energy bands, a new method has recently been developed for estimating total energies of structural relaxations and defects [2.130–136]. Computing the creation energies of defects or of various Si–H bonds estimates the likelihood of their formation. A measure of the total energy of a system is given by summing over the occupied one electron eigenvalues to get the electronic energy

$$E_{el} = \sum_i E_i = \int_{-\infty}^{E_F} EN(E)\,dE \ . \tag{2.8}$$

This term, however, doubly counts electron-electron interactions and overlooks the coulomb repulsion between atomic cores. Incorporating these corrections into a term U, the total energy may be written

$$E_{tot} = E_{el} + U \ . \tag{2.9}$$

Electronic screening forces U to be short ranged so that U may be described by a short-range force constant model [2.130]. Thus, we associate contributions to U with nearest-neighbor bonds in the system

$$U = \sum_b \Delta U_b \tag{2.10}$$

and expand ΔU_b about the equilibrium bond length:

$$\Delta U_b(d) = U_0 + (1/2) \sum_{\langle ij \rangle} (U_1 \delta_{ij} + U_2 \delta_{ij}^2 + U_3 \delta_{ij}^3) . \qquad (2.11)$$

Here δ_{ij} is the fractional change in bond length between nearest neighbor atoms i and j. In Fig. 2.5 we sketch the contributions ΔE_{el} and ΔU_b to the total energy per bond ΔE_{tot} as a function of bond length d. Here we have taken the atom as the zero of energy. This corresponds to taking the bond length d to infinity.

The correction U is difficult to calculate from first principles (it cannot be obtained purely from the ETB model). The terms which express variation with bond length (U_1, U_2, and U_3) can be fit to bulk crystalline values for the lattice constant, bulk modulus and thermal expansion coefficient. The equilibrium term U_0, which contributes when bonds are cleaved or rebonded, is easily estimated empirically by comparison with solid and molecule dissociation energies (or bond strengths) [2.135, 136].

Error estimation for this total energy computational method is difficult. The method has compared favorably with self-consistent pseudopotential estimation of the relaxation energy of dimerization on the Si(100) surface [2.133]. Since the ETB method usually gives good valence bands, reproducing $N(E)$ and hence reproducing E_{el} well, the use of these ETB parameters is not a large source of error. *Lee* and *Joannopoulos* recently used this total energy method to obtain force constants for SiC, enabling a calculation of the phonon bands throughout the Brillouin zone [2.137]. They obtained outstanding agreement with known experimental frequencies. The reliability of the method and the estimation of U_0 are discussed in [2.136] where the method is applied to the a-Si:H system.

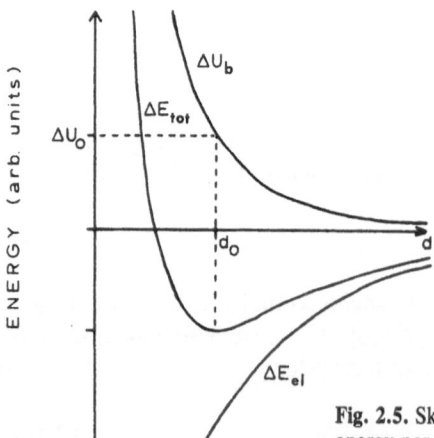

Fig. 2.5. Sketch of contributions ΔE_{el} and ΔU_b to the total energy per bond ΔE_{tot} as a function of bond length d

b) Ab Initio Tight-Binding or LCAO Methods

In this approach the localized wave functions ϕ_a are known either analytically or numerically and the matrix elements and overlap integrals are computed explicitly for a given model one-electron Hamiltonian. In one implementation [2.10, 138, 139] the so-called chemical or localized pseudopotential method [2.122–124] is employed. This is not a pseudopotential in the usual sense (described in the next section) of projecting out the influence of core states. Rather, the influence of neighboring atoms is projected out. In this approach [2.122–124] one develops localized atomic-like orbitals which are eigenstates of an effective Hamiltonian. The effective Hamiltonian includes a projection operator which removes most of the perturbing influence of surrounding atoms near a given atom under consideration. Hence, the localized basis is only slightly distorted from the atomic orbitals of the isolated atoms. Since the projection operator is defined in terms of the atomic-like orbitals on other atoms in the system, a self-consistent solution is appropriate. This localized pseudopotential method has not yet been used extensively on the a-Si:H system, but it does represent an ab initio and potentially self-consistent localized orbital theory which should be useful for performing realistic calculations. Further approximations are often necessary to render calculations tractable, however, such as disregard of multicenter integrals and lack of self-consistency in the potential. While the results thus obtained are still qualitatively useful, they represent little improvement in accuracy over the simpler ETB method. In practice the method provides a justification for the success of empirical parameterizations, and with further developments it may provide a quantitatively accurate ab initio procedure for obtaining tight-binding matrix elements. A discussion of this method and its application to various systems can be found in [2.10] by *Bullett*.

Another version of the ab initio LCAO method utilizes an expansion of both the potential and the localized basis functions in Gaussian-type orbitals [2.140–143]. In this scheme the atomic orbitals (e.g., Herman-Skillman [2.144] wave functions) are expanded in terms of Gaussians for each atom in the structure. The atomic potential for a free atom (e.g., Hartree-Fock-Slater self-consistent-field [2.144] is fitted to the form

$$V_{\text{atom}}(r) = -(Z/r)\,e^{-ar^2} + \sum_i c_i e^{-b_i r^2}\,. \qquad (2.12)$$

The overall system potential is a non self-consistent superposition of the atomic potentials. A Hamiltonian matrix element may be decomposed into a series of three-center integrals containing a Gaussian at site A, another at B and $V_{\text{atom}}(r)$ centered at C. Integrals associated with $(Z/r)\exp(-ar^2)$ expressed in terms of the error function and those associated with $\exp(-br^2)$ can be done analytically [2.140–143]. Thus, it becomes practicable on a given geometric structure to compute all multicenter integrals exactly which is the great advantage of this method. Two-center and nearest-neighbor approximations are not needed. Core states can be removed from the problem (reducing the size of the secular equation and greatly facilitating solution) by (i) orthogonalizing the valence manifold to the core manifold and deleting core states from the basis set, or (ii) using at the outset atomic pseudopotentials and working with only the valence electrons.

c) Pseudopotential Methods

A pseudopotential is an effective one-electron potential which acts only on the valence electrons of atoms while it models the influence of the core (inner electrons plus nucleus). Since it is the valence electrons which participate in chemical bonding while the core electron wave functions remain largely inert and unchanged as atoms bond to form a solid, this approach reproduces most properties of interest. Elimination of the core states usually simplifies the representation of eigenstates or reduces the size of the secular determinant. This can render an intractable problem tractable. Pseudopotentials with no high momentum components, which are typically useful in conjunction with plane-wave expansions in periodic systems, come in a great many varieties. No complete review will be given; see [2.145] by *Cohen* and *Heine* for a general introduction and also the references given below. The basic idea is that one can write a one-electron Schrodinger equation for a crystal in the form

$$T|\phi_k\rangle + \left(1 - \sum_{\text{cores}\, c} |\phi_c\rangle\langle\phi_c| \right) V|\phi_k\rangle = E_k|\phi_k\rangle , \qquad (2.13)$$

where the summation extends over all the core states. Most of the strength of the potential V in the core region is projected out onto the core states (obviously all the strength would be projected out if the set of core states constituted a complete set of states). Since the remaining potential, formally

$$\left(1 - \sum_{\text{cores}\, c} |\phi_c\rangle\langle\phi_c| \right) V , \qquad (2.14)$$

is weak, the solutions $|\phi_k\rangle$ are nearly eigenstates of the kinetic energy operator T, i.e., plane waves. Thus, a basis of a fairly small number of plane waves, typically about 200, can give excellent convergence for representing the pseudoeigenstates of the crystal. The pseudopotential V_{ps} simulates the influence on the valence electrons of the core electrons and the nucleus. In the empirical pseudopotential method, the potential may be Fourier expanded and its Fourier components fit to experimental data such as optical transitions, or an analytic form of atomic pseudopotential may be fit to similar data from atomic spectroscopy [2.145]. In another approach ab initio atomic pseudopotentials are obtained from atomic structure calculations. Recent advances, reviewed by *Yin* and *Cohen* [2.146] (and see references therein), make this latter method a highly accurate model of the atomic core. Also see [2.147] by *Bachelet* et al., for a detailed description and table of Gaussian fitting parameters for "norm conserving" pseudopotentials from H to Pu. When the atomic cores are modeled by an atomic (sometimes called ionic) pseudopotential, then the valence electrons are typically allowed to adjust self-consistently as the atoms are brought together to form a solid. It is assumed that the pseudopotential model of the core developed for the isolated atom will remain valid in the interacting solid. This approximation of transferability can be theoretically justified in the ab initio methods [2.146] and is amply verified by the agreement of these calculations with experiment. The feature of self-consistent adjustment of the charge may prove important in the study of defect structures and bonding anomalies which give rise to localized states in the gap and may also provide reliable charge density maps for the study of bonding. Self-consistent adjustment of charge has proved crucial, for example, in considering charge transfer between Si atoms in the buckling model of the Si(111) surface [2.40]. In this case electronic correlation energy greatly reduces the amount of charge transfer. The self-consistent calculation [2.40] shows that the charge transfer and buckling predicted by empirical tight-binding models [2.43, 44] are greatly diminished, to the extent that buckling is not a stable relaxation. The restrictions of periodicity (or symmetry in the case of point defect [2.148–151] calculations) in applying these methods to amorphous systems will be discussed in Sect. 2.4.2.

d) Self-Consistent Quantum Chemical Methods

Some recently developed methods of quantum chemistry (e.g., the generalized valence bond method) can provide highly realistic solutions of the Schrödinger equation, sometimes including electron correlation to a large degree. The complexity of the problem (i.e., the computational expense) grows so rapidly with the number of orbitals in the basis set, however, that these methods can only treat molecular clusters of up to 12–15 atoms. One would like to model certain aspects of the bulk amorphous system with molecular clusters. Since nearly every atom of such a cluster is a surface atom, there is often difficulty separating real bulk electronic information from intrinsically surface information. Certain properties, for example, of defect states which are believed to be very highly localized can be investigated by these methods. One models a given defect or bonding conformation by considering a molecule or atom cluster which includes the desired conformation. We will discuss three methods which have been applied to cluster models of a-Si:H. These are the (i) self-consistent field X_α scattered wave (SCF-X_α-SW), (ii) Hartree-Fock (HF) and (iii) generalized valence bond (GVB) methods.

The SCF-X_α-SW method [2.152] (and references therein) can be considered a bound-state version of the KKR method [2.153], familiar in the theory of periodic solids. One forms a muffin-tin type potential in which an electron within a muffin-tin radius of a given atomic center sees the potential of only that atom (actually the muffin-tin spheres are allowed to overlap somewhat), and the potential in the interstitial region is constant. The atomic potential is the Slater X_α self-consistent field potential with local exchange [2.152]. The atomic wave functions of each muffin-tin region are matched at the muffin-tin boundary to spherical functions which are equivalent to outgoing and incoming spherical waves. The outgoing spherical waves from other atomic centers in the cluster must be matched self-consistently with the incoming waves of any given center. Energy eigenvalues are discrete energies at which such a matching is possible. The resulting one-electron energies and wave functions can yield an approximate description of energy states and charge densities associated with given Si–H bonding conformations in a-Si:H. The use of the X_α potential is an advantage for molecular cluster calculations. By utilizing the "transition state" method one may calculate optical excitation energies which include the effects of orbital relaxation, avoiding the poor approximation of Koopmans' theorem [2.152]. Electron correlation effects are supposedly incorporated to some degree in the choice of the parameter α for each atomic type. This approximation is not well justified on theoretical grounds. This particular method is, therefore, less reliable than other quantum chemical methods. The GVB method (see below), for example, does include electron correlation in an effective and controlled way.

Unrestricted Hartree-Fock (UHF) methods have been applied to small clusters of Si and H atoms [2.154]. The basic idea of HF is the variational solution of the many electron problem in which the many-electron wave function is approximated by a single Slater determinant made up of one-electron eigenfunctions. In [2.154] the further approximation of an effective core potential (pseudopotential) is used for the Si atoms to reduce the number of electrons in the problem. See [2.154] and references therein for details of this method.

The GVB method has certain similarities with HF. In this approach [2.155–157] (and references therein) one forms valence bond orbitals using one orbital per electron and then solves self-consistently to optimize the orbitals as in HF. This models electron correlation effects which are absent from the HF treatment. The GVB wave function is the generalization of the HF wave function in which electron correlation in included and in which occupied and correlating orbitals are computed self-consistently [2.157]. One feature of GVB theory is that it describes most effectively the variation of bond energy with changing bond length. The GVB method is used particularly for charged defects where electron correlation may be a significant part of the total energy [2.155].

2.4.2 Models of Hydrogenated Amorphous Silicon

In this section we present specific theoretical models, including models of the geometrical structure and approximations used in specifying and solving the Hamiltonian, used to understand electronic states in a-Si:H. All results are presented in Sect. 2.5.

a) Finite Cluster Models

The use of small molecules in conjunction with quantum chemical methods was discussed in the last section. In another approach, a much larger continuous random network (CRN) cluster of about 200 atoms is built by hand and then relaxed by computer, based on bond stretching and bending force constants [2.158–162]. The defect or configuration of interest is placed near the center of this cluster. An illustrative sketch is presented in Fig. 2.6. The secular equation is developed by using the ab initio LCAO method in which both the (SCF) potential and the atomic orbitals are expanded in a Gaussian form [2.140–143], allowing exact evaluation of all multicenter integrals. In forming the basis for the Hamiltonian, atomic-like orbitals from only a small number (about 35) of atoms near the center of the 200 atom cluster are used. The 35 atom cluster is chosen to include the defect or configuration of interest. The valence orbitals of these atoms are orthogonalized to the core orbitals; then the core orbitals are deleted from the basis set. The resulting basis is called an OLCAO (orthogonalized LCAO) basis. This approximation greatly reduces the size of the secular determinant, yet gives reasonably good valence and conduction bands [2.142]. Note that the Hamiltonian operator is that of the full cluster of about 200 atoms, while the basis includes orbitals from only the central 35 atoms. It is argued that because the basis functions do not overlap the boundary, they see the Hamiltonian of an infinite system and no spurious surface states arise [2.140–143]. A simple example [2.163] shows, however, that this is not entirely correct. In an ETB calculation for an "ideal surface", the "infinite system Hamiltonian" is retained in the sense

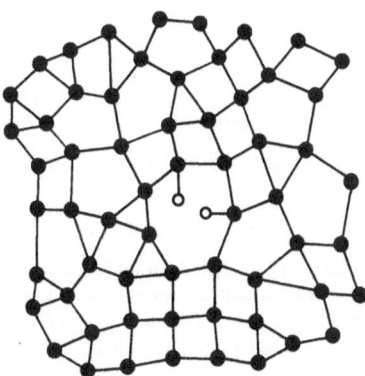

Fig. 2.6. Sketch of a monohydride embedded in a large cluster model of a random network

that interaction parameters appropriate to the bulk are used for the surface. Surface states nevertheless arise from the truncation of the Hamiltonian matrix by truncating the basis set [2.163]. The basis truncation used in OLCAO should give rise to surface states, but the additional terms in the Hamiltonian apparently shift the energies into the bands by changing the effective self-energies of these states. It is not clear whether this feature could cause problems in interpreting the resulting spectrum of states in the bands. After the wave functions $|\psi_i\rangle$ are obtained by solving the secular equation, the local density of states $N_a(E)$, (2.3), is evaluated on atoms of interest in the center of the cluster.

b) Repeated Unit Cell (Supercell) Models

It has been a popular and powerful technique to model an amorphous system in terms of a crystal with a large disordered unit cell. A schematic representation of such a unit cell is shown in Fig. 2.7. Early work [2.10, 15] utilized empirical pseudopotential Hamiltonians in k-space on either (i) arbitrarily built disordered unit cells or (ii) large crystalline unit cells which are natural polymorphs of Si or Ge. Recently, large (about 54 atoms) computer-grown and relaxed periodic cells have been developed which incorporate hydrogen bonded to Si [2.164, 165]. A smaller unit cell of 8 atoms has also been used to study the hydrogenated vacancy defect in which one Si atom is removed from a crystalline-like region and an H atom is bonded to each of the four broken bonds [2.166]. Periodicity retains the convenience of k-space which facilitates fully self-consistent and accurate calculations in a plane wave basis, but produces certain spurious effects due to LRO. Another benefit of periodicity is that the system is infinite and no surface states arise. The OLCAO method utilizing Gaussian expansions for orbitals and potential [2.140, 142, 167–169] and the chemical pseudopotential approach [2.139] have both been applied to periodic CRN structures.

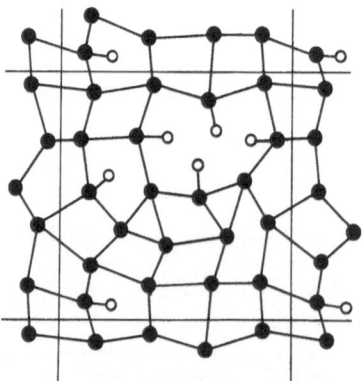

Fig. 2.7. Sketch of a large disordered unit cell containing several monohydrides in a disordered Si environment. This provides a model of the amorphous system but retains the advantages of periodicity

c) Effective Crystal Models

Effective crystal models incorporate disorder into an effective or renormalized Hamiltonian which is then solved on a crystalline lattice. Periodicity is usually obtained in a statistical sense by averaging over potential variations [2.27]. These methods are especially useful and popular in the theory of random alloys [2.9–11, 16], which has some overlap with the present subject. We will discuss three models which have been applied to either a-Si or a-Si:H. These are (i) the virtual crystal approximation (VCA), (ii) the coherent potential approximation (CPA) and (iii) the complex-band structure (CBS) model. Typically one wants to calculate an averaged Green's function $\langle G \rangle$ where the averaging is done over random potential fluctuations induced by structural changes or by random atomic substitution. Recall that the Green's function is related to $N(E)$ by

$$N(E) = - (1/\pi)\text{Im}\{\text{Tr}[G(E_+)]\} , \tag{2.15}$$

and to the projected density of states on orbital $|\phi_a\rangle$ by

$$N_a(E) = - (1/\pi)\text{Im}\{\text{Tr}[G_{aa}(E_+)]\} , \tag{2.16}$$

where $E_+ = E + i\delta$ and δ is a positive infinitesimal [2.27].

In a recent series of papers [2.98–101], the electronic structure of a-Si was modeled by examining the influences of bond length variations, bond angle variations, dihedral angle disorder and topological disorder on the band structure of c-Si. To estimate the quantitative disorder, a deformation potential model was used in conjunction with an empirical tight-binding scheme [2.98]. In the model calculation of *Tanaka* and *Tsu* [2.97], a realistic ETB Hamiltonian is used in conjunction with a type of VCA. The Hamiltonian of a weakly distorted diamond structure of identical (Si or Ge) atoms is transformed to a Hamiltonian describing nonidentical atoms in identical sites of a perfect diamond structure. The resulting Hamiltonian is made periodic by averaging its matrix elements over all configurations. The band structure and resulting $N(E)$ for this Hamiltonian are easily computed [2.97]. The effect on $N(E)$ of various specific distortions such as bond extensions and contractions, bond angle and dihedral angle variations have been studied.

In the CPA one develops an effective Hamiltonian with a self-energy $\Sigma(E)$ adjusted so that on the average the t-matrix for scattering off of an impurity site is zero [2.9–12, 27, 170]. This is usually implemented with an ETB Hamiltonian, as in the presently considered calculations [2.171–173]. The CPA does not correctly treat multiple-scattering processes from clusters of a fixed number of sites [2.27], hence it is expected to artificially smooth the local $N(E)$ associated with localized defects. Some attempts have been made to generalize the CPA by exactly treating clusters up to some given size in order to include the desired correlation effects [2.174].

The CBS method is an implementation of the VCA with a *k*-space pseudopotential [2.175–178]. The long-range order of the crystalline system is relaxed by performing a configurational average of the Green's function using a two-particle correlation function which is Gaussian-like near the crystalline atomic positions. The poles of the averaged Green's function give complex energy bands $E(k) = \varepsilon(k) + i\Gamma(k)$, where the imaginary part $\Gamma(k)$ is a measure of the lifetime of the state at k. This method has allowed computation of the imaginary part of the dielectric function $\varepsilon_2(\omega)$. This particular reponse function, however, is not very sensitive to the microscopic structural aspects of the amorphous phase [2.15].

d) Infinite Effective Medium – Cluster Models

In this section we describe a class of models which accurately treat some cluster of atoms, e.g., a defect configuration or simply a disordered bonding conformation, and then embed the cluster in an infinite effective medium which represents the influence of the rest of the solid. We first consider the case where the effective medium is periodic, i.e., a defect structure embedded in a crystal.

Recent work in the theory of point defects in semiconductors [2.125] has yielded a powerful technique which can treat the point-defect problem with the same accuracy available for crystalline band structure calculations [2.148–151]. The use of fully self-consistent (sometimes ab initio) pseudopotentials makes this type of calculation a standard with which more approximate methods can be compared. The point defects or bonding clusters of the amorphous system, such as the hydrogenated vacancy, can be modeled by this method [2.179, 180]. The basic principle of the method is the following. One obtains the perfect crystal Green's function $G_0(E)$ from a crystalline pseudopotential calculation. Writing the perturbation (potential change) induced by the defect as U, one forms the quantity

$$D(E) = \det[1 - G_0(E)U] = |D(E)|e^{i\delta(E)} . \tag{2.17}$$

Then the change in the density of states, $\Delta N(E)$, can be shown to be [2.148–151]

$$\Delta N(E) = d[\delta(E)]/dE . \tag{2.18}$$

As a model of the amorphous system, one assumes that local cluster properties are not too strongly influenced by the strictly crystalline surroundings vis à vis the expected variety of random network-like surroundings in the real material. Placing the defect in crystalline surroundings limits the flexibility of this approach. For example, (i) one cannot model an isolated monohydride in an otherwise fully bonded network, and (ii) one cannot model fivefold or sevenfold rings of bonds. The very large cluster of atoms which would be needed to allow enough distortion of the crystalline environment to model these configurations renders the problem intractable.

We now present an infinite nonperiodic effective medium theory, the cluster Bethe lattice method (CBLM) [2.181, 182]. The CBLM is a very flexible method which has been applied extensively to a-Si and a-Si:H [2.104, 136, 183–187]. The Bethe lattice [2.188] is an infinite, nonperiodic, fully connected system of atoms, where every atom is bonded in the same configuration but rings of bonds are disallowed. The system branches out like a tree to infinity as shown in Fig. 2.8c. This network is in some ways an ideal mathematical model of the random network discussed in Sect. 2.3.2. The electronic states of an infinite random network are generally quite similar to those of a Bethe lattice. The reasons for this similarity are (i) both systems are infinite and nonperiodic, (ii) the local bonding configurations of atoms in each model is similar, and (iii) the topological nature of the random network is such that many rings of bonds of various sizes can exist in the system. These rings impose boundary conditions on the wave function which tend to quantize states and put peaks in $N(E)$. With many different ring sizes, however, the peaks shift in energy and the resulting $N(E)$ tends to appear smooth and featureless. $N(E)$ for the Bethe lattice is smooth and featureless because of the lack of rings [2.15]. The Bethe lattice is useful mathematically because the Green's function for a Bethe lattice can be solved exactly for any tight-binding Hamiltonian [2.104, 136]. All these properties make the Bethe lattice an excellent candidate for an effective medium in which to study the electronic states at local structurally inhomogeneous or disordered sites of an amorphous material, which is precisely the basis of the CBLM. The basic idea is depicted schematically in Fig. 2.8. Within the infinite random network, a particular set of atoms is chosen as a reference. A cluster of atoms surrounding and including the reference set is removed from the system. A Bethe lattice whose atoms are in the same bonding configuration as in the

CLUSTER BETHE-LATTICE METHOD

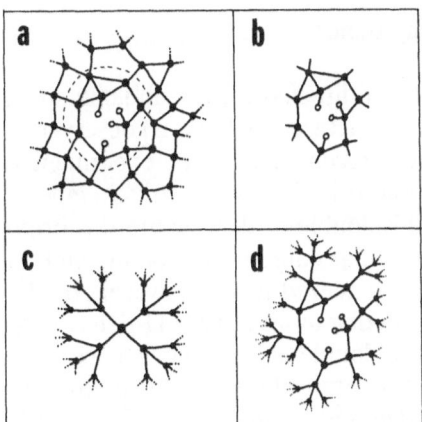

Fig. 2.8a–d. The cluster Bethe lattice method involves (**a**) taking an infinite system of atoms and choosing one or more atoms as a reference set, (**b**) removing from the system a cluster of atoms surrounding the reference set, (**c**) using a Bethe lattice to model the connected infinite network, and (**d**) attaching one Bethe lattice tree to each dangling bond on the surface of the cluster

original system (e.g., tetrahedral for Si) is attached to every dangling bond on the surface of the cluster. Now the new system in Fig. 2.8 d is infinite and nonperiodic, the local environment of the reference atoms is the same as in the original structure, and the local density of states $N_a(E)$ of each central atom may be solved exactly. Since the $N(E)$ structure of the Bethe lattice is smooth and featureless, sharp structures obtained in $N_a(E)$ can immediately be associated with the bonding in the reference cluster. Used in conjunction with a realistic ETB Hamiltonian the method is very flexible, allowing investigation of a variety of bonding conformations and defects [2.104, 136, 183–187].

2.5 Results of Theoretical Work

In this section we present the results of many theoretical calculations and the interpretations of myriad models. It should be clear from preceeding sections that a-Si and its alloys with H may possibly exist in a variety of more or less defective structures, none of which are completely understood or characterized. In addition, many of the assumptions and methods employed in theoretical work in this area are not yet well tested, although the field has seen improvement in recent years. One is therefore cautioned not to accept agreement between theory and experiment too uncritically. Because the problem is so difficult one tends to overlook what may be important structural differences among films produced in different laboratories or under slightly different preparation conditions. One must always consider whether the model used is too flexible and the experiment too poor in information for a meaningful comparison, or whether the theory truly models an essential feature of the material which the experiment probes. With this caveat, we proceed to give theoretical results.

2.5.1 Electronic States Within the Energy Bands

Much recent effort on band states has been directed toward understanding what type of bonding could give rise to the structure (peaks A–E) observed in the original (Fig. 7.3) [2.30, 31] and more recent [2.32] UPS experiments. We direct our attention first to peaks A and B in the low-substrate temperature (poorer quality) film (Fig. 2.3 a). This double-peak structure was originally attributed to SiH_3 groups based on similar data for SiH_3 on the Si(111) surface [2.30, 31]. The interpretation was rendered moot, however, by the results of theoretical calculations indicating that at least three H-rich configurations, SiH_2, SiH_3, and $(SiH_2)_2$, have the double-peak in the valence band with about the right splitting [2.187]. The theoretical results of two different groups (*Ching* et al. [2.159, 160] and *Allan* and *Joannopoulos* [2.189, 190]

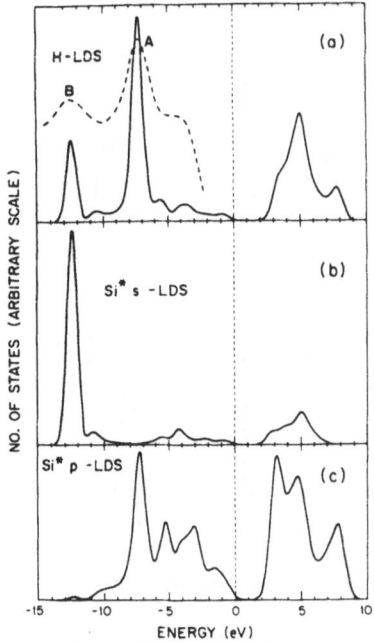

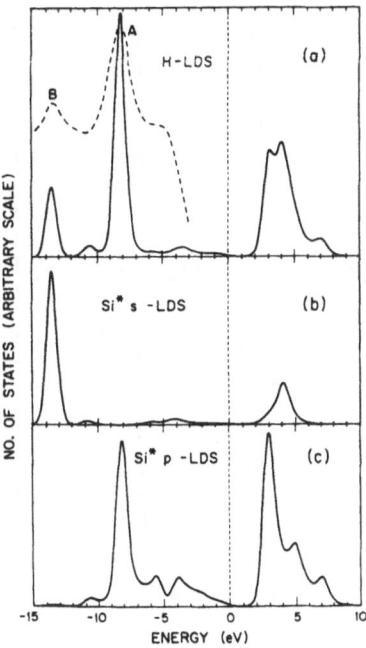

Fig. 2.9 a–c. Projected local densities of states (LDS) for a dihydride (SiH₂) embedded in a continuous random network, computed by OLCAO methods [2.159, 160]. The dashed line is a sketch of the UPS spectrum shown in Fig. 2.3 a, shifted by about − 2.5 eV. Compare with cluster Bethe lattice results of Fig. 2.11

Fig. 2.10 a–c. Projected local densities of states (LDS) for a trihydride (SiH₃) embedded in a continuous random network, computed by OLCAO methods [2.159, 160]. The dashed line is a sketch of the UPS spectrum shown in Fig. 7.3 a, shifted by about − 3.0 eV. Compare with the cluster Bethe lattice results of Fig. 2.12

are compared in Figs. 2.9–12. H(s), Si($3p$) and Si($3s$) local densities of states are presented in panels a, b and c, respectively. It is apparent that rather different theoretical approaches (ab initio OLCAO and ETB CBLM) agree fairly well, even on the projection of the local density of states $N_a(E)$ on different orbitals. This is seen as follows. The OLCAO local density of states for H(s) and nearby Si(s) and Si(p) for an SiH₂ embedded in a random network is given in Fig. 2.9. The CBLM results for the same conformation are provided in Fig. 2.11. Concentrating on the valence bands, the main spectral features are reproduced at close to the same energies and with the same orbital character. (The conduction bands arising from the ETB Hamiltonian used in the CBLM are not very reliable. They are known, for example, to be too narrow [2.191, 192], as seen by comparing Figs. 2.9, 11.) Note, in particular, that peaks A and B observed in UPS (Fig. 2.2 a) are reproduced in these calculations. The corresponding results for the SiH₃ conformation are presented in Figs. 2.10, 12 (OLCAO and CBLM, respectively). Again there is agreement between the two methods. In particular, both

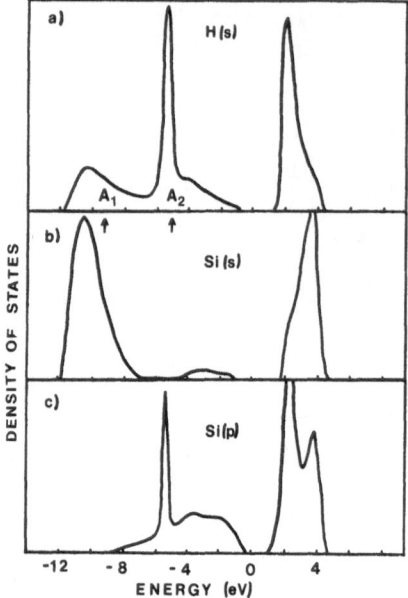

Fig. 2.11. Density of states for a dihydride (SiH_2) embedded in a Bethe lattice [2.189]. (a) Local density of states on the $H(1s)$ orbital; (b) local density of states on the $Si(3s)$ orbital; (c) local density of states on the $Si(3p)$ orbital. Compare with OLCAO results of Fig. 2.9

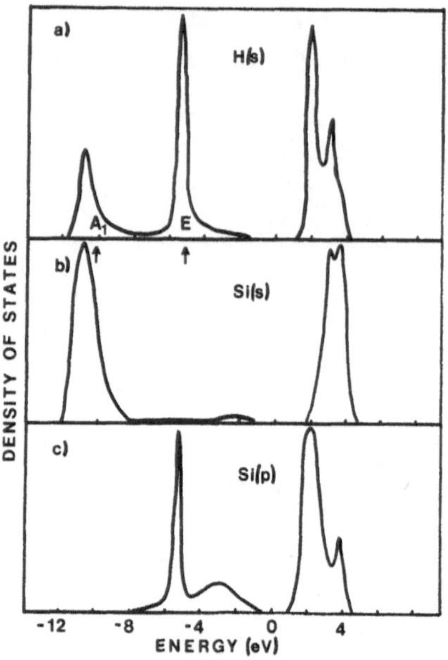

Fig. 2.12 a–c. Density of states for a trihydride (SiH_3) embedded in a Bethe lattice [2.189]. (a) Local density of states on the $H(1s)$ orbital; (b) local density of states on the $Si(3s)$ orbital; (c) local density of states on the $Si(3p)$ orbital. Compare with OLCAO results of Fig. 2.10

methods predict that both SiH_2 and SiH_3 bonding are consistent with peaks A and B. Note that the two-peak structure is also seen in the small cluster $(SiH_3)Si_4H_9$ SCF-X_α-SW model calculation for the SiH_3 (Fig. 2.13, right panel) in which an "optical density of states" is obtained from transition-state calculations [2.152]. (The right panel of this figure refers to a cluster calculation which models the SiH_3 conformation. The middle panel refers to a cluster model of an isolated monohydride, while the left panel refers to a model of "bulk" a-Si [2.152]. The latter two calculations will be discussed later in this section.) Within the accuracy of UPS and theory, a unique bonding assignment is not possible.

Recent UPS work on c-Si surfaces indicates that the purportedly observed $Si(111):SiH_3$ spectrum [2.193] may well have arisen from oxygen contamination [2.194]. Recently the same A and B peaks have been carefully checked in a-Si:H by taking UPS for hydrogenated and oxidated films [2.195]. The polyhydride identification of these peaks [2.30, 31] is reinforced on the basis of significant small distinctions between the oxide and hydride spectra [2.195].

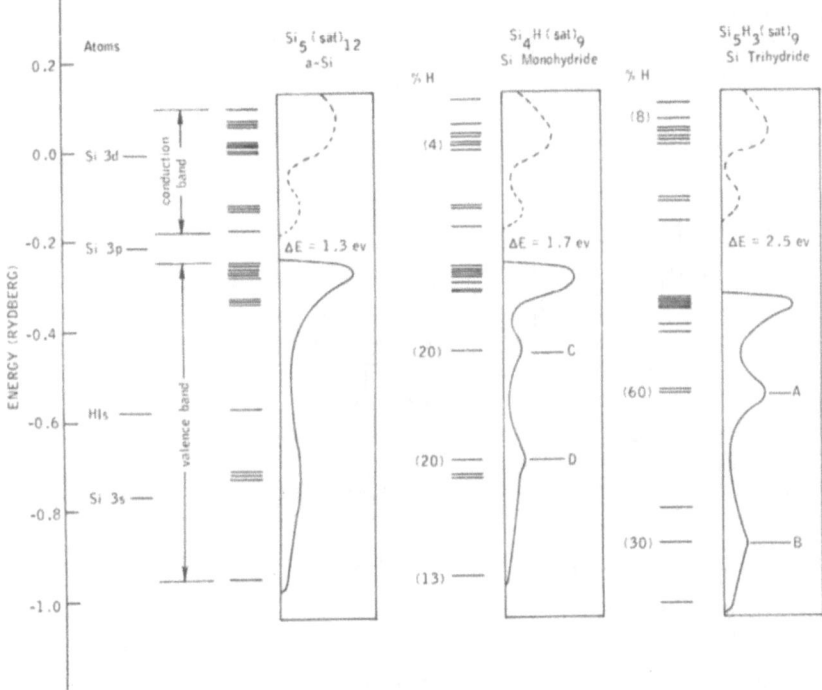

Fig. 2.13. Self-consistent field X_α molecular orbital calculations for several clusters of atoms modeling hydrogenated Si [2.152]. Each smooth curve is a broadened version of the discrete eigenenergies which are indicated as lines below. (*right panel*) trihydride (SiH₃) bonded to a Si atom which is then bonded to three more Si atoms in tetrahedral fashion; the dangling bonds of the Si atoms are saturated by pseudohydrogens. (*middle panel*) monohydride (SiH) bonded to three Si atoms; the dangling bonds of the Si atoms are saturated by bonding to pseudohydrogens. (*left panel*) a Si atom bonded tetrahedrally to four other Si atoms; dangling Si bonds are saturated by pseudohydrogens. The diagonal energies of the respective atomic orbitals are also shown

There is general agreement that monohydride (SiH) bonding dominates the high-temperature spectrum (peaks C, D, E), although simultaneous UPS and annealing experiments [2.32] and theoretical results [2.136, 183–187] have complicated the interpretation. A variety of theoretical models of the Si–H bond in a-Si : H [2.136, 152, 159, 160, 165, 166, 171–173, 179, 180, 183–187] have recently been investigated. Local density-of-states results are presented in Figs. 2.13–2.21 for the various models indicated in the captions. These results will be discussed individually as we develop a framework for interpretation. Our thinking is first guided by the results of Fig. 2.14. Figure 2.14a is a sketch of UPS for a-Si (dashed line) and high quality a-Si : H (solid line) [2.31]. From the featureless spectrum of the nonhydrogenated sample, the three bumps C, D, and E emerge upon hydrogenation. Panel (b) shows the theoretical local density of states for H on the monohy-

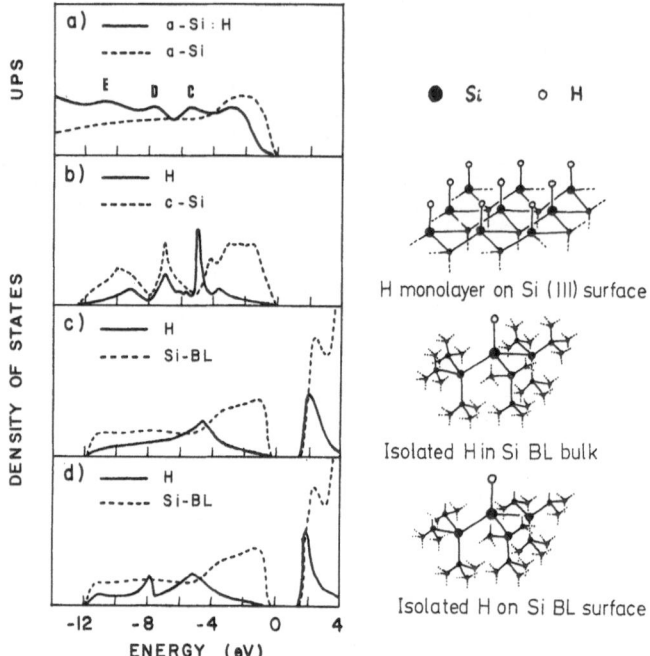

Fig. 2.14 a–d. Cluster Bethe lattice calculations investigating the Si–H bond [2.136, 183–185]. (a) UPS measurements [2.130, 131] for a-Si (– – –) and a-Si : H (——). (b–d) Local densities on H atoms in configurations shown. The dashed curves in (b–d) are densities of states of the underlying Si crystal (b) or the underlying Si Bethe lattice (c, d), respectively. Results of panel (b) are from [2.192] and the normalization of the H density of states is 1/4 that of the Si density of states

drated Si(111) : SiH surface (solid line) and for bulk crystalline Si (dashed line) [2.192]. Note that the three features of the solid curves line up rather well. It was suggested [2.31] on the basis of this agreement that the C, D, E structure arises directly from the Si–H bond. Before accepting such a hypothesis, however, the following must be considered. (a) What influence does the underlying Si lattice have on the local $N_a(E)$ near a Si–H bond? Note, for example, that the two lower bumps in the H local density of states (Fig. 2.14 b, solid curve) seem to line up with corresponding bumps in the density of states of the underlying Si (dashed curve). (b) Do interactions among the monohydrides play an important role in $N_a(E)$? We begin answering these questions by considering Fig. 2.14 c. The influences of underlying Si bonding topology and SiH–SiH interactions have been removed by embedding a single SiH in a Si Bethe lattice (BL) [2.136, 183–187]. Note the smoothness of the density of states for the Si BL (Fig. 2.14 c, dashed curve), a consequence of the absence of rings of bonds. The resulting H local density of states (Fig. 2.14 c, solid curve) retains only a single peak, corresponding to peak C. We therefore take this peak to be the intrinsic signature of the Si–H bond. Every calculation involving monohydride bonds exhibits this peak in the local density of states on the hydrogen atom, including very different theoretical techniques and models. It is seen, for example, on Si(111) : SiH (Fig. 2.14 b, solid curve), where it is the only peak which does not line up with peaks in the underlying Si $N(E)$. It is therefore either the interactions among monohydrides or the underlying Si topology which produces the two peaks below -6 eV in Fig. 2.14 b (solid line). The distinction is important, for it may lead to an understanding of related bonding structures in a-Si : H [2.136, 183–187].

We continue the investigation of the influence of interactions by placing a monohydride in a surface-like environment (within the BL model), surrounded by dangling bonds at second neighbors. Rings of bonds are absent. The resulting local density of states (Fig. 2.14 d, solid line) shows the C peak and another interaction-induced peak lower in the valence band. Evidently the interacting environment of a monohydride can strongly influence its local density of states in the lower half of the valence band.

To further demonstrate the influences of interactions on the SiH local density of states, the results for several interacting invironments are presented in Fig. 2.15 [2.136]. In this figure the solid curves refer to $N_a(E)$ on the H atom, while the dashed curves refer to $N_a(E)$ on the nearest Si atom. By utilizing the BL model, the influences of rings of bonds are absent from the following comparisons. Note first that the C peak (characteristic Si–H

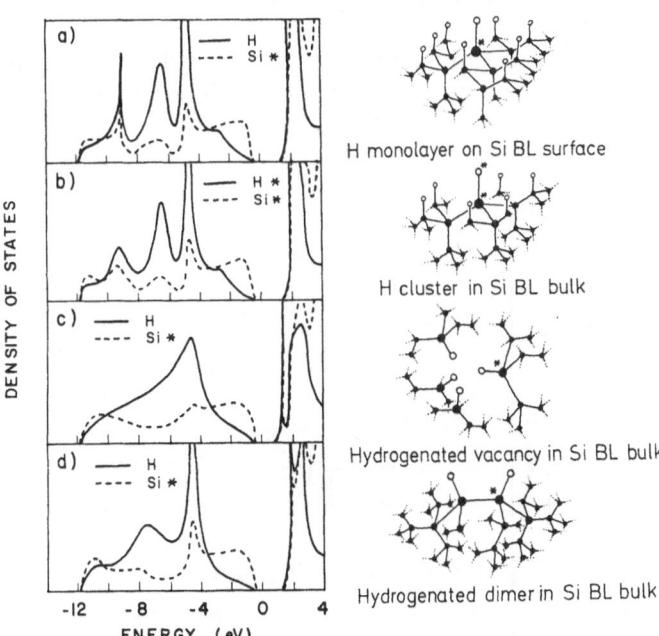

H monolayer on Si BL surface

H cluster in Si BL bulk

Hydrogenated vacancy in Si BL bulk

Hydrogenated dimer in Si BL bulk

Fig. 2.15 a–d. Local densities of states for the particular atoms in the configurations shown [2.184]

bond signature) is present in each spectrum, just below -4 eV. In the top panel we consider an infinite set of monohydrides interacting in a surface-like geometry. This calculation was performed by defining surface transfer matrices for a "surface BL" after *Yndurain* and *Louis* [2.196]. A very similar calculation is presented in Fig. 2.15b. Here only the nearest six monohydrides around a central reference monohydride are kept – the rest of the bonds are completed with BL trees [2.136]. The similarity of the spectra indicate that only neighboring monohydrides have much influence. As a consequence, small patches of adsorbed monohydrides, for example, on the internal surfaces of voids in a-Si:H, would give rise to a three-peak C, D, E type of density of states. Thus, monohydride interactions could be responsible for the observed UPS spectrum (Fig. 2.3b, c).

As a second example of interacting monohydrides, we consider in Fig. 2.15c a hydrogenated vacancy in the bulk of a-Si:H. In this case the SiH–SiH interactions are not strong enough to significantly perturb the density of states from that of an isolated SiH in the BL (compare with Fig. 2.14c, ignoring arbitrary normalization). Finally we place a pair of monohydrides at nearest-neighbor separation, a strongly interacting geometry which may be found in bulk a-Si:H. In this case the interactions give rise to the D peak localized mostly on the H atom and a hint of the E peak on the neighboring Si atom (Fig. 2.15d, solid and dashed curves, respectively). In another geometry not shown here, two monohydrides at second neighbors do not interact enough to produce D or E peaks in the lower valence band [2.197].

Having investigated influences of interactions, we proceed to investigate the influences of the bonding topology of the underlying Si. We first demon-

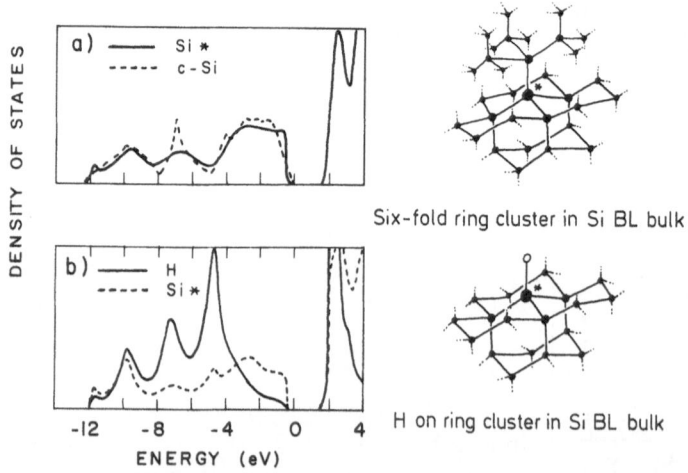

Six-fold ring cluster in Si BL bulk

H on ring cluster in Si BL bulk

Fig. 2.16a, b. Local densities of states for particular atoms in the configurations shown [2.184]. The density of states of crystalline Si is given as a dashed line in (**a**) [2.192]. The normalization of the H local density of states is chosen to be the same as that of the Si in this figure

strate that the main features in the density of states of crystalline Si (sketched in Fig. 2.16a, dashed curve) can be understood completely in terms of the sixfold rings of bonds of the diamond lattice. A cluster of Si atoms including one reference atom (denoted by Si*) and a surrounding set of atoms completing six sixfold rings through Si* is embedded in a Si BL [2.136, 183–185]. The local density of states on Si* (Fig. 2.16a, solid line) reproduces with excellent agreement the three-peak spectrum of c-Si. (The small bump at the bottom of the valence band is purely an artifact of the BL model which has unphysical long-range order which affects states at the band edges). It is clear that the three peaks arise entirely from rings, since in the absence of rings the $N(E)$ for the BL is completely smooth (recall Fig. 2.14b or c, dashed line). Now to investigate how this underlying structure will influence the monohydride $N_a(E)$, we bond an H atom to Si* (there are still six sixfold rings of bonds). The local density of states (Fig. 2.16b, solid curve) shows clearly the very strong influence of the underlying Si. The characteristic C peak occurs as expected, while the two lower peaks line up perfectly with the ring-induced Si peaks displayed in Fig. 2.16a. In other words, when hydrogen atoms bond to Si atoms which are participating in several sixfold rings of bonds, the characteristic local density-of-states structure induced by the rings induces peaks in the hydrogen local density of states.

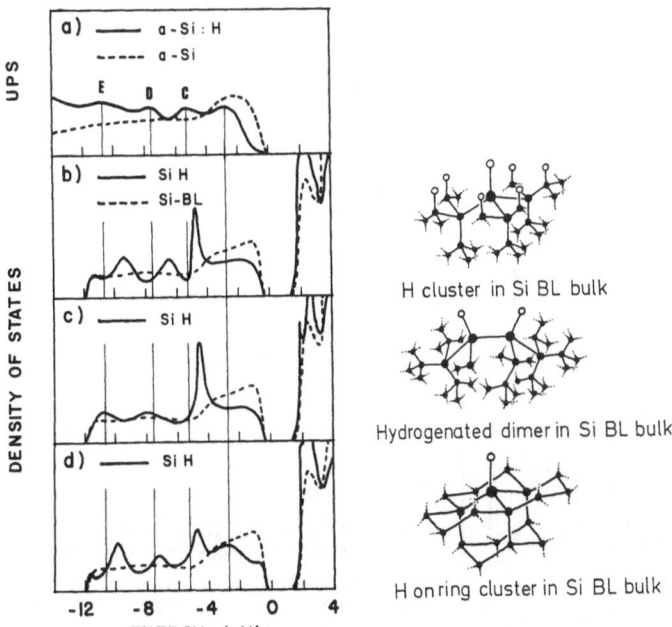

Fig. 2.17. (a) UPS spectra from [2.130, 131]. **(b–d)** Average densities of states over H and neighboring Si atoms in the configurations shown. The Si Bethe lattice density of states is shown as a dashed curve in (b–d). Note comparison of theoretical results with main peaks of experimental spectrum

In Fig. 2.17 we summarize our results thus far and compare again with the UPS results. Monohydrides clustered together at second-neighbor separation give adequate agreement with the observed peak separations among C, D, and E (panel b). The nearest-neighbor monohydrides (hydrogenated dimer) agree slightly better with the experiment (panel c). Finally, the occurrence of H atoms bonded to Si atoms with sixfold ring topology (as would be found in microcrystalline regions in a-Si : H) may also be the source of the experimental spectrum. We will return later in this section to the question of how to distinguish these possibilities. If sixfold rings indicative of microcrystalline regions are indeed present, then we may begin to understand substitutional doping in a-Si : H [2.136, 183–185].

We return now to Fig. 2.13 (second curve), the small cluster model of the monohydride [2.152]. There is a clustering of eigenvalues which is associated with the usual C peak and there is also a clustering which appears to reproduce the D peak. Since H atoms are used to saturate all the dangling Si bonds on the "surface" of this small cluster of 13 atoms, the resulting spectrum evidently must include the effects of H–H interactions propagating through the Si–Si bonds. The presence of an additional peak in the lower valence band as a consequence of these interactions is consistent with our foregoing theoretical interpretation.

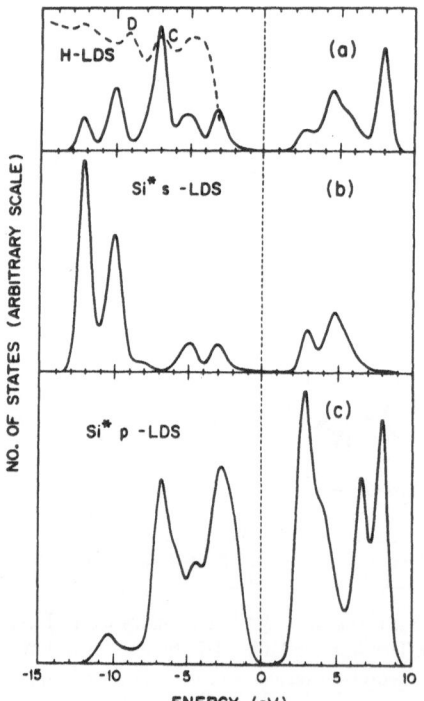

Fig. 2.18 a–c. Projected local density of states (LDS) for a monohydride (SiH) embedded in a continuous random network, computed by OLCAO methods [2.159, 160]. The dashed line is a sketch of the UPS spectrum shown in Fig. 2.3 c, shifted by about − 2 eV. Compare with the cluster Bethe lattice results of Fig. 2.14 c

We now consider the OLCAO result [2.159, 160] for an isolated SiH in a CRN (Fig. 2.18). After shifting the energy scale the largest peak is interpreted as the C peak. There is a lot of structure in this spectrum, however, which we contend is a direct consequence of the specific local bonding environment in which the SiH is placed in this one example calculation. The CBLM result and interpretation predicts that an averaging over many local bonding topologies will considerably smear out all but the central large peak. *Ching* et al. [2.160] did check to a limited extent the persistence of features upon ensemble averaging over various CRN models by constructing two different models for the SiH. The differences in peak position were considered to be small [2.160], so no further testing was done. It is not clear what aspects of local topology (i.e., ring statistics), if any, were varied.

In another CRN calculation, computer models were developed containing monohydrides in a large (about 60 atoms) unit cell [2.164]. The electronic structure is solved self-consistently using an empirical ionic pseudopotential for Si^{+4} and H^+ [2.165]. Averaging three examples and smoothing the results provides the $N(E)$ spectrum shown in Fig. 2.19. The line drawn through the theoretically computed circles seems to identify four subbands in the local density of states on the H atoms (this line is merely to guide the eye). The Hamiltonian used and the self-consistent solution with an adequate plane-wave basis should render the results reliable; the amount of scatter in the theoretical data, however, makes comparison with experiment difficult. These computations are being extended by constructing more examples and sampling more k-points in the Brillouin zone [2.165]. The consideration of self-consistent charge densities available from this method should eventually provide useful information about bonding and charge transfer in this mate-

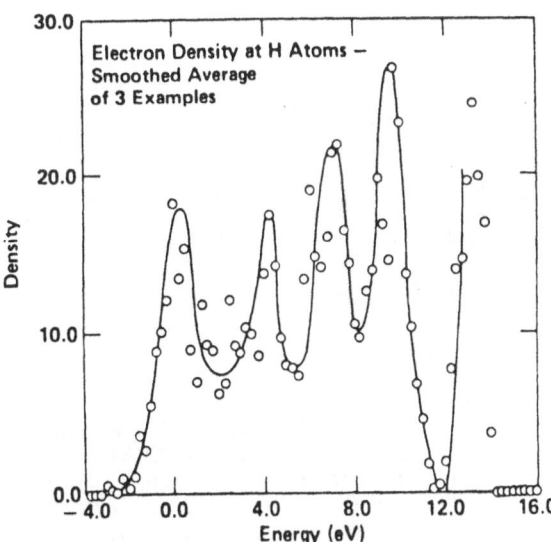

Fig. 2.19. Local density of states near H atoms bonded as monohydrides in a periodic continuous random network model of a-Si:H [2.165]. Electronic states are obtained using an empirical pseudopotential. The theoretical results are given as dots; the lines are merely to guide the eye

rial. At this time too few examples have been considered for unambiguous trends to emerge. This approach may play an important role in future investigations.

Another probe of the Si–H bond, in the context of the hydrogenated vacancy in crystalline Si, is presented in Figs. 2.20, 2.21. Figure 2.20 shows results of CPA calculations [2.171–173] in which 5% of Si atoms in the lattice were removed (Fig. 2.20 a) and then all dangling bonds saturated by H (Fig. 2.20 b–d). In the top panel, dangling bond states fill the energy gap. With H bonded to all these dangling bonds, the gap states are removed (panels b–d). The gap region will be discussed in more detail in Sect. 2.5.3. Focusing on the valence band, note in particular that the intrinsic Si–H "C" peak is again present (panel d), near − 5 eV. A second pronounced feature near − 7 eV in the H local density of states is apparently induced by the density of states of the underlying Si (compare with panel c), consistent with earlier results. The feature lying below the Si valence band arises from interactions among hydrogenated vacancies. This feature is also seen in a supercell model of the hydrogenated vacancy, from the interactions among the cells [2.166]. Turning to Fig. 2.21, we consider an isolated hydrogenated vacancy treated by self-consistent pseudopotential Green's function scattering theoretic methods [2.179, 180]. The density of states in this figure is

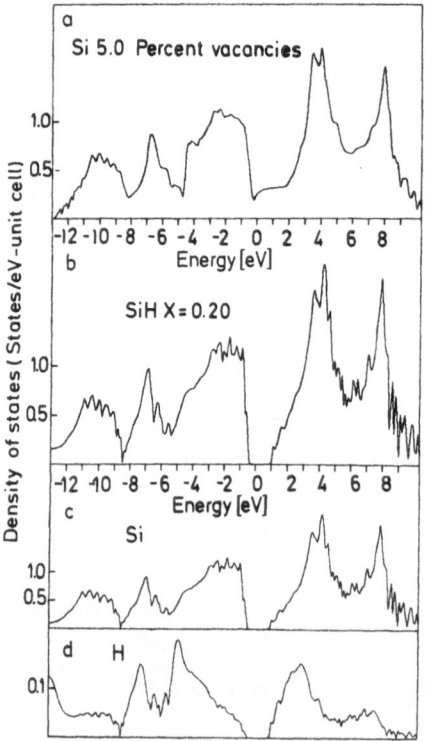

Fig. 2.20. Total (**a, b**) and local (**c, d**) densities of states obtained by CPA [2.171–173]. Panel (a) refers to 5% vacancies and no H, and (b–d) refer to the case with all the dangling bonds terminated by H

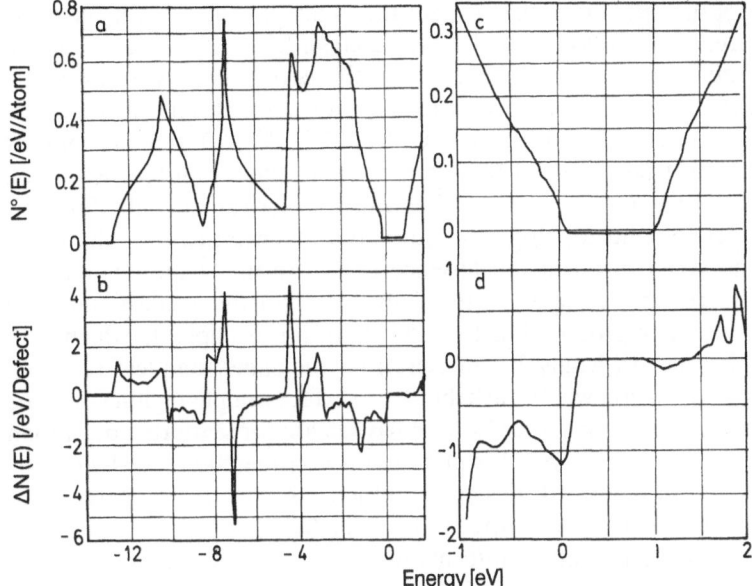

Fig. 2.21 a–d. Density of states for crystalline Si (**a**, **c**) and the change $\Delta N(E)$ due to one hydrogenated vacancy (**b**, **d**). These results are obtained by self-consistent pseudopotential Green's function scattering theoretic methods [2.179, 180]

presented in terms of (a) the unperturbed crystalline $N(E)$, and (b) the change $\Delta N(E)$ brought about by the presence of the hydrogenated vacancy. Panels (c) and (d) expand the gap regions of (a) and (b), respectively. One sees in Fig. 2.21 b the C peak just below -4 eV in the change $\Delta N(E)$. In a reciprocal space description, this feature is attributed to the existence of a critical point (W_2) in the crystalline Si band structure [2.180], although the foregoing random network results seem to show that the C peak is better understood in terms of local bonding and not in terms of LRO. For the single hydrogenated vacancy, note that no states are found below the c-Si valence band.

In the recent experiments of *Smith* and *Strongin* [2.32], UPS spectra were observed while heating the H out of a-Si:H. Since the peak at higher-binding energy (corresponding to the D peak) persists to higher temperatures, where mostly monohydride bonding is expected, *Smith* and *Strongin* suggest that the D peak is actually the intrinsic signature of the Si–H bond and the C peak arises from SiH–SiH interactions. They suggest that the same type of annealing behavior may be observed with hydrogenation of the Si(100) : 2X1 surface [2.198]. This interpretation, however, is misleading. In both cases reconstruction should play an important role in determining the UPS spectrum. Since the "D" peak is present even on the clean Si(100) : 2X1 surface [2.198], it obviously can arise purely from dangling bonds and their interactions in the absence of H. In addition, as H is driven out in the *Smith* and

Strongin experiment, one would expect the peak associated with the Si–H bond to decrease, whereas the peaks associated with the production of dangling bonds interacting with each other and with nearby monohydrides should increase. Thus, the experimental result need not be inconsistent with theoretical assignments of the C and D peaks. UPS at 21.2 eV excitation probes only the outer 10 angstroms or so of the sample, where significant reconstruction may occur and may not be completely indicative of the sample bulk. Furthermore, interactions and topology can strongly influence peaks in the lower valence band $N(E)$, even in the absence of H. Further experimental and theoretical work can be expected to resolve some of these questions. One experiment has been proposed [2.136, 183, 184], for example, in which both XPS and UPS spectra would be measured on the same samples. By using glancing angle detection with x-rays the same surface sample region would be probed, but the x-rays have a far lower cross section for scattering by H than for scattering by Si. This type of experiment is considered especially important for the case of monohydrides because theory has made it apparent that the local environment strongly affects the local $N_a(E)$. We now explain the important microscopic bonding information that might be obtained by such an experiment.

If the peaks D and E observed in UPS on a-Si:H actually arise from sixfold rings of bonds through Si atoms to which H atoms are bonded, then these peaks should persist in the XPS. On the other hand, if the D and E peaks arise predominantly from monohydride interactions, then these peaks

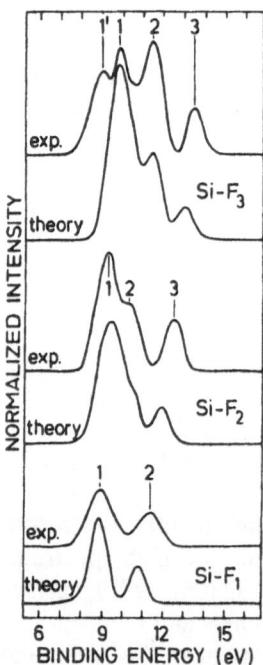

Fig. 2.22. Experimental [2.199, 200] and theoretical [2.161] densities of states for several Si–F bonding conformations

should be absent from XPS. (In any case, the C peak should be diminished or absent in XPS.) The former case is consistent with the formation of microcrystalline regions upon hydrogenation. Recall that the smooth XPS of nonhydrogenated a-Si (Fig. 2.2 c) indicates absence of microcrystallinity. The steric constraints of microcrystallinity might explain how the material is doped, since substitutional bonding would be facilitated. The latter case would indicate that rings of bonds are of random sizes, that the system is truly a random network. In this case the question would remain as to why impurity atoms would bond substitutionally, even in the presence of presumed steric freedom. These questions remain of fundamental importance. We note that some very recent XPS measurements seem to indicate that the latter case is correct (see e.g. Fig. 3.11 in Chap. 3). In any case it is quite clear that hydrogen plays an important role for both doping mechanisms (microcrystalline order or not) by removing states from the gap.

We turn our attention briefly to a-Si : F, where F atoms are used in place of H atoms to tie off dangling bonds. In this case the theoretical $N_a(E)$ for the SiF, SiF_2 and SiF_3 complexes in a CRN, computed by OLCAO, were available [2.161] in advance of the experimental UPS. After normalization of the experimental results and after empirically shifting the theoretical curve by -3.0 eV, theory and experiment are brought into agreement (Fig. 2.22). Thus, a combination of theory and UPS experiments has identified certain bonding conformations of Si and F in a-Si : F films prepared under various conditions [2.199, 200].

2.5.2 Electronic States Within the Band Tails

Early theoretical work [2.187] suggested that addition of H to the a-Si matrix would remove states from the top of the valence band. This work suggested, moreover, that the observed widening of the gap with increasing H [2.130, 131] occurs primarily by depletion of states from the valence band edge [2.187]. This suggestion has been confirmed theoretically within the CBLM by comparison of the densities of states for structures which contain increasing amounts of H [2.136, 185]. In Fig. 2.23 we compare calculations (using a first nearest-neighbor ETB Hamiltonian) which include increasingly more H: (i) a pair of nearest-neighbor monohydrides (solid line) in a Si BL (dashed line), (ii) a BL comprising SiH units bonded together, and (iii) polysilane. Note the valence band edge recedes but the conduction band edge remains the same as H is added. This trend can be understood heuristically as follows. The gap should widen as Si–Si bonds are replaced by stronger Si–H bonds and the bonding-antibonding splitting becomes larger. In the conduction band the effect is compensated, however, by the lower self-energy of the H(1s) orbital with respect to the Si(sp^3) hybrid.

Variation in the H concentration throughout a sample of a-Si : H causes variations in the valence band maximum. We therefore expect hole trap

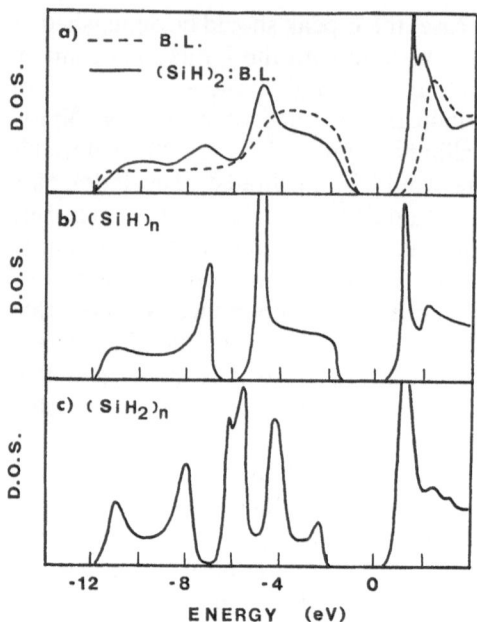

Fig. 2.23 a–c. Comparison of Bethe lattice results for structures which incorporate varying concentrations of H atoms [2.185]. Note particularly that the valence band edge recedes with increasing H concentration, while the conduction band edge remains constant

states at the top of the valence band, localized in regions of pure a-Si and bounded by regions rich in Si–H bonds [2.136, 183–185]. This may explain the large asymmetry in hole and electron mobilities [2.17, 18], and may also explain other observations which indicate a concentration of states near the valence band edge.

Recession of the valence band edge with hydrogenation is seen in many theoretical calculations [2.152, 159, 160, 171–173, 179, 180, 185]. Consider, for example, the hydrogenated vacancy calculation results in Fig. 2.21. Panel (c) shows the band edges and gap region for perfect crystalline Si, while panel (d) shows the change $\Delta N(E)$ in the density of states upon forming a hydrogenated vacancy. Observe in panel (d) the large removal of states at the top of the valence band; states are depleted to a far lesser extent from the conduction band edge [2.179, 180]. The same trend is observed in CPA calculations in which the dangling bonds of a set of Si vacancies are randomly bonded with varying concentrations of H [2.171–173]. CPA calculations clearly show a stationary conduction band edge E_c and receding valence band edge E_v as the H concentration is continuously raised, as shown in Fig. 2.24 [2.172]. The picture of spatially varying band edges led *Brodsky* [2.201] to propose a model in which carriers near the band gap are spatially correlated. The model suggests that spatial coincidence of states near the valence and conduction band edges can account for observed optical absorption, conductivity, photoconductivity and photoluminescence. It is important to notice that the number of such localized states per unit volume of a sample should be on the order of the density of band states, which is orders of magnitude larger than

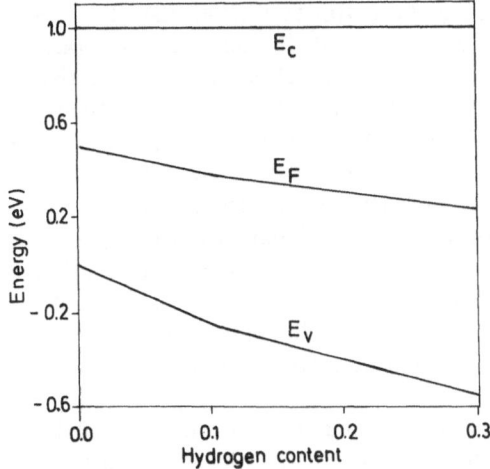

Fig. 2.24. Variation of the conduction band edge (E_c), the Fermi energy (E_F) and the valence band edge (E_v) with hydrogen concentration. This result is obtained using the CPA [2.172]

the expected number defect-related gap states. Estimates of the band edge localized region energy width arising from this type of H-induced quantitative disorder indicate a mobility edge lying at least 0.1 eV into the top of the valence band [2.179, 180].

We noted in Sect. 2.2 that, in the absence of H, the valence band edge steepens in a-Si compared with c-Si. Figure 2.2a compared with Fig. 2.2c shows the steepening and the movement of the *p*-like hump to higher energies. Theoretical results presented in Fig. 2.25 show that both bond-length

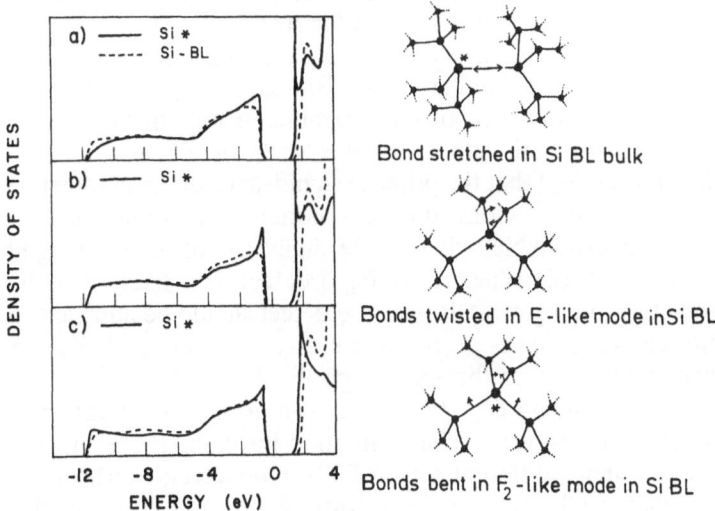

Fig. 2.25 a–c. Local densities of states on the atoms designated by an asterisk, undergoing the indicated distortions [2.184, 186]

and bond-angle variations will cause these trends [2.104, 184, 186, 187]. Figure 2.25 a shows that, as a bond is stretched, bonding and antibonding levels begin to rejoin from the valence and conduction bands, respectively. (The dashed line in each panel is $N(E)$ for the undistorted Si BL.) If the bond were completely broken, two degenerate half-filled dangling bond states would occur at the Si(sp^3) hybrid energy, which lies in the gap. Effects of bond-angle distortions are shown in panels (b) and (c). The only independent distortions of a tetrahedron which preserve bond length are distortions of E and F_2 symmetry [2.104], presented in Fig. 2.21 b, c, respectively. Both distortions have essentially the same effect on the band edges. Since bond lengths change little but bond angles vary over a 20 degree range [2.28], the steepening of the valence band edge might be a consequence of bond-angle variations [2.104].

Virtual crystal approximation (VCA) calculations using realistic ETB Hamiltonians have shown that dihedral angle variations may cause band tailing at the top of the valence band [2.97–101]. Based on VCA calculations, a band of localized states of about 0.3 eV width is estimated to arise from dihedral angle disorder [2.98]. There are also some indications that topological disorder can cause band tailing [2.102]. Any realistic model or picture must incorporate all these effects.

2.5.3 Electronic States Within the Band Gap

The origin and nature of gap states in a-Si : H is of primary concern for technological applications. This is also an area in which, unfortunately, reliable theoretical methods have not yet been developed. The ETB methods which succeed in the study of band states suffer several drawbacks for gap states: (i) To reproduce the energies of these states, which mix wave functions from valence and conduction bands, the Hamiltonian must reproduce both energy regions well. Most sets of ETB parameters only fit the valence band well [2.191]; fitting both bands requires such an extension of the number of parameters [2.202] that the physical significance of the parameters is obscured. Transferability among different bonding environments, for example, becomes questionable when large numbers of longer-ranged interactions are included. (ii) When modeling a defect, it is likely that the interaction parameters among orbitals near the defect should be adjusted to reflect the altered environment. In particular, *Krieger* and *Laufer* [2.128, 129] have shown that as the set of tight-binding orbitals per atom tends toward completeness, no gap states at all can arise without changing the parameters. There is not yet a reliable method for doing the adjusting. Because of these problems, different sets of ETB parameters, which fit the valence band equally well, can predict a variety of gap state energies [2.128, 129]. In spite of these drawbacks, however, the qualitative trends of gap state energies may be reproduced by ETB calculations. It is preferable, whenever possible, to compare ETB results with results from electrostatically self-consistent calculations such as self-consistent pseudopotential calculations. In some cases the comparison is favorable [2.133]. Note that self-consistency is already included in the ETB parameterizations of the bulk bands, but it is not necessarily adequate for the defect configuration. The more reliable methods such as self-consistent pseudopotential or generalized valence bond (GVB) are usually so cumbersome that relatively few defects of interest can be investigated.

A number of defects have been studied which place one-electron states in the gap. The chemical bonding arguments of *Adler* [2.14, 53, 115–117] have suggested that a twofold coordinated Si atom defect may introduce gap states which act as electron and hole traps. (In *Adler's* [2.117] notation, T_2 refers to a twofold coordinated tetrahedrally bonding atom. The charge state of the defect is given as a superscript, as T_2^0 for a neutral twofold coordinated Si.) CBLM calculations predict two gap states (the upper of which is empty) for the T_2^0 [2.136, 183–186]. Two gap states are similarly found for the relaxed dimer (nearest-neighbor dangling bond) defect, which is most likely a lower energy defect [2.136, 183–186]. This would result from bonding a pair of T_2^0 defects. The common feature of these defects is that they would pair spins in the lower state, and yet they would provide a light induced ESR (LESR) signal. Some such defect might explain the low equilibrium ESR in conjunction with higher LESR observed in a-Si : H [2.50]. Spin-pairing defects may also be the cause of the large H infusion needed to completely quench the spin signal [2.33–37, 53].

Some results from a CPA study of the gap region [2.171–173] are given in Fig. 2.20. Panel (a) shows a continuous band of gap states associated with vacancy dangling bonds. Absence of correlations in the CPA prevents any sharp structure in $N(E)$ [2.170]. A generalization of the CPA, exactly treating scattering from larger atomic clusters, gives fairly sharp dangling bond states in the gap [2.174]. Figure 2.20 a, b shows, furthermore (within the CPA), how saturation of all the dangling bonds with H atoms removes the dangling bond states from the gap. This particular result is quite straightforward. Relaxation and charging effects of dangling bonds are also investigated, in the context of the Si-SiO$_2$ interface, in model calculations of *Ngai* and *White* [2.203]. We next describe investigations of these effects using the GVB method.

Photoluminescence (PL), ESR, and LESR experiments indicate the possible existence of a spin-pairing defect which may have a negative effective correlation energy U_{eff}. It has been suggested [2.115, 117] that relaxations associated with charged states of the dangling bond may provide a negative U_{eff}. *Redondo* et al. [2.155] have used the GVB method to compute the total energies of a small molecule model of a Si dangling bond on a surface. In this calculation they consider the neutral and two charged states of the dangling bond, allowing relaxation of the threefold coordinated atom along the direction of the broken bond. Based on these results, we have constructed

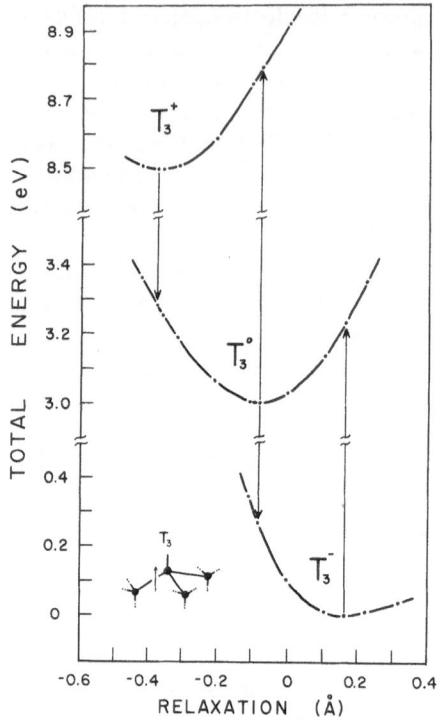

Fig. 2.26. Configurational coordinate diagram for the three charge states of a dangling bond. The coordinate represents deviation of the dangling bond atom along the $\langle 111 \rangle$ direction from its "ideal" tetrahedral position. This result is based on GVB calculations for a small molecule model of the dangling bond [2.155]

[2.136, 184] the configurational coordinate diagram for the three charge states (T_3^+, T_3^0 and T_3^-), as shown in Fig. 2.26.

Two important results follow from Fig. 2.26. Firstly, the transition

$$2\,T_3^0 \;\to\; T_3^+ + T_3^- \tag{2.19}$$

is endoergic. Hence, U_{eff}, the energy of this reaction, is positive. From Fig. 2.26 one obtains $U_{\text{eff}} \sim 2.5$ eV. Secondly, there are significant relaxations associated with charging the dangling bond, although these relaxations are not as large as the complete dehybridization assumed by *Adler* [2.115]. The Stokes shifts for the transitions $T_3^0 \to T_3^+$ and $T_3^- \to T_3^0$ are about 0.6 eV and 0.5 eV, respectively. These relaxation energies are consistent with interpreting 0.9 eV PL as de-excitation of T_3^-, and 1.4 eV absorption as the transition $T_3^- \to T_3^0$ [2.90, 92, 136, 184]. This links both these observations to the T_3^0 spin density, as is observed [2.88–90, 92].

The configurational coordinate diagram also provides an estimate of the one-electron levels associated with charged dangling bonds. The ionization energy of the T_3^0 is about 5.8 eV, in close agreement with the ionization energy of a surface dangling bond [2.155]. The ionization energy of the T_3^-, however, is only about 3.2 eV. This would place the T_3^- state in the conduction band, about 2.6 eV above the T_3^0 level. These results, however, only pertain to dangling bonds on a surface; furthermore, they only pertain to noninteracting dangling bonds. When broken bonds occur in the bulk, additional polarization of the Si medium brings the ionization energies of the T_3^0 and the T_3^- closer together. A rough calculation for a dangling bond in a small void of radius approximately equal to the second neighbor distance in Si, for example, gives a difference between T_3^0 and T_3^- ionization energies of less than 1.0 eV [2.190]. Interactions with other dangling bonds should further reduce this energy. Thus, the effective correlation energy U_{eff} may be of the order of 1.0 eV or slightly less, closer to the range (0.2–0.4 eV) estimated from earlier [2.67] and more recent [2.64] spin-resonance experiments. The GVB method, unfortunately, can only treat a small molecule including only a single dangling bond. Influences of bulk surroundings and of other dangling bonds are hard to estimate reliably. While it seems clear that an isolated dangling bond on a surface has a positive U_{eff} of about 2.5 eV, the consequences for dangling bonds occurring in the bulk of a-Si : H are still not completely understood.

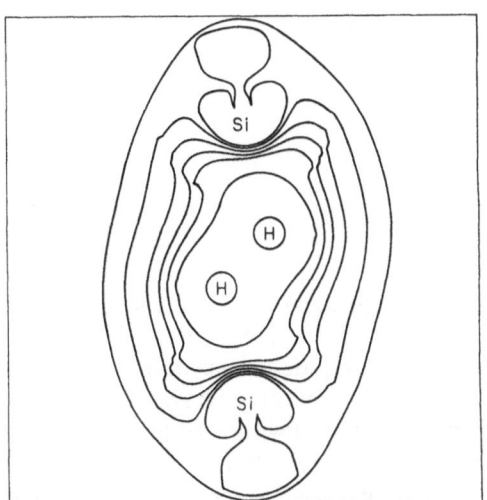

Fig. 2.27 Self-consistent field X_α scattered wave molecular orbital wave function of lowest energy for a small molecule model of molecular hydrogen three center bond [2.121]. The molecule comprises two Si atoms, two H atoms and six "saturator" effective H atoms on the Si back bonds. In this example the bond lengths are H–H, Si–H and Si–Si distances = 0.8, 1.35, and 3.2 Å, respectively. The orbital shown is bonding between the H atoms

A very recent SCF-X_α-SW calculation [2.121] suggests that molecular H_2 may possibly enter strained Si-Si bonds and would tend to stabilize the bonds over a range of bond lengths. The predicted geometry and charge density are given in Fig. 2.27. It is suggested that this novel bonding conformation can compensate stretched bonds and remove states from the conduction and valence band tails. The authors further suggest that this type of bonding could be responsible for the unusual low energy peak observed in H effusion studies.

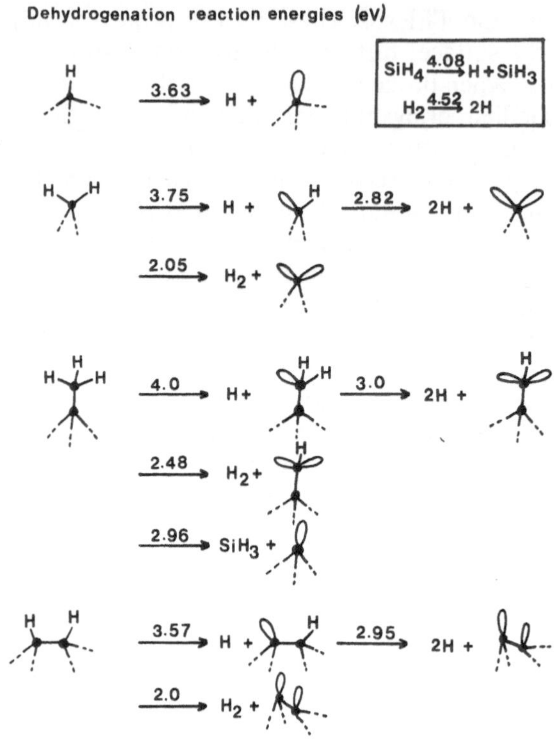

Dehydrogenation reaction energies (eV)

Fig. 2.28. Energies required for the indicated dehydrogenation reactions, derived from semi-empirical energies [2.136, 184]

2.5.4 Hydrogen Desorption Energies

The energetics of Si–H bonding has been investigated using the semi-empirical total energy scheme described in Sect. 2.4.1a. Numerical results are shown in Fig. 2.28 for a number of endoergic dehydrogenation reactions [2.136, 184]. To compare with experimentally measured activation energies, one should be aware that these processes may have an additional energy barrier whose value is not yet estimated. Notice that the Si medium reduces the first Si–H bond strength from its value of 4.08 eV in SiH_4 [2.204, 205] to the range 3.6–4.0 eV. The removal of H from SiH or from $(SiH)_2$ requires 3.6 eV, near the observed [2.206] value of about 3.4 eV. The removal of a second H atom requires in each case less energy than removal of the first. Figure 2.28 also shows that formation of H_2 molecules in the bond-breaking process, in cases where H atoms are proximal at the outset, reduces the energy required to about 2–2.5 eV. The lower value is close to most of the measured values obtained from H effusion studies, although the interpretation remains controversial [2.206–214]. Finally, it can be seen that the energy needed to remove neutral SiH_3 from a surface is ~ 3.0 eV, making SiH_3 units slightly more surface mobile than H atoms.

2.6 Conclusions and Future Directions

We have presented a number of experimental observations and puzzles, discussed in more detail throughout the other chapters of this book, and we have presented a variety of recently developed theoretical models and methods which attempt to interpret and predict experimental results for electronic structure. The theoretical methods used range from simple parameterized chemical bonding models to fully self-consistent pseudopotential and all-electron calculations. The simpler models have introduced some important concepts and have indicated where interesting physics may lie. Because of uncertainties in these models, however, they must give way to the more sophisticated calculations becoming available. The empirical tight-binding (ETB) Hamiltonian has been used to implement a variety of approximation schemes. The tremendous flexibility of this approach, as well as its quantitative success in many cases, suggests that it will remain a popular approach in future work. The chemical or localized pseudopotential scheme is a potentially self-consistent and ab initio method utilizing a localized orbital basis. This scheme shows promise, but has not yet been applied extensively to a-Si : H. The self-consistent Green's function scattering theoretical approach to point-defect calculations has been successfully applied to the hydrogenated vacancy in a crystalline silicon environment. This method, however, has limited applicability to disordered systems. Self-consistent pseudopotential calculations for both electronic states and total energies of defects in disordered unit cells are just becoming available [2.215]. This method promises to be an important tool for future investigations. Hopefully, we will see a trend in which the sophisticated methods becoming available will provide unambiguous statements about electronic states, local atomic order and bonding for many kinds of disordered systems.

The valence band region of a-Si : H is reasonably well understood, but the band tails and especially the gap region remain poorly understood theoretically. Moreover, a realistic determination of the mobility edge is still lacking. Quantum chemical cluster calculations would be ideally suited for predicting the energy states and relaxations expected for defects, but the influence on these states of the rest of the infinite medium in the real system is still poorly understood. There is much work to be done in this area, which is an area ripe for new theoretical advances.

Throughout this chapter we have investigated the relationship between the atomic (or geometric) structure of a-Si : H and its electronic states. Various signatures of monohydride, dihydride and trihydride bonding conformations have been determined. It has been observed that the Si environment of a monohydride has a strong influence on its electronic structure. In particular, it is the ring *topology* of the underlying Si lattice that is most important. This is not true, however, for di and trihydride bonding configurations. Interactions among monohydrides, at nearest-neighbor separation or for many of them at second-neighbor separation, can also be very important. In addition, interactions between monohydrides and nearby intrinsic Si defects can also play an important role in the nature of the electronic states.

We have also investigated a number of defects which place states in the gap. There is no theoretical evidence that either the Si–H–Si three center bond or an isolated dangling bond has a negative effective electron correlation energy. However, positively charged vacancies and interstitial B impurities will exhibit a negative effective correlation energy. This has been observed experimentally in crystalline Si [2.216] and ought to be the same in amorphous Si. The relaxations of the isolated dangling bond defect have been studied using a quantum chemical cluster model. The results predict a large Stokes shift of about 0.5 eV accompanying electron capture. This fits well with the observed increase of both 0.9 eV PL and 1.4 eV absorption with dangling bond spin density, since the absorption could correspond to excitation of the doubly occupied dangling bond and the 0.9 eV defect PL could correspond to de-excitation of the same defect.

By studying the consequences of bond length, bond angle and dihedral angle distortions, it was shown that all of these forms of disorder principally lead to band tailing.

The energetics of the Si–H bond in various environments in a-Si : H was studied using a semi-empirical total energy method. Experimental studies of H effusion during annealing can probe these bonds and provide important information about the atomic structure of a-Si : H. Presently, there are some conflicting experimental results for the observed activation energies and the theoretical method is not sufficiently tested and proven to make detailed predictions. Observations seem to indicate (i) the breaking of an isolated Si–H bond with an activation energy near 3.4 eV, and (ii) the release of molecular H_2 from nearby monohydrides with an activation energy in the range of 2.0 eV. The theoretical results are consistent with both of these possibilities.

Optical and photoemission experiments have shown that additional H widens the gap in a-Si : H. The theoretical results show that this occurs mainly by depletion of states from the top of the valence band. The bonding of H, furthermore, removes states from the gap, placing states with H character at the conduction band edge. The removal of gap states, which has been reproduced in theoretical calculations, indicates that a movement of the Fermi level could be observed *if* there were a mechanism for introducing a population of additional electrons or holes.

Experimentally, a-Si : H can be doped with impurities with the creation of additional electrons or holes. The fundamental theoretical problem, however, is why the valency of the impurity atoms is not completely satisfied. If the underlying Si network is microcrystalline, then there is no problem. Doping can be understood precisely as it is in crystalline systems. If the underlying Si lattice is a random network structure, however, it is not clear why Mott's argument regarding the flexibility of disordered systems to satisfy the valency of impurity atoms is not valid. Perhaps the answer lies in that disordered solids which form network structures intrinsically resist having their local bonding configuration changed by considerable amounts. These systems do not have the steric freedom of a liquid. Consequently, satisfying the valency of all impurity atoms could create strains in the disordered network that are too large. Developing a detailed understanding of doping remains an important area for future investigation.

Recently, *Street* [2.217] has proposed a model of doping in amorphous semiconductors which differs from that of crystalline semiconductors. He proposes that fourfold coordinated substitutional defects are only *stable* when *charged*, satisfying a generalized 8–N rule reflecting not the ideal atomic valency but the actual number of valence electrons of the bonded atom. Thus the stability of the substitutional sites depends directly on the location of the Fermi energy. This predicts that in the doped material E_F can never lie in a large density of states. The structure will adjust to effectively open a gap at E_F. *Robertson* [2.218], using an empirical tight-binding parameterized Hamiltonian, has carried out a systematic study of shallow and deep levels produced by various group III and group V atoms bonded in a-Si : H. His variety of numerical and qualitative predictions should help guide investigations of doping and associated gap states. In the search for novel models of nonperiodic systems, several investigators [2.219–221] have introduced the idea of studying regular structures in curved three-dimensional space whose projections on Euclidean three-dimensional space resemble regions of real disordered materials. By discovering exact symmetries in the curved-space models, approximate symmetries in the actual material are suggested. (For example, a remnant of the crystalline k selection rule is proposed near the 1.1 eV absorption edge [2.220].) Such theories provide a promising new and systematic approach to the study of order in amorphous materials.

Acknowledgement. This review was made possible by Grant No. NSF DMR-76-80895 from the U.S. National Science Foundation.

References

2.1   W. E. Spear, P. G. LeComber: Solid State Commun. **17**, 1193 (1975)
2.2   N. F. Mott: Phil. Mag. **19**, 835 (1969)
2.3   J. Tauc: Phys. Today 23 (Oct. 1976)
2.4   M. H. Cohen: Phys. Today 26 (May 1971)
2.5   J. Keller, J. M. Ziman: J. Non-Cryst. Solids **8**, 111 (1972)
2.6   M. H. Cohen: J. Non-Cryst. Solids **4**, 391 (1970)
2.7   D. Adler: *Amorphous Semiconductors* (CRC Press, Cleveland, OH 1971)

2.8   M. H. Brodsky (ed.): *Amorphous Semiconductors;* Topics Appl. Phys., Vol. 36
       (Springer, Berlin, Heidelberg, New York 1981), Chap. 2 by B. Kramer and D. Weaire,
       Chap. 6 by R. Fischer, Chap. 7 by I. Solomon, Chap. 8 by G. Lucovsky and T. M.
       Hayes, Chap. 9 by P. G. LeComber and W. E. Spear
2.9   V. Heine: *Solid State Physics*, Vol. 35 (Academic, New York 1980), p. 1
2.10   D. W. Bullett: ibid. p. 129
2.11   R. Haydock: ibid. p. 216
2.12   M. J. Kelly: ibid. p. 296
2.13   N. F. Mott: J. Phys. C **13**, 5433 (1980)
2.14   D. Adler: Sol. Cells **2**, 199 (1980)
2.15   J. D. Joannopoulos, M. L. Cohen: Solid State Phys. **31**, 71 (1976)
2.16   R. J. Elliott, J. A. Krumhansl, P. L. Leath: Rev. Mod. Phys. **46**, 465 (1974)
2.17   W. Paul, D. A. Anderson: Sol. Energy Mat. **5**, 229 (1981)
2.18   H. Fritsche: Sol. Energy Mat. **3**, 447 (1980)
2.19   9th Conf. on Amorphous and Liquid Semiconductors, Grenoble, 1981; J. Phys. **42**,
       Suppl. 10, (Paris) (1981)
2.20   *Tetrahedrally Bonded Amorphous Semiconductors,* Carefree, AZ, 1981, ed. by R. A.
       Street, D. K. Biegelsen, J. C. Knights, AIP. Conf. Proc. **73** (1981)
2.21   *Fundamental Physics of Amorphous Semiconductors,* Proc. 1980 Kyoto Summer Institute,
       ed. by F. Yonezawa, Springer Ser. Solid-State Sci., Vol. 25 (Springer, Berlin, Heidel-
       berg, New York 1981), p. 1 by H. Fritzsche and p. 119 by F. Yonezawa and M. H. Cohen
2.22   15th Intern. Conf. on the Physics of Amorphous Semiconductors, Kyoto, 1980, ed. by S.
       Tanaka, Y. Toyozawa; J. Phys. Soc. Jpn. **49**, Suppl. A (1980)
2.23   Photovoltaic Material and Device Measurements Workshop, San Diego, 1980, ed. by D.
       L. Stone; Sol. Cells **2** (1980)
2.24   8th Conf. on Amorphous and Liquid Semiconductors, Cambridge, 1979, ed. by W. Paul,
       M. Kastner; J. Non-Cryst. Solids **35/36** (1980)
2.25   7th Conf. on Amorphous and Liquid Semiconductors, Edinburgh, 1977, ed. by W. E.
       Spear (CICL, University of Edinburgh 1977)
2.26   *Tetrahedrally Bonded Amorphous Semiconductors,* Yorktown Heights, 1974, ed. by M.
       H. Brodsky, S. Kirkpatrick, D. Weaire, AIP. Conf. Proc. **20** (1974)
2.27   E. N. Economou: *Green's Functions in Quantum Physics*, 2nd ed., Springer Ser. Solid-
       State Sci., Vol. 7 (Springer, Berlin, Heidelberg, New York 1983)
2.28   S. C. Moss, J. F. Graczyk: Proc. 10th Intern. Conf. on the Physics of Semiconductors,
       Cambridge, 1970, ed. by J. C. Hensel, F. Stern (US Atomic Energy Commission,
       Washington (1970), p. 658)
2.29   L. Ley, S. Kowalczyk, R. Pollack, D. A. Shirley: Phys. Rev. Lett. **29**, 1088 (1972)
2.30   B. von Roedern, L. Ley, M. Cordona: Phys. Rev. Lett. **39**, 1576 (1977)
2.31   B. von Roedern, L. Ley, M. Cardona, F. W. Smith: Phil. Mag. B **40**, 433 (1980)
2.32   R. J. Smith, M. Strongin: Phys. Rev. B **24**, 5863 (1981)
2.33   W. E. Spear: Adv. Phys. **26**, 811 (1977)
2.34   G. A. N. Connell, J. R. Pawlick: Phys. Rev. B **13**, 787 (1976)
2.35   P. Thomas, M. H. Brodsky, D. Kaplan, D. Lepine: Phys. Rev. B **18**, 3059 (1978)
2.36   D. Kaplan, N. Sol, G. Velasco, P. Thomas: Appl. Phys. Lett. **35**, 440 (1978)
2.37   J. I. Pankove, M. A. Lampert, M. L. Tarng: Appl. Phys. Lett. **32**, 439 (1978)
2.38   J. C. Phillips: Phys. Rev. Lett. **23**, 1151 (1979)
2.39   K. C. Pandey: Phys. Rev. Lett. **47**, 1913 (1981)
2.40   K. C. Pandey: Phys. Rev. Lett. **49**, 223 (1982)
2.41   J. E. Northrup, J. Ihm, M. L. Cohen: Phys. Rev. Lett. **47**, 1910 (1981)
2.42   J. E. Northrup, M. L. Cohen: J. Vac. Sci. Technol. **21**, 333 (1982)
2.43   R. Del Sole, D. J. Chadi: Phys. Rev. B **24**, 7431 (1981)
2.44   D. J. Chadi, R. Del Sole: J. Vac. Sci. Technol. **21**, 319 (1982)
2.45   A. Redondo, W. A. Goddard III, T. C. McGill: J. Vac. Sci. Technol. **21**, 649 (1982)
2.46   J. C. Knights, T. Hayes, J. Mikkelsen: Phys. Rev. Lett. **39**, 712 (1977)
2.47   J. A. Reimer, T. M. Duncan: Phys. Rev. B **27**, 4895 (1983)

2.48 C. Tsang, R. Street: Phil. Mag. B **37**, 601 (1978)
2.49 I. Austin, T. Nashashibi, T. Searle, P. G. LeComber, W. Spear: J. Non-Cryst. Solids **32**, 373 (1979)
2.50 D. K. Biegelsen: Sol. Energy Mat. **2**, 421 (1980) (review)
2.51 R. A. Street, D. K. Biegelsen: J. Non-Cryst. Solids **36**, 651 (1980)
2.52 J. Knights, D. Biegelsen, I. Solomon: Solid State Commun. **22**, 133 (1977)
2.53 D. Adler: J. Phys. **42**, Suppl. 10, (Paris) 3 (1981)
2.54 A. Friederich, D. Kaplan: J. Phys. Soc. Jpn. **49**, Suppl. A, 1233 (1980)
2.55 S. Hasegawa, T. Shimizu, M. Hirose: J. Phys. Soc. Jpn. **49**, Suppl. A, 1237 (1980)
2.56 D. Biegelsen, J. Knights, R. Street, C. Tsang, R. White: Phil. Mag. B **37**, 477 (1978)
2.57 R. Street, D. Biegelsen: Solid State Commun. **33**, 1159 (1980)
2.58 U. Voget-Grote, W. Kummerle, R. Fischer, J. Stuke: Phil. Mag. B **41**, 127 (1980)
2.59 S. Hudgens: Phys. Rev. B **14**, 1547 (1976)
2.60 J. Stuke: 7th Conf. on Amorphous and Liquid Semiconductors, Edinburgh, 1977, ed. by W. E. Spear (CICL, University of Edinburgh 1977) p. 406
2.61 H. Fritsche, C. C. Tsai, P. Persans: Solid State Technol. **21**, 55 (1978)
2.62 D. Biegelsen, R. Street, C. C. Tsai, J. Knights: Phys. Rev. B **20**, 4839 (1979)
2.63 R. Street, D. Biegelsen, J. Stuke: Phil. Mag. B **40**, 4511 (1979)
2.64 J. D. Cohen, J. P. Harbison, K. W. Wecht: Phys. Rev. Lett. **48**, 109 (1982)
2.65 D. L. Staebler, C. R. Wronski: Appl. Phys. Lett. **31**, 292 (1977)
2.66 D. L. Staebler, C. R. Wronski: J. Appl. Phys. **51**, 3262 (1980)
2.67 H. Dersch, J. Stuke, J. Beichler: Appl. Phys. Lett. **38**, 456 (1981)
2.68 A. Madan, P. G. LeComber, W. Spear: J. Non-Cryst. Solids **20**, 239 (1976)
2.69 W. E. Spear, P. G. LeComber, A. J. Snell: Phil. Mag. B **38**, 303 (1978)
2.70 Z. Jan, R. Bube, J. Knights: J. Appl. Phys. **51**, 3278 (1980)
2.71 N. B. Goodman, H. Fritsche, H. Ozaka: J. Non-Cryst. Solids **35**, 599 (1980)
2.72 D. V. Lang: In *Thermally Stimulated Relaxation in Solids,* ed. by P. Bräunlich, Topics Appl. Phys., Vol. 37 (Springer, Berlin, Heidelberg, New York 1979)
      M. Lannoo, J. Bourgoin: *Point Defects in Semiconductors* II, Springer Ser. Solid-State Sci., Vol. 35 (Springer, Berlin, Heidelberg, New York 1983)
2.73 D. V. Lang, J. D. Cohen, J. P. Harbison: Phys. Rev. B **25**, 5285 (1982)
2.74 J. D. Cohen, D. V. Lang, J. P. Harbison: AIP. Conf. Proc. **73**, 217 (1981)
2.75 J. D. Cohen, D. V. Lang, J. P. Harbison: Phys. Rev. Lett. **45**, 197 (1980)
2.76 B. von Roedern, G. Moddel: Solid State Commun. **35**, 467 (1980)
2.77 L. Ley, M. Cardona, R. A. Pollak: In *Photoemission in Solids* II, ed. by L. Ley and M. Cardona, Topics Appl. Phys., Vol. 27 (Springer, Berlin, Heidelberg, New York 1979) Chap. 2
2.78 D. A. Anderson, G. Moddel, W. Paul: J. Non-Cryst. Solids **35**, 345 (1980)
2.79 R. A. Street: Adv. Phys. **30**, 593 (1981) (review)
2.80 R. Street, D. K. Biegelsen, J. C. Knights: Phys. Rev. B **24**, 969 (1981)
2.81 C. Tsang, R. Street: Phys. Rev. B **19**, 3027 (1979)
2.82 C. Tsang, R. Street: Phil. Mag. B **37**, 601 (1978)
2.83 R. Street: Phil. Mag. B **37**, 43 (1978)
2.84 R. Street, J. Knights, D. K. Biegelsen: Phys. Rev. B **18**, 1880 (1978)
2.85 R. Street: Phys. Rev. B **17**, 3984 (1978)
2.86 J. I. Pankove, F. H. Pollack, C. Schnabolk: J. Non-Cryst. Solids **35**, 459 (1980)
2.87 R. W. Collins, M. A. Paesler, G. Moddel, W. Paul: J. Non-Cryst. Solids **36**, 681 (1980)
2.88 W. B. Jackson, N. B. Amer: AIP. Conf. Proc. **73**, 263 (1981)
2.89 W. B. Jackson, N. B. Amer: Phys. Rev. B **25**, 5559 (1982)
2.90 S. G. Bishop, U. Strom, P. C. Taylor, W. Paul: AIP. Conf. Proc. **73**, 278 (1981)
2.91 B. A. Wilson, T. P. Kerwin: Phys. Rev. B **25**, 5276 (1982)
2.92 R. Street: Phys. Rev. B **21**, 5775 (1980)
2.93 A. F. Ioffe, A. R. Regel: Progr. Semicond. **4**, 239 (1960)
2.94 D. Weaire, M. F. Thorpe: Phys. Rev. B **4**, 2508 (1971)
2.95 M. F. Thorpe, D. Weaire: Phys. Rev. B **4**, 3518 (1971)

2.96   M. F. Thorpe, D. Weaire: Phys. Rev. Lett. **27**, 1581 (1971)
2.97   K. Tanaka, R. Tsu: Phys. Rev. B **24**, 2038 (1981)
2.98   J. Singh: Phys. Rev. B **23**, 4156 (1981)
2.99   M. H. Cohen, H. Fritsche, J. Singh, F. Yonezawa: J. Phys. Soc. Jpn. **49**, Suppl. A, 1175 (1980)
2.100 M. H. Cohen, J. Singh, F. Yonezawa: Solid State Commun. **36**, 923 (1980)
2.101 M. H. Cohen, J. Singh, F. Yonezawa: J. Non-Cryst. Solids **35**, 55 (1980)
2.102 J. A. Blackman, M. F. Thorpe: Phys. Rev. **23**, 2871 (1981)
2.103 T. C. McGill, J. Klima: Phys. Rev. B **5**, 1517 (1972)
2.104 J. D. Joannopoulos: Phys. Rev. B **16**, 2764 (1977)
2.105 J. D. Joannopoulos, G. Lucovsky (eds.): *The Physics of Hydrogenated Amorphous Silicon I,* Topics in Appl. Phys., Vol. 55 (Springer, Berlin, Heidelberg, New York 1984)
2.106 C. P. Slichter: *Principles of Magnetic Resonance,* Springer Ser. Solid-State Sci., Vol. 1 (Springer, Berlin, Heidelberg, New York 1980)
2.107 J. A. Reimer, R. W. Vaughan, J. C. Knights: Phys. Rev. B **24**, 3360 (1981)
2.108 W. E. Carlos, P. C. Taylor, S. Oguz, W. Paul: AIP. Conf. Proc. **73**, 67 (1981)
2.109 J. A. Reimer, J. C. Knights: AIP. Conf. Proc. **73**, 78 (1981)
2.110 F. R. Jeffrey, M. E. Lowry, M. L. S. Garcia, R. G. Barnes, D. E. Torgeson: AIP. Conf. Proc. **73**, 83 (1981)
2.111 J. A. Reimer, R. W. Vaughan, J. C. Knights: Solid State Commun. **37**, 161 (1981)
2.112 J. A. Reimer, R. W. Vaughan, J. C. Knights: Phys. Rev. Lett. **44**, 193 (1980)
2.113 M. S. Conradi, R. E. Norberg: Phys. Rev. B **24**, 2285 (1981)
2.114 W. E. Carlos, P. C. Taylor: Phys. Rev. B **25**, 1435 (1982)
2.115 D. Adler, R. C. Frye: AIP. Conf. Proc. **73**, 146 (1981)
2.116 D. Adler: J. Non-Cryst. Solids **36**, 819 (1980)
2.117 D. Adler: Phys. Rev. Lett. **41**, 1755 (1978)
2.118 S. R. Elliott: Phil. Mag. B **38**, 325 (1978)
2.119 S. R. Ovshinsky, D. Adler: Contemp. Phys. **19**, 109 (1978)
2.120 R. Fisch, D. Licciardello: Phys. Rev. Lett. **41**, 889 (1978)
2.121 M. E. Eberhart, K. H. Johnson, D. Adler: Phys. Rev. B **26**, 3138 (1982)
2.122 P. W. Anderson: Phys. Rev. **181**, 25 (1969)
2.123 J. D. Weeks, P. W. Anderson, A. G. H. Davidson: J. Chem. Phys. **58**, 1388 (1973)
2.124 D. W. Bullett: J. Phys. C **8**, 2695 (1975)
2.125 S. T. Pantelides, J. Pollman: J. Vac. Sci. Technol. **16**, 1349 (1979)
2.126 J. C. Slater, G. F. Koster: Phys. Rev. **94**, 1498 (1954)
2.127 W. A. Harrison: *Electronic Structure and the Properties of Solids* (Freeman, San Francisco 1980)
2.128 J. B. Krieger, P. M. Laufer: Phys. Rev. B **23**, 4063 (1981)
2.129 J. B. Krieger, P. M. Laufer: J. Vac. Sci. Technol. **19**, 307 (1981)
2.130 D. J. Chadi: Phys. Rev. Lett. **41**, 1062 (1978)
2.131 D. J. Chadi: J. Vac. Sci. Technol. **16**, 1290 (1979)
2.132 D. J. Chadi: Phys. Rev. Lett. **43**, 43 (1980)
2.133 J. Ihm, M. L. Cohen, D. J. Chadi: Phys. Rev. B **21**, 4592 (1980)
2.134 D. J. Chadi, J. R. Chelikowsky: Phys. Rev. B **24**, 4892 (1981)
2.135 D. Vanderbilt, J. D. Joannopoulos: Phys. Rev. B **22**, 2927 (1980); see especially Appendix B
2.136 D. C. Allan, J. D. Joannopoulos, W. B. Pollard: Phys. Rev. B **25**, 1065 (1982); B **26**, 3475 (1982)
2.137 D. H. Lee, J. D. Joannopoulos: Phys. Rev. Lett. **48**, 1846 (1982)
2.138 D. W. Bullett, M. J. Kelly: J. Non-Cryst. Solids **32**, 225 (1979)
2.139 M. J. Kelly: AIP. Conf. Proc. **20**, 174 (1974)
2.140 W. Y. Ching, C. C. Lin: Phys. Rev. Lett. **34**, 1223 (1975)
2.141 W. P. Menzel, K. Mednick, C. C. Lin, C. F. Dorman: J. Chem. Phys. **63**, 4708 (1975)
2.142 W. Y. Ching, C. C. Lin: Phys. Rev. B **12**, 5536 (1975)
2.143 R. C. Chaney, T. K. Tung, C. C. Lin, E. E. Lafon: J. Chem. Phys. **52**, 361 (1970)

2.144 F. Herman, S. Skillman: *Atomic Structure Calculations* (Prentice-Hall, Englewood Cliffs, NJ 1963)
2.145 M. L. Cohen, V. Heine: *Solid State Physics* **24**, 38 (Academic, New York 1970)
2.146 M. T. Yin, M. L. Cohen: Phys. Rev. B **25**, 7403 (1982)
2.147 G. B. Bachelet, D. R. Hamann, M. Schluter: Phys. Rev. B**26,** 4199 (1982)
2.148 G. A. Baraff, M. Schlueter: Phys. Rev. Lett. **41**, 892 (1978)
2.149 J. Bernholc, N. O. Lipari, S. T. Pantelides: ibid. p. 895
2.150 J. Bernholc, S. T. Pantelides: Phys. Rev. B **18**, 1780 (1978)
2.151 G. A. Baraff, M. Schlueter: Phys. Rev. B **19**, 4965 (1979)
2.152 K. H. Johnson, H. J. Kolari, J. P. deNeufville, D. L. Morel: Phys. Rev. B **21**, 643 (1980)
2.153 J. Korringa: Physica **13**, 392 (1947);
      W. Kohn, N. Rostoker: Phys. Rev. **94**, 1111 (1954)
2.154 A. C. Kenton, M. W. Ribarsky: Phys. Rev. B **23**, 2897 (1981)
2.155 A. Redondo, W. A. Goddard III, T. C. McGill, G. T. Surrat: Solid State Commun. **20**, 733 (1976)
2.156 W. A. Goddard III, T. H. Dunning, Jr., W. J. Hunt, P. J. Hay: Accounts Chem. Res. **6**, 368 (1963)
2.157 W. A. Goddard III, J. J. Barton, A. Redondo, T. C. McGill: J. Vac. Sci. Technol. **15**, 1274 (1978)
2.158 W. Y. Ching: Phys. Rev. B **22**, 2816 (1980)
2.159 W. Y. Ching, D. J. Lam, C. C. Lin: Phys. Rev. Lett. **42**, 805 (1979)
2.160 W. Y. Ching, D. J. Lam, C. C. Lin: Phys. Rev. B **21**, 2378 (1980)
2.161 W. Y. Ching: J. Non-Cryst. Solids **35**, 61 (1980)
2.162 W. Y. Ching: AIP. Conf. Proc. **73**, 151 (1981)
2.163 S. T. Pantelides: Rev. Mod. Phys. **50**, 797 (1978); see p. 849
2.164 L. Guttman: Phys. Rev. B **23**, 1866 (1981)
2.165 C.-Y. Fong, L. Guttman: AIP. Conf. Proc. **73**, 125 (1981); L. Guttman, C.-Y. Fong: Phys. Rev. B**26,** 6756 (1982)
2.166 W. E. Pickett: Phys. Rev. B **23**, 6603 (1981)
2.167 W. Y. Ching, C. C. Lin: Phys. Rev. B **18**, 6829 (1978)
2.168 W. Y. Ching, C. C. Lin, D. L. Huber: Phys. Rev. B **14**, 620 (1976)
2.169 W. Y. Ching, C. C. Lin, L. Guttman: Phys. Rev. B **16**, 5488 (1977)
2.170 P. Soven: Phys. Rev. **156**, 809 (1967)
2.171 E. N. Economou, D. A. Papaconstantopoulos: Phys .Rev. B **23**, 2042 (1981)
2.172 D. A. Papaconstantopoulos, E. N. Economou: Phys. Rev. B **24**, 7233 (1981)
2.173 D. A. Papaconstantopoulos, E. N. Economou: AIP. Conf. Proc. **73**, 130 (1981); W. E. Pickett, D. A. Papaconstantopoulos, E. N. Economou: Phys. Rev. B**28,** 2232 (1983)
2.174 C. T. White, W. E. Carlos: Phys. Rev. B **24**, 3380 (1981)
2.175 B. Kramer, J. Treusch: In Proc. 12th Intern. Conf. on the Physics of Semiconductors, ed. by M. H. Pilkuhn (Teubner, Stuttgart 1974)
2.176 J. Treusch, B. Kramer: Solid State Commun. **14**, 169 (1974)
2.177 B. Kramer, K. Maschke, P. Thomas: Phys. Stat. Sol. B **48**, 635 (1971)
2.178 B. Kramer: Phys. Stat. Sol. **41**, 649 (1970)
2.179 D. P. DiVincenzo, J. Bernholc, M. H. Brodsky, N. O. Lipari, S. T. Pantelides: AIP. Conf. Proc. **73**, 156 (1981)
2.180 D. P. DiVincenzo, J. Bernholc, M. H. Brodsky: J. Phys. (Paris) C4 **42**, Suppl. 10, 137 (1981); Phys. Rev. B**28,** 3246 (1983)
2.181 J. D. Joannopoulos, F. Yndurain: Phys. Rev. B **10**, 5164 (1974)
2.182 F. Yndurain, J. D. Joannopoulos: Phys. Rev. B **11**, 2957 (1974)
2.183 D. C. Allan, J. D. Joannopoulos: AIP. Conf. Proc. **73**, 136 (1981)
2.184 J. D. Joannopoulos, D. C. Allan: *Adv. in Solid State Phys.,* Vol. 21, ed. by J. Treusch (Vieweg, Braunschweig 1981) p. 167
2.185 D. C. Allan, J. D. Joannopoulos: Phys. Rev. Lett. **44**, 43 (1980)
2.186 J. D. Joannopoulos: J. Non-Cryst. Solids **36**, 781 (1980)
2.187 J. D. Joannopoulos: J. Non-Cryst. Solids **32**, 241 (1979)

2.188 C. Domb: Adv. Phys. **9**, 145 (1960)
2.189 D. C. Allan: Ph. D. Thesis, Massachusetts Institute of Technology, Cambridge, MA (1982) (unpublished)
2.190 J. D. Joannopoulos: unpublished
2.191 K. C. Pandey, J. C. Phillips: Phys. Rev. B **13**, 750 (1976)
2.192 K. C. Pandey: Phys. Rev. B **14**, 1557 (1976)
2.193 K. C. Pandey, T. Sakuri, H. D. Hagstrum: Phys. Rev. Lett. **35**, 1728 (1975)
2.194 R. Butz, R. Memeo, H. Wagner: Phys. Rev. B **25**, 4327 (1982)
2.195 R. Karcher, L. Ley: Solid State Commun. **43**, 415 (1982)
2.196 F. Yndurain, E. Louis: Solid State Commun. **25**, 439 (1978)
2.197 D. C. Allan: unpublished
2.198 T. Sakuri, H. D. Hagstrum: Phys. Rev. B **14**, 1593 (1976)
2.199 K. J. Gruntz, L. Ley, R. L. Johnson: Phys. Rev. B **24**, 2069 (1981)
2.200 L. Ley, K. J. Gruntz, R. L. Johnson: AIP. Conf. Proc. **73**, 161 (1981)
2.201 M. H. Brodsky: Solid State Commun. **36**, 55 (1980)
2.202 D. A. Papaconstantopoulos, E. N. Economou: Phys. Rev. B **22**, 2903 (1980)
2.203 K. L. Ngai, C. T. White: J. Appl. Phys. **52**, 320 (1981)
2.204 *Handbook of Chemistry and Physics,* 61st ed., ed. by R. C. Weast (CRC, Boca Raton, FL 1980)
2.205 M. A. Ring, M. J. Puentes, H. E. O'Neal: J. Am. Chem. Soc. **92**, 4845 (1970)
2.206 K. Zellama, P. Germain, S. Squelard, B. Bourbon, J. Fontenille, R. Danielou: Phys. Rev. B **23**, 6648 (1981)
2.207 K. Zellama, P. Germain, C. Picard, B. Bourbon: J. Phys. (Paris) C4 **42**, Suppl. 10, 815 (1981)
2.208 M. H. Brodsky, M. A. Frisch, J. Z. Ziegler: Appl. Phys. Lett. **30**, 561 (1977)
2.209 D. E. Carlson, C. W. Magee: Appl. Phys. Lett. **33**, 81 (1978)
2.210 J. A. McMillan, E. M. Peterson: Thin Solid Films **63**, 189 (1979)
2.211 S. Oguz, R. W. Collins, M. A. Paesler, W. Paul: J. Non-Cryst. Solids **35**, 231 (1980)
2.212 S. Oguz, M. A. Paesler: Phys. Rev. B **22**, 6213 (1980)
2.213 D. K. Biegelsen, R. A. Street, C. C. Tsai, J. C. Knights: Phys. Rev. B **20**, 4839 (1979)
2.214 D. K. Biegelsen, R. A. Street, C. C. Tsai, J. C. Knights: J. Non-Cryst. Solids **35**, 285 (1980)
2.215 D. Vanderbilt, J. D. Joannopoulos: Rev. Lett. **49**, 823 (1982)
2.216 G. Watkins, J. Troxell: Phys. Rev. Lett. **44**, 593 (1980)
2.217 R. A. Street: Phys. Rev. Lett. **49**, 1187 (1982)
2.218 J. Robertson: Phys. Rev. B **28**, 4647, 4658, 4666 (1983)
2.219 J. F. Sadoc, R. Mosseri: Phil. Mag. B **45**, 467 (1982)
2.220 M. H. Brodsky, D. P. Di Vincenzo: 10th Conf. on Amorphous and Liquid Semiconductors, Tokyo (1983)
2.221 D. P. Di Vincenzo, R. Mosseri, M. H. Brodsky, J. F. Sadoc: "Long range structural and electronic coherence in amorphous semiconductors," unpublished

3. Photoemission and Optical Properties

Lothar Ley

With 61 Figures

This chapter deals with the information obtained from photoelectron spectroscopy (PES or photoemission) and optical spectroscopy about the electronic structure of amorphous silicon (a-Si).

3.1 Connection Between Photoemission and Optical Spectroscopy

The two techniques are complementary. Photoemission probes the energy distribution of occupied states and optical spectra are related to the transitions between occupied and empty state densities.

In a particular form of optical spectroscopy the initial state is a sharp core level. This soft x-ray absorption spectroscopy yields, therefore, a reasonable (within the resolution given by the core-level width) picture of the density of states of the conduction bands. In a special form of x-ray absorption spectroscopy, the so-called yield spectroscopy, the absorption coefficient is measured through the secondary electron emission that follows the filling of the core hole after the absorption process has taken place. The inverse process is x-ray emission spectroscopy in which a transition takes place between filled valence bands and an empty core level. Finally, the energy losses encountered by fast electrons (1 to several hundred $\times 10^3$ eV) are used to measure the plasmon energy which corresponds to the collective excitation of the valence electrons in a solid.

There are a number of excellent books and review articles that cover the optical properties of semiconductors in general [3.1–3] and those of amorphous semiconductors in particular [3.4–8]. The various techniques of soft x-ray spectroscopies have been reviewed by *Brown* [3.9] and *Kunz* [3.10]. Two volumes published in the same series as the present book deal with most aspects of photoemission [3.11, 12]. One chapter in [3.12] is devoted, in particular, to semiconductors and it contains a section on photoemission work in amorphous semiconductors [3.13].

Under these circumstances only a minimum of theoretical background is given in the next section. We shall then proceed to discuss the photoemission spectra and the results obtained from related spectroscopies such as yield spectroscopy (Sect. 3.3). Section 3.4 will be devoted to the optical spectra in the visible and near ultraviolett (UV).

Emphasis will always be on hydrogenated amorphous silicon (a-Si:H), though results for a-Si, a-Ge and amorphous alloys of silicon with fluorine are included to complete the picture where necessary. Since photoemission is a very surface sensitive method (sampling depth $\sim 4 \cdots 20$ Å), a section on the surface properties of amorphous silicon as far as it has been explored with PES is included (Sect. 3.3.16).

3.2 The Dielectric Constant

The linear response of a solid to electromagnetic radiation of frequency ω is determined by the complex dielectric function $\tilde{\varepsilon}(\omega) = \varepsilon_1(\omega) + i\varepsilon_2(\omega)$. $\tilde{\varepsilon}$, usually a tensor, reduces to a scalar in cubic and amorphous solids. The dielectric function is related to the refractive index n and the extinction coefficient κ through

$$\varepsilon_1 = n^2 - \kappa^2 \quad \text{and} \tag{3.1a}$$

$$\varepsilon_2 = 2n\kappa . \tag{3.1b}$$

The absorption coefficient $\alpha(\omega)$ is

$$\alpha(\omega) = \frac{2\pi}{n\lambda} \varepsilon_2(\omega) = \frac{\omega}{nc} \varepsilon_2(\omega) , \tag{3.1c}$$

where λ is the wavelength of light in vacuum. The real and imaginary parts of the dielectric constant are related to each other through the Kramers-Kronig dispersion relations [3.2, 3]

$$\varepsilon_1(\omega) - 1 = \frac{2}{\pi} \oint_0^\infty \frac{\varepsilon_2(\omega')\,\omega'}{\omega'^2 - \omega^2} d\omega' \tag{3.2a}$$

$$\varepsilon_2(\omega) = \frac{2}{\pi} \oint_0^\infty \frac{\varepsilon_1(\omega')\,\omega}{\omega^2 - \omega'^2} d\omega' , \tag{3.2b}$$

where $\oint$ represents the Cauchy principal of the integral.

Two sum rules follow from the Kramers-Kronig relations [3.14]. For $\omega \to 0$, (3.2a) yields

$$\varepsilon_1(0) - 1 \equiv \varepsilon_0 - 1 = n^2 - 1 = \frac{2}{\pi} \int_0^\infty \frac{\varepsilon_2(\omega')}{\omega'} d\omega' \tag{3.3}$$

where $n \equiv n(0)$ is the static refractive index.

For sufficiently high frequencies the electrons may be considered free and $\varepsilon_1(\omega)$ is given by the Drude formula

$$\varepsilon_1(\omega) = 1 + \Omega_p^2/\omega^2 \quad \text{for} \quad \omega \to \infty \, , \tag{3.4}$$

with Ω_p the plasmon frequency. Inserting (3.4) into (3.2b), we obtain the second sum rule:

$$\Omega_p^2 = \frac{2}{\pi} \int\limits_0^\infty \omega' \, \varepsilon_2(\omega') \, d\omega' \, . \tag{3.5}$$

The quantity Ω_p^2 is proportional to the *total* (core + valence) number of electrons N_T per unit volume V of the solid:

$$\Omega_p^2 = 4 \pi e^2 \cdot N_T/m \, , \tag{3.6}$$

where e and m are the charge and mass of an electron.

In the random phase approximation, the electronic contribution to $\varepsilon_2(\omega)$ is proportional to the sum of all optical transitions between occupied and empty states separated in energy by $\hbar\omega$:

$$\varepsilon_2(\omega) = \frac{4 \pi^2 e^2}{m^2 \omega^2} \frac{1}{V} \sum_{v,c} |P_{vc}^\mu|^2 \, \delta \, (E_c - E_v - \hbar\omega) \, , \tag{3.7}$$

where V is the volume of the specimen. For the optical response in the visible and near *UV* which is of interest here, the sum extends over occupied valence states $|v\rangle$ with energy E_v and empty conduction states $|c\rangle$ (energy E_c) which are separated by an energy $\hbar\omega$. P_{vc}^μ is the matrix element of one component μ of the momentum operator $\boldsymbol{p} = -i\hbar \nabla$ between states $\langle v|$ and $|c\rangle$:

$$P_{vc}^\mu = \langle v|e_\mu \cdot \boldsymbol{p}|c\rangle \, , \tag{3.8}$$

where e_μ is the unit vector in the direction μ. Unpolarized light averaging over all directions yields $|\langle P^\mu \rangle|^2 = \frac{1}{3}|\langle P \rangle|^2 \equiv \frac{1}{3} P^2$.

In a crystal, $|v\rangle$ and $|c\rangle$ are Bloch states and therefore eigenstates of the wave vector $\boldsymbol{k}$. The momentum matrix element (3.8) vanishes between states of different $\boldsymbol{k}$ (in the absence of phonons) and only direct or $\boldsymbol{k}$ conserving transitions contribute to $\varepsilon_2(\omega)$. Replacing the sum over v and c in (3.7) by an integral over $\boldsymbol{k}$ restricted to the Brillouin zone (BZ) and a summation over band indices i and j, we obtain

$$\varepsilon_2(\omega) = \frac{4 \pi^2 e^2}{m^2 \omega^2} \frac{1}{3} \sum_{i,j} \int\limits_{BZ} \frac{2}{(2\pi)^3} d^3k \, P_{ij}(\boldsymbol{k})^2 \, \delta(\omega_{ij}(\boldsymbol{k}) - \omega) \, , \tag{3.9}$$

where $\hbar\omega_{ij} = E_j - E_i$ and it is understood that bands with index i are occupied and those with index j are empty. In a commonly made approximation, the variation of $P_{ij}(\boldsymbol{k})$ with $\boldsymbol{k}$ and the band indices i, j are neglected and $P_{ij}(\boldsymbol{k})$ is replaced by an average $P(\omega)$ which is a function of transition frequency ω only. $P(\omega)$ can then be taken out of the integral:

$$\varepsilon_2(\omega) = \left(\frac{2\pi e}{m\omega}\right)^2 \frac{1}{3} P^2(\omega) \sum_{i,j} \int_{BZ} \frac{2}{(2\pi)^3} d^3k\, \delta(\omega_j(\boldsymbol{k}) - \omega_i(\boldsymbol{k}) - \omega) \qquad (3.10\,\text{a})$$

$$= \left(\frac{2\pi e}{m\omega}\right)^2 \frac{1}{3} P^2(\omega) \sum_{i,j} \int_{\omega_j - \omega_i = \omega} \frac{2}{(2\pi)^3} \frac{dS_k}{|\nabla_k(\omega_j - \omega_i)|} \qquad (3.10\,\text{b})$$

$$= \left(\frac{2\pi e}{m\omega}\right)^2 \frac{1}{3} P^2(\omega) J_{vc}(\omega) . \qquad (3.10\,\text{c})$$

In (3.10 b) we have replaced the volume integral by an integral over the surface of constant energy difference $\omega_{ji} = \omega$ in $\boldsymbol{k}$-space. Equation 3.10 a–c define $J_{vc}(\omega)$, the optical or joint density of states (JDOS). It represents the number of states which can undergo energy and $\boldsymbol{k}$-conserving transitions for photon frequencies between ω and $\omega + d\omega$. The joint density of states exhibits singularities whenever the denominator in (3.10 b) is zero and these

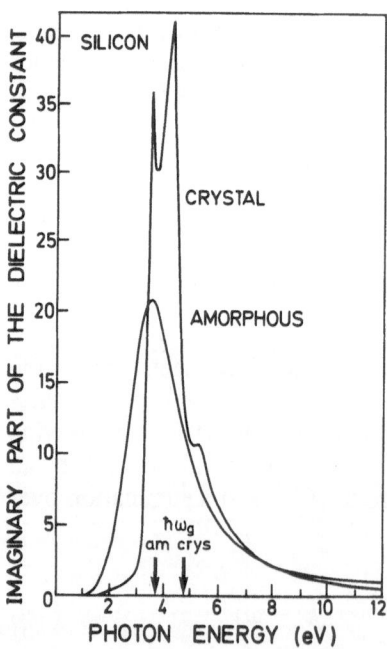

Fig. 3.1. The ε_2 spectrum of crystalline and unhydrogenated amorphous silicon [3.15]. The arrows indicate the Penn gaps $\hbar\omega_g$

van Hove singularities [3.1, 3] account for the structure in $\varepsilon_2(\omega)$ of crystalline semiconductors as illustrated for c-Si in Fig. 3.1 [3.15]. The identification of *structures* (peaks and edges) in the experimental ε_2 spectrum with the corresponding features in $J_{vc}(\omega)$ obtained from band structure calculations is indeed one of the main sources of information about the electronic structure of crystalline semiconductors [3.1–3]. The magnitude and energy dependence of $P(\omega)$ is, by comparison, of secondary importance as long as $P(\omega)$ varies slowly with ω compared to $J_{vc}(\omega)$.

Through the loss of translational symmetry in an amorphous semiconductor, the wave vector k is no longer a good quantum number that needs to be conserved in an optical transition. The appropriate expression for $\varepsilon_2(\omega)$ is therefore (3.7). The evaluation of the momentum matrix element requires, however, the knowledge of all wave functions in a disordered system – a problem that has yet to be solved.

Instead, an expression analogous to (3.10c) is used [3.16, 17]:

$$\varepsilon_2(\omega) = \left(\frac{2\pi e}{m\omega}\right)^2 P_{am}^2(\omega)\, a^3 \times \int dE\, N_v(E)\, N_c(E + \hbar\omega) \tag{3.11}$$

in which the joint density of states of the crystalline ε_2 (ω) has been replaced by a simple convolution of occupied valence and empty conduction densities of states $N_v(E)$ and $N_c(E)$, respectively. Expression (3.11) is what one intuitively expects instead of (3.10) for transitions that are subject only to energy conservation but no longer to k conservation.

The state densities $N_v(E)$ and $N_c(E)$ remain to be well-defined quantities even in cases without translational symmetry. They have been calculated for various models of the amorphous structure as described in Chapter 2. $N_v(E)$ and $N_c(E)$ are also obtained directly from photoemission and yield spectroscopy (Sects. 3.3.7–3.3.10) and we shall calculate the ε_2 spectrum of a-Si using (3.11) and the experimental state densities in Sect. 3.4.2. The relaxation of k conservation accounts for the main differences in the optical spectra of amorphous and crystalline Si: the enhanced absorption below 3 eV and the loss of fine structure in $\varepsilon_2(\omega)$ of a-Si (Fig. 3.1). In c-Si, transitions between 1.1 and 3.4 eV are indirect, i.e., they take place between states of different k vector and only the simultaneous absorption or emission of a phonon allows for a weak absorption in that region [3.1]. The lack of k conservation in a-Si makes these transitions quasi-allowed. The van Hove singularities are absent in a-Si and the convolution of N_v and N_c in (3.11) smears out whatever structure is left in the state densities. It is, however, apparent from Fig. 3.1 that $\varepsilon_2(\omega)$ of a-Si is not just a broadened version of the dielectric function of c-Si. The difference can be traced to differences in the density of valence states (Sect. 3.3.7) which also account for the red shift in the maximum of ε_2 of a-Si (Sect. 3.4.2a).

A derivation of (3.11) has been given by *Tauc* et al. [3.4, 16] under the assumption that the states of the amorphous solid can be expanded in terms of crystal wave functions. Under these circumstances, (3.11) holds for transitions between delocalized states with $|P_{am}(\omega)|^2 \simeq |P_{cryst}(\omega)|^2$.

If one of the states in (3.11) is localized over a volume $V(E)$, Tauc obtained in his derivation an enhancement of the transition probability proportional to $V(E)/a^3$, where a is the nearest-neighbor distance. This view has been challenged by *Davis* and *Mott* [3.18] who argue that the factor $V(E)/a^3$ is exactly offset by the inverse factor entering in $|P_{am}(\omega)|^2$ through the phase randomization of the delocalized states. The arguments of *Davis* and *Mott* appear to be generally accepted.

Expression (3.11) will therefore be used to evaluate $\varepsilon_2(\omega)$ in amorphous semiconductors if at least one of the states is delocalized, despite the fact that a proof of (3.11) under less restricting assumptions than those made by Tauc is still lacking.

If initial *and* final states are both localized in the tails of the state densities, the transition matrix element will depend on the spatial separation and the degree of localization of the two states. In the extreme case that the spatial extension of the wave functions is small compared to their average separation, $\varepsilon_2(\omega)$ will vanish. The form of the optical constant in the presence of spatial correlation has been considered by a number of authors [3.19–22]. We shall return to this point in connection with the discussion of the Urbach tail in Sect. 3.4.4.

3.2.1 Sum Rules and Plasmons

With little modifications, sum rules are derived from (3.3, 5) which involve only integrations over the fundamental part of the absorption spectrum, i.e. transitions between valence and conduction electrons [3.23]. Equation (3.3) is replaced by

$$\varepsilon_{1,\infty} \equiv \varepsilon_\infty = n_\infty^2 = 1 + \frac{2}{\pi} \int\limits_{\omega_t}^{\omega_m} \frac{\varepsilon_2(\omega')}{\omega'} d\omega' \; . \tag{3.12}$$

Integration extends from ω_t, the threshold of interband transitions to an energy $\hbar\omega_m$ that is large compared to the width of the $\varepsilon_2(\omega)$ spectrum. Equation 3.12 defines the long wavelength limit of the dielectric constant ε_∞ which differs from ε_0 by the contributions of infrared active lattice vibrations to $\varepsilon_2(\omega)$ that are excluded from the integral in (3.12). Lattice modes are not infrared active in crystalline homopolar semiconductors such as Si and Ge. In a-Si, disorder induces static charges on the Si atoms (see Sect. 3.3.14) and the lattice modes become infrared active [3.24]. The long wavelength refractive index $n_\infty = \sqrt{\varepsilon_\infty}$ is therefore determined at infrared frequencies that are just above the lattice modes ($\gtrsim 500$ cm^{-1}) (see Sect. 3.4). The sum rule (3.12) is exhausted for $\hbar\omega_m \simeq 20$ eV in most semiconductors [3.23]. The contribution from core levels to $\varepsilon_\infty(\omega_m \to \infty)$ is negligible in Si where the threshold for core level excitation is ~ 100 eV. In Ge, transitions from the $3d$ levels set in at ~ 30 eV but their contribution to ε_∞ amounts to a mere 3% [3.23] due to the energy denominator in (3.12).

If the integral (3.5) is evaluated to a frequency ω,

$$\frac{2}{\pi} \int_0^\omega d\omega' \omega' \varepsilon_2(\omega') = \frac{4\pi e^2 N_{\text{eff}}(\omega)}{m} = \frac{4\pi e^2 \varrho}{mM} n_{\text{eff}}(\omega) , \tag{3.13}$$

one can express the result of the integration in terms of n_{eff}, the effective number of valence electrons per atom that contribute to ε_2 in the frequency interval from zero to ω. The relationship between n_{eff} and the electron density N_{eff} is given through the atom density ϱ/M according to $N_{\text{eff}} = (\varrho/M) \cdot n_{\text{eff}} = \varrho \cdot (L_A/A) \cdot n_{\text{eff}}$, where ϱ is the density of the material, M the atomic mass, A the atomic weight, and L_A Avogadro's number (6.022×10^{23} mol^{-1}).

In c-Si, the integral (3.13) reaches a plateau for $\hbar\omega \simeq 20$–25 eV with a value of $n_{\text{eff}} = n_v = 4$, the number of valence electrons per atom [3.23]. In semiconductors with low-lying core levels such as Ge, no plateau is observed and n_{eff} increases smoothly beyond n_v, even for $\hbar\omega$ smaller than the actual onset of core-level transitions. This is due to the so-called transfer of oscillator strength [3.2].

Inserting N_v, the valence electron density, into (3.6) we obtain the valence plasmon energy $\hbar\omega_p$ which is usually referred to as simply *the* plasmon energy:

$$\omega_p^2 = \frac{4\pi e^2}{m} \frac{\varrho L_A}{A} \cdot n_v . \tag{3.14}$$

The calculated values for $\hbar\omega_p$ using (3.14) are 16.5 eV for c-Si, 15.7 eV for c-Ge, 31 eV for diamond and 12.6 eV for graphite.

Plasmons are collective excitations of the electron gas. Electrons interact strongly with these excitations and they give rise to maxima in the electron energy loss spectrum of fast electrons traversing a sample [3.25]. They show also up as loss peaks on the low energy side of sharp structures in photoelectron spectra. The probability that an electron suffers an energy loss $\hbar\omega$ is proportional to $-\text{Im}\{1/\tilde{\varepsilon}(\omega)\} = \varepsilon_2(\omega)/[\varepsilon_1^2(\omega) + \varepsilon_2^2(\omega)]$. The loss function has a strong peak at an energy E_p which occurs when plasmons are excited (see Fig. 3.2). E_p coincides with the plasmon energy of (3.14) only in the free-electron limit (Drude limit), i.e., in the absence of interference from nearby interband transitions [$\varepsilon_2(\omega_p) \simeq 0$]. The comparison in Table 3.1 shows that Si fulfills this requirement quite well. In Ge, E_p is slightly higher (by ~ 0.5 eV) than $\hbar\omega_p$ calculated using (3.14). In this case $\hbar\omega_p$ can be made to agree with E_p by replacing the free electron mass m in (3.14) by an optical mass m^*.

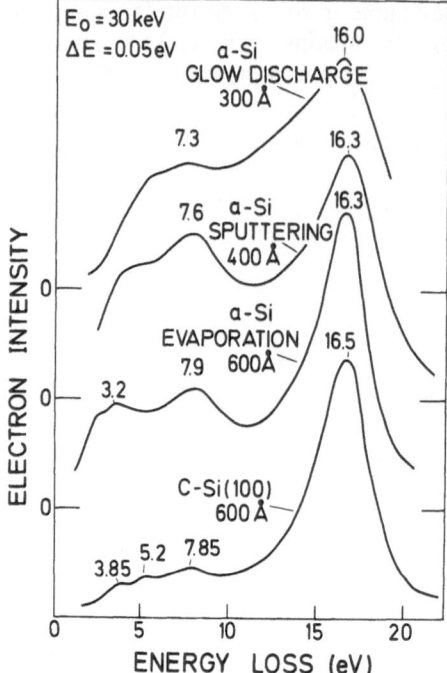

Fig. 3.2. High energy electron loss spectra of three amorphous Si films and of c-Si. The dominant feature is the plasmon loss around 16.4 eV [B. Schröder: private communication]

In amorphous Si und Ge we expect rather drastic reductions in the value of $\hbar\omega_p$ and therefore E_p, depending on the densities of the amorphous films which can be as low as 75% of the crystalline density (for films evaporated at room temperature) [3.26–31]. Instead, the mere 2% reduction observed in a-Si is rather independent of preparation conditions and would correspond to a density deficit of only ~4% according to (3.14) (see Table 3.1).

In a-Ge the situation is similar. In one case [3.30] the density of the a-Ge film was measured and $\hbar\omega_p$ (a-Ge) calculated to be 14.8 eV compared to 15.7 eV for c-Ge. The measured values E_p of 16.3 (c-Ge) and 15.9 eV (a-Ge) reflect only half the density loss. We must therefore assume that the plasmons are mainly localized in the dense regions of the amorphous network and penetrate only little into the voids which are mainly responsible for the density deficit in a-Si and a-Ge [3.32–33] and also in a-Si:H [3.34].

The plasmon energy may thus serve as a measure of the *microscopic* density of an amorphous material. *Katayama* et al. have analyzed plasmon energies in a-Si_xC_{1-x}:H alloys in this sense [3.35]. They were able to deduce from the variation in plasmon energy with carbon concentration a change from 3-fold (graphitic) to 4-fold (diamond) coordination as the Si concentration approaches $x \simeq 0.6$.

Table 3.1. Loss energies E_p measured in electron loss experiments compared with free electron plasmon energies $\hbar\omega_p$ in silicon and germanium

Sample	Preparation			E_p[eV]		$\hbar\omega_p$[eV]	Ref.
c-Si				16.45 ± 0.1		16.5	a
c-Si				16.5 ± 0.1			b
a-Si	evaporated,	T_D = RT		16.1 ± 0.1			a
	evaporated,	T_D = RT		16.4 ± 0.1			b
	evaporated,	T_D = RT		16.4 ± 0.3			c
				16.3 ± 0.2			f
	sputtered,	T_D = RT		16.3 ± 0.2			f
	sputtered,	T_D = RT		17.0 ± 0.1			g
a-Si:H	gd	T_D = RT		16.0 ± 0.2			f
a-Si:F	sputtered,	T_D = RT		17.0 ± 0.2			g
c-Ge				16.0 ± 0.2		15.7	c
c-Ge				16.2 ± 0.1			e
c-Ge				16.5			d
a-Ge	evaporated,	T_D = RT		15.9		14.8	d
	evaporated,	T_D = RT		15.5 ± 0.1			b
	evaporated,	T_D = RT		15.8 ± 0.2			c

[a] K. Zeppenfeld, H. Raether: Z. Phys. **193**, 471 (1966)
[b] T. Aiyama, K. Yada: J. Phys. Soc. Jpn. **36**, 1554 (1974)
[c] O. Sueoka: J. Phys. Soc. Jpn. **20**, 2203 (1965)
[d] H. Richter, A. Rukwied: Z. Phys. **160**, 473 (1960)
[e] H. J. Hintz, H. Raether: Thin Solid Films **58**, 281 (1979)
[f] B. Schröder: (private communication)
[g] K. Gruntz: Dissertation, University of Stuttgart (1981)

3.2.2 The Penn Model and Other Parametrization Schemes

The sum rules (3.12, 13) provide convenient checks on the consistency of measured ε_2 spectra. They relate moments M_r of the $\varepsilon_2(\omega)$ spectrum to macroscopic quantities like the density ϱ or the refractive index n_∞ and to microscopic quantities like n_v:

$$\varepsilon_\infty = 1 + M_{-1} \quad \text{and} \tag{3.15}$$

$$\omega_p^2 = M_1 \, , \quad \text{where} \tag{3.16}$$

$$M_r = \frac{2}{\pi} \int_{\omega_t}^{\omega_m} \omega^r \varepsilon_2(\omega)\, d\omega \, . \tag{3.17}$$

These integral relationships do not depend on the details of the shape of $\varepsilon_2(\omega)$ and on the presence or absence of critical points and they are, therefore, particularly useful for the comparison of crystalline and amorphous modifications of the same material. Equations 3.15, 16 constitute thus parametrization schemes for $\varepsilon_2(\omega)$.

Another widely used parameter is the Penn gap $\hbar\omega_g$ defined through [3.36, 37]

$$\varepsilon_\infty = 1 + \omega_p^2/\omega_g^2 \, , \tag{3.18}$$

Equation (3.18) represents the single oscillator approximation to the dielectric response of a solid. The distribution of transition energies has been replaced by one energy $\hbar\omega_g$ which represents the *average* separation between valence and conduction electrons, i.e., between bonding and antibonding states. Again, ω_g may be expressed in terms of moments M_r of ε_2 using (3.15, 16):

$$\omega_g^2 = M_1/M_{-1} \, . \tag{3.19}$$

The quantity ω_g was introduced by *Penn* in an isotropic model for the band structure of a semiconductor [3.38]. It is assumed that the solid has a spherical Brillouin zone with an isotropic gap $\hbar\omega_g$ at its boundary. The states are supposed to be free electron-like except for the existence of this gap which separates valence from conduction states. The dielectric constant $\varepsilon_2(\omega)$ starts with a singularity at ω_g and falls off proportional to ω^{-2} for $\omega > \omega_g$. It fulfills the sum rule (3.6) [3.2].

While the state densities N_v and N_c of the Penn model are a rather poor representation of the valence and conduction bands in a-Si and a-Ge, it turns out that a slightly broadened version of the Penn dielectric constant is actually a fair representation of the singly-peaked dielectric function of a-Si (Fig. 3.1) with its steep leading edge and its tail towards higher energy [3.39]. To a good approximation, $\hbar\omega_g$ coincides with the maximum of the $\varepsilon_2(\omega)$ spectrum.

The Penn gap $\hbar\omega_g$ plays the pivotal role in the theory of ionicity due to *Phillips* and *van Vechten* [5.36, 37] where it is a measure of the average bond strength that can be sensibly divided into a covalent and an ionic contribution. The Penn gaps of c-Si and c-Ge derived from (3.18) are 4.8 and 4.3 eV, respectively. The position for c-Si is marked in Fig. 3.1. The reduction in $\hbar\omega_g$ to ~ 3.7 eV in the a-Si spectrum of Fig. 3.1 must be attributed, according to the model, to a weakening of the Si–Si bonds in a-Si (see also Sect. 3.4.2).

Another parametrization scheme of $\varepsilon_2(\omega)$ introduced by *Wemple* and *DiDominico* [3.40, 41] and applied to a-Si and a-Si : H [3.42] considers two energies. The oscillator energy $\hbar\omega_0$ defined through

$$\omega_0^2 = M_{-1}/M_{-3} \tag{3.20}$$

takes the place of $\hbar\omega_g$. The definition of ω_g and ω_0 as *ratios* of two momenta of $\varepsilon_2(\omega)$ of equal power makes these quantities independent of the amplitude of $\varepsilon_2(\omega)$ which is determined by the magnitude of the transition matrix element. The second energy $\hbar\omega_d$ defined by *Wemple* and *DiDominico*

$$\omega_d^2 = M_{-1}^3/M_{-3} \tag{3.21}$$

takes such variations in the oscillator strength of interband transitions into account.

3.3 Photoemission

3.3.1 The Three-Step Model

In a photoemission experiment, a beam of monochromatic light of energy $\hbar\omega$ impinges on the sample surface and the energy distribution $I(E, \omega)$ of the emerging photoelectrons is measured by means of an electrostatic energy analyzer. The spectra so obtained will be interpreted throughout this chapter in terms of the three-step model first proposed by *Spicer* [3.43]. Rigorous theories of photoemission which may be cast into the three-step model are found in [3.44–46].

In the three-step model, photoemission is treated as a sequence of

i)   the optical excitation of an electron,
ii) its transport through the solid which includes the possibility for inelastic scattering by other electrons, and finally,
iii) the escape through the sample surface into the vacuum.

The energy distribution curve (EDC) of the photoemitted electrons $I(E, \omega)$ is consequently a sum of a primary distribution of electrons $I_p(E, \omega)$ that have not suffered an inelastic collision and a background of secondary electrons $I_s(E, \omega)$ due to electrons that have undergone one or more inelastic collisions:

$$I(E, \omega) = I_p(E, \omega) + I_s(E, \omega) . \tag{3.22}$$

The primary distribution is factorized according to the three-step model into a distribution of photoexcited electrons $J(E, \omega)$, a transmission function $T(E)$, and an escape function $D(E)$:

$$I_p(E, \omega) = J(E, \omega) \cdot T(E) \cdot D(E) . \tag{3.23}$$

Under the assumption that the inelastic scattering probability can be characterized by an isotropic electron mean free path $\lambda_e(E)$, $T(E)$ is given by

$$T(E) = \frac{\lambda_e(E)/\lambda_{ph}(\omega)}{1 + \lambda_e(E)/\lambda_{ph}(\omega)} . \tag{3.24}$$

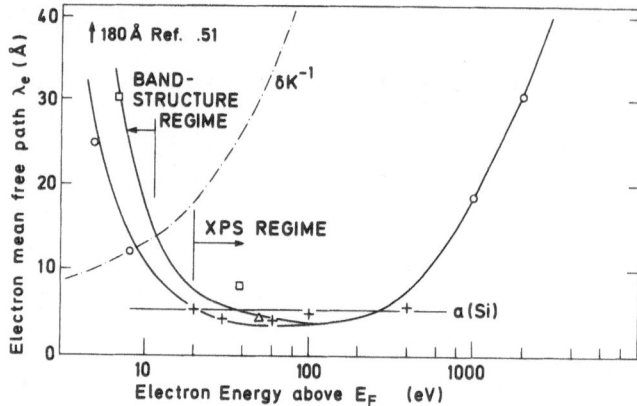

Fig. 3.3. The electron mean free path λ_e in silicon as a function of electron energy. Data points are from [3.47–51]. The high energy portion of λ_e ($E \gtrsim 500$ eV) is from [3.52]. The two curves below 100 eV reflect the scatter in measured values of λ_e. The line labelled a(Si) marks the lattice constant of c-Si. We indicate in the curve two important regimes of photoemission. The XPS regime, in which the density of valence states is basically investigated, and the band structure regime, where the effects of k conservation in the excitation process are dominant. The meaning of δk^{-1} is explained in the text

The photon penetration depth $\lambda_{ph}(\omega)$ exceeds hundred Ångströms, whereas $\lambda_e(E)$ is given by the graph in Fig. 3.3. Figure 3.3 is based on $\lambda_e(E)$ values measured for Si by a number of authors [3.47–52]. The same energy dependence of λ_e holds with little deviations for most materials and Fig. 3.3 thus represents what is termed the "universal" curve of $\lambda_e(E)$ [3.52]. For the electron energies of interest here (10 eV $\lesssim E \lesssim$ 1000 eV), λ_e is seen to vary between about 20 Å and 4 Å so that $T(E)$ is $\lesssim 0.05$. The pronounced minimum in λ_e around 50 to 200 eV is due to the onset of plasmon excitations which constitute the main loss mechanism for photoelectrons with energies above the plasmon excitation threshold of 15–20 eV. Below the plasmon onset, electron-hole scattering determines λ_e and the diminishing phase space reduces the scattering probability with decreasing E [3.51].

The short mean free path of the electrons effectively limits the sampling depth in PES to ~4 to 20 atomic layers. It should be kept in mind, however, that the genuine surface electronic structure extends no more than 2 to 3 atomic layers into the bulk so that EDC's will normally represent largely bulk properties [3.53, 54]. On the other hand, with the proper choice of E around the minimum in λ_e, PES can be turned into a sensitive probe of surface electronic structure.

The factor $D(E)$ takes into account the fact that the photoelectrons have to overcome a surface barrier before they emerge into the vacuum. This confines the emerging electrons to an escape cone with an opening angle 2θ given by

$$\cos \theta = \sqrt{B/E} \; , \tag{3.25}$$

where B, the barrier, is equal to the work function ϕ of the sample plus the Fermi energy E_F measured relative to the inner potential of the specimen.

The gist of the three-step model is that $T(E)$ and $D(E)$ are slowly-varying functions of E so that the *structure* in the primary photocurrent $I_\mathrm{p}(E, \omega)$ reflects the structure in $J(E, \omega)$. It is the latter quantity, therefore, which we shall consider exclusively when we refer to the EDC or photoemission spectrum in what follows.

3.3.2 The Energy Distribution Curve (EDC) and the Density of States (DOS)

Photoemission and optical absorption are closely related. In an optical absorption experiment we ask for the probability that an electron makes a transition between two states separated by an energy $\hbar\omega$ and then sum over all possible transitions with the same energy difference. In photoemission we choose one energy $\hbar\omega$ and pick from all possible transitions only those that lead to final states with an energy E, where E is the energy set by our electron analyzer. We thus recognize in the formula for the primary photocurrent

$$J(E, \omega) \propto \sum_{n, n'} |P_{nn'}|^2 \, \delta(E_n - E_{n'} - \hbar\omega) \, \delta(E - E_{n'}) \tag{3.26}$$

an expression analogous to $\varepsilon_2(\omega)$ (3.7) augmented by a δ-function that selects the proper final state energy E. In the absence of this δ-function and apart from constants, (3.26) reduces to $\omega^2 \varepsilon_2(\omega)$.

Following the argument given in Sect. 3.2.2, we arrive at the following expression for $J(E, \omega)$ in amorphous solids:

$$J(E, \omega) \propto |P(\omega)|^2 \cdot N_\mathrm{v}(E - \hbar\omega) \cdot N_\mathrm{c}(E) \; . \tag{3.27}$$

That is, the photoelectron spectrum $J(E)$ for a fixed ω is a direct measure of the density of valence states $N_\mathrm{v}(E - \omega)$ weighted by $|P(\omega)|^2$ and the density of conduction states at energy E.

For crystalline materials the corresponding expression is

$$J(E, \omega) \propto \sum_{n, n} |P_{n, n'}(\omega)|^2 \int_{BZ} d^3k \, \delta(E_n(k) - E_{n'}(k) - \hbar\omega)$$
$$\times \delta(E_{n'}(k) - E) \; , \tag{3.28}$$

or evaluating the δ-functions and assuming that P^2 is independent of n and n',

$$J(E, \omega) \propto |P(\omega)|^2 \sum_{n, n'} \int_L \frac{dl_{nn'}}{|\nabla_k E_{n'}(k) \cdot \nabla_k E_n(k)|} \; . \tag{3.29}$$

The integration is performed along that line L in k-space where the two surfaces given by $[E_{n'}(k) - E_n(k)] = \hbar\omega$ and $E_{n'}(k) = E$ intersect. The sum on the right-hand side of (3.29) without the matrix element defines the so-called energy distribution of the joint density of states (EDJDOS). Momentum conservation and the two δ-functions in (3.28) impose rather stringent conditions on the photoemission spectrum of crystalline materials. A peak in $J(E, \omega)$ requires that a direct transition with energy $\hbar\omega$ is allowed which leads to a final state with energy E. These conditions which constitute the band-structure regime of photoemission [3.55] (Fig. 3.3) prevail only as long as k conservation holds, i.e., as long as optical transitions take place in an effectively infinite crystal.

The finite $\lambda_e(E)$ produces a smearing of the component of k perpendicular to the surface, $k_\perp$, which can be interpreted as an imaginary part in $k_\perp = k_\perp^{(1)} + ik_\perp^{(2)}$ with $k_\perp^{(2)} \simeq \lambda_e^{-1}$. The component $k_\perp$ will therefore cease to be conserved if $k_\perp^{(2)}$ approaches the linear dimension of the Brillouin zone. The contribution to $J(E, \omega)$ in (3.28, 29) for a given direction $k_{||}$ yields under the assumption of a constant $|P|^2$ the one-dimensional density of states ($k_\perp$ varies along $k_{||}$) [3.56]. Averaging over all directions of k produces the density of valence states. The averaging is automatically performed in polycrystalline samples so that under these circumstances we are dealing, in effect, with indirect transitions just as in amorphous materials. This regime is called the XPS (x-ray photoemission spectroscopy) regime for historical reasons and it starts according to Fig. 3.3 at about 20 eV photon energy where $\lambda_e(E)$ becomes comparable to the lattice constant of c-Si. The band-structure regime holds for $\hbar\omega \lesssim 11$ eV. The region between 11 and 20 eV is to be considered a transition region.

Beyond ~ 200 eV, where the electron mean free path starts to rise again and $k_\perp^{(2)}$ decreases proportionally, another mechanism ensures that we can describe the photoemission spectrum in terms of the indirect transition model [3.57]. At high electron energies, the quasi-free electron bands form a tight mesh in k-space as they are folded over into the reduced zone. The average separation δk in k-space of energy-degenerate final states decreases rapidly with energy. Using as an example a spherical-zone empty lattice model with a lattice constant a, Grobman et al. [3.57] obtained (in atomic units)

$$\delta k \simeq \frac{1}{E} \frac{(2\pi)^3}{3a^3} \left(\frac{3}{\pi}\right)^{3/2}. \tag{3.30}$$

The energy dependence of $(\delta k)^{-1}$ so estimated for Si is plotted as the dash-dot line in Fig. 3.3. The relation $(\delta k)^{-1} \gg \lambda_e$ is seen to hold also for $\hbar\omega > 200$ eV and there will, therefore, always be a final state of energy E that has the correct k. This means, in effect, that we can drop the requirement of k-conservation, and $J(E, \omega)$ for polycrystalline samples is given for $\hbar\omega \gtrsim 20$ eV by

$$J(E, \omega) \propto |P|^2 \, N_v(E - \hbar\omega) \, N_c(\omega) \ , \tag{3.31}$$

i.e., an expression identical to that for amorphous semiconductors. We are thus in a position to compare the density of valence states of amorphous and crystalline materials obtained in photoemission as long as we use photon energies in excess of about 20 eV.

For higher photon energies, core levels will also be excited and contribute to the photoemission spectrum in the form of sharp peaks. Core levels, as localized states, are, of course, without dispersion and their spectral shape is – aside from experimental contributions – given by a Lorentzian with a width Γ determined by the lifetime τ of the hole left behind after photoexcitation according to $\Gamma = \hbar/\tau$ [3.58, 59].

3.3.3 Photoemission Cross Sections

We have so far simplified our discussion of $J(E, \omega)$ by assuming that the matrix elements $|P_{n'n}(\omega)|^2$ can be considered constant and taken out of the summation in (3.28). This will not be the case in general and the EDC's will represent the density of valence states modulated by a factor which corresponds to an average $|P_{n'n}(\omega)|^2$ where the average is taken over all initial states n with the same energy. For the photon energies normally used in photoemission ($\hbar\omega > 20$ eV), the final state electrons have wavelengths which are short compared to interatomic distances. The matrix elements are therefore mainly determined by the rapidly varying segments of the initial state wave functions close to the atom core rather than the smooth interatomic portions. The inner parts are little influenced by the bonding environment and the matrix elements depend, consequently, on the atomic character ($2s$, $3s$, $3p$, etc.) of the wave functions [3.60, 61]. Imagine, therefore, that the density of valence states of Si is partitioned into its partial densities $N_s(E)$ and $N_p(E)$ due to the $3s$ and $3p$ electrons, respectively. Then

$$J(E, \omega) \propto \sigma_s(\omega) \, N_s(E) + \sigma_p(\omega) \, N_p(E) \ , \tag{3.32}$$

where $\sigma_{s,p}(\omega)$ are the energy dependent photoionization cross sections for the $3s$ and $3p$ atomic levels of silicon. Equation 3.32 can be generalized to more than two atomic orbitals. When the σ's are known as a function of ω, (3.32) may be used to obtain the partial densities of states from measurements of the EDC's for a series of values of $\hbar\omega$. This has been done for a-Ge:H, where the partial cross sections are reasonably well known, to deduce the relative contribution of H $1s$ and Ge $4p$ states to hydrogen induced peaks in the valence bands [3.62].

The atomic cross sections σ for valence and core electrons of the elements have been calculated [3.63–65]. They have also been obtained from measurements on solids, atoms, or simple molecules [3.61, 66, 67]. For the valence

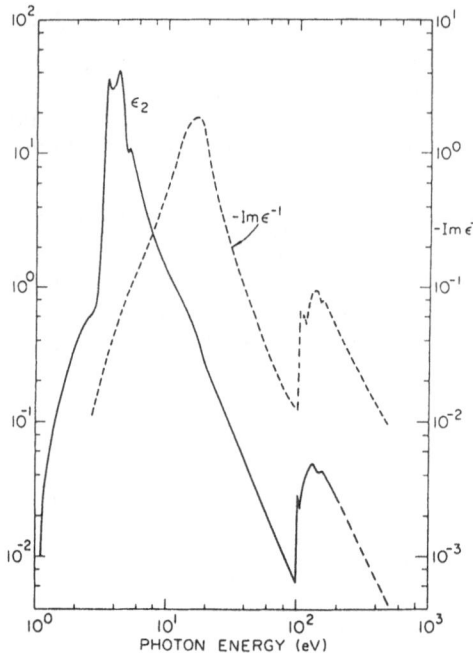

Fig. 3.4. Imaginary part of the dielectric function (ε_2) and the loss function $-$ Im $\{\bar{\varepsilon}^{-1}\}$ for silicon from 1 to 1000 eV. Notice the difference between $\varepsilon_2(\omega)$ and the loss function for energies below 50 eV and their similarity above that energy [3.9]

electrons of Si and Ge, the ratios $\sigma_{3s}/\sigma_{3p} = 3.4$ and $\sigma_{4s}/\sigma_{4p} = 1.0$ at $\hbar\omega = 1486.6$ eV were so obtained [3.67].

The integrated intensity of the EDC is, of course, proportional to $\varepsilon_2(\omega)$. Figure 3.4 shows $\varepsilon_2(\omega)$ for Si over a wide energy range of photon energies. The dielectric constant drops rapidly ($\sim \omega^{-2}$) with energy, except for the onset of transitions from the Si 2p and 2s core lines at 100 and 140 eV, respectively [3.68, 69] (see Sect. 3.3.10).

3.3.4 Binding Energies, Work Function, and Photoemission Threshold

The results of the previous sections are schematically summarized in the energy level diagram of Fig. 3.5. In an EDC, energies are given as binding energies E_B. The initial state or binding energy E_B is related to the measured kinetic energy E_{kin} through

$$E_B = \hbar\omega - E_{kin} - \phi_{sp} , \tag{3.33}$$

where ϕ_{sp} is the work function of the spectrometer. With this choice of energy scale, the zero of binding energy corresponds to the Fermi level E_F of the spectrometer. For conducting samples, the Fermi levels of sample and spectrometer line up and the zero of energy always coincides with E_F (sample). The position of E_F is easily determined from the sharp cut-off observed

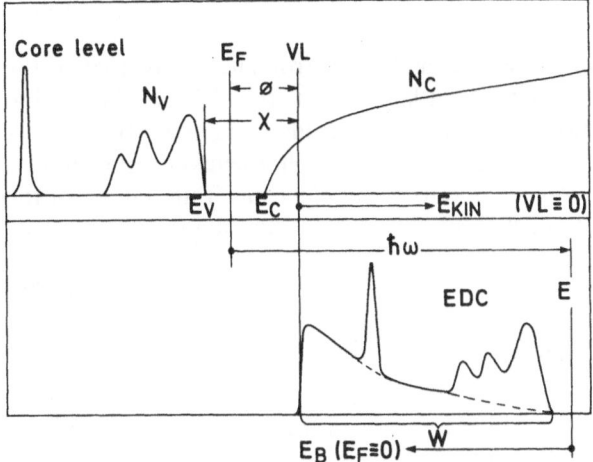

Fig. 3.5. Schematic representation of the photoemission process and the energy levels involved. The upper half represents the level scheme of the semiconductor including one core level and the lower panel depicts the energy distribution of the photoelectrons (EDC). (N_v, N_c) valence and conduction densities of states; (E_v, E_c) valence and conduction band edges; (χ) photoemission threshold; (ϕ) work function; (E_F) Fermi level; (VL) vacuum level; $(\hbar\omega)$ photon energy; (W) width of EDC (it is assumed that $\phi_{sample} > \phi_{analyzer}$). The dashed line indicates the background of inelastically scattered electrons

at the high kinetic energy end of a metal spectrum. For nonconducting samples the Fermi levels of sample and spectrometer differ by the potential difference

$$\Delta V = i_e \cdot R , \qquad (3.34)$$

where i_e is the photoelectron current and R the resistance of the sample measured between the illuminated surface and the back contact with the spectrometer. The small electron current ($\lesssim 10^{-11}$ A) insures that ΔV remains negligible for all but the best insulators. Amorphous Si samples prepared as thin films pose no problem in this respect. In semiconductors E_F is free to move within the gap and its position depends on doping, surface states and for amorphous specimens, on the gap state distribution. With E_F as the fixed reference energy, shifts in E_F are reflected in equal but opposite shifts of the whole photoemission spectrum. Using the sharp core levels or distinct features in the valence bands, such movements of E_F are determined with considerable accuracy (see Sect. 3.3.13). It is, of course, also possible to place E_F relative to the spectroscopically determined top of the valence bands. The low energy cut-off of a spectrum is reached when the kinetic energy of the electrons is zero. This corresponds, according to Fig. 3.5, to a binding energy

$$E_B^0 = \hbar\omega - \phi \, , \tag{3.35}$$

where ϕ is either the work function of the spectrometer ϕ_{sp} or that of the sample ϕ_{sa}, depending on which one is larger. Equation 3.35 can be used to determine the work function of the sample, provided it is larger than ϕ_{sp}. A more useful quantity for semiconductors is the photoemission threshold χ which is the minimum energy needed to excite electrons from the top of the valence bands E_v to the vacuum level just outside the sample:

$$\chi = \phi + E_F - E_v \, . \tag{3.36}$$

With reference to Fig. 3.5 and $\phi_{sa} > \phi_{sp}$,

$$\chi = \hbar\omega - W \, , \tag{3.37}$$

where W is the total width of the EDC from the lower energy cut-off to the top of the valence bands.

3.3.5 Core Levels

The binding energies (to within a few eV) of core levels are characteristic of each element. Core level spectra are therefore used to identify the composition of specimens within the sampling depth of photoemission. A list of up-to-date binding energies accessible with Al K_α x-rays is given in [3.11, 12]. The intensity of a core level j from an element Z, $I(j, Z)$, is related to the concentration $c(Z)$ of this element according to

$$I(j, Z) = F \cdot \sigma(j, Z, \omega) \cdot C(Z) \cdot T(E) \cdot D(E) \cdot A(E) \, , \tag{3.38}$$

where σ is the photoelectron cross section of the core level and $T(E)$ and $D(E)$ are the transmission and escape functions, respectively. $A(E)$ is the energy-dependent transmission function of the analyzer and F is an energy-independent scale factor that takes photon intensity and geometrical factors into account. For the determination of relative concentrations of elements Z_1 and Z_2, $c(Z_1)/c(Z_2)$, (3.38) may be written as

$$\frac{c(Z_1)}{c(Z_2)} = \frac{I(j_1 Z_1)}{I(j_2 Z_2)} \frac{\sigma(j_2, Z_2, \omega)}{\sigma(j_1, Z_1, \omega)} \frac{\lambda_e(j_2, Z_2)}{\lambda_e(j_1, Z_1)} \frac{A(E_2)}{A(E_1)} \, . \tag{3.39}$$

Here we have made the following simplifications. $D(E)$ has been dropped since it varies by less than $\pm 5\%$ for electron energies between 500 and 1500 eV. $T(E_2)/T(E_1)$ has been replaced by the ratio of the electron mean free paths λ_e, an approximation that is justified by the magnitude of $\lambda_{ph}/\lambda_e \gtrsim 100$, c.f. (3.24).

Table 3.2. Factors entering the calculation of atomic concentrations from core level intensities using Al K_α excitation ($h\nu$ = 1486.6 eV) [3.70]

Level	E_B^a [eV]	E_{kin} [eV]	$E_{kin}^{-0.3}$	$\sigma/10^{4\,b}$ [barn]	K^c
Si $2p_{3/2}$	98.9	1388	0.114	0.74	844
Ge $3p_{3/2}$	121.6	1366	0.115	3.2	3860
Ge $3s$	181.2	1306	0.116	1.7	1972
O $1s$	531.7	955	0.128	4.0	5120
C $2s$	284.7	1202	0.119	1.4	1666
F $1s$	686.2	800	0.135	6.0	8100
N $1s$	397.6	1089	0.123	2.45	3014
B $1s$	188.1	1299	0.116	0.66	766
P $2p_{3/2}$	129.0	1358	0.115	1.1	1265
P $2s$	186.6	1300	0.116	1.6	1856
Ar $2p_{3/2}$	241.6	1245	0.118	2.75	3245
Mo $3d_{5/2}$	227.5	1259	0.117	7.7	9009

[a] relative to E_F [b] [63, 64]   [c] $K = \sigma \cdot \lambda \cdot A$

For the commonly used electron energy analyzer with fixed pass energy and preretardation, $A(E) \propto E^{-1}$. The high energy part of $\lambda_e(E)$ ($E > 500$ eV) in Fig. 3.3 is reasonably well approximated by $\lambda_e(E) \propto E^{0.7}$. We may therefore replace $\lambda_e \cdot A$ in (3.39) by $E^{-0.3}$.

In Table 3.2 we have collected the relevant parameters of (3.39) with the simplifications just discussed for the core levels of a number of elements that are of interest in work on a-Si, a-Ge and its alloys [3.70]. The entries of Table 3.2 are appropriate for Al K_α excitation. The accuracy of concentrations obtainable using (3.39) and this table is estimated to be better than 20%. The lowest concentration that can be detected with photoemission is about 1 at.% averaged over the sampling depth. The application of (3.39) is considerably facilitated if the factors on the right-hand side can be determined from a sample with known composition, as done for a-Si$_x$C$_{1-x}$ alloys [3.71].

The exact binding energy of a core level depends on the valence charge q. For a change Δq, the binding energy changes by

$$\Delta E_B = \beta \, \Delta q \ . \tag{3.40}$$

In the simplest approximation for this so-called chemical shift, the q's are treated as screening charges localized on the valence shell with radius r_v. The corresponding screening potential shifts all inner core levels with radii smaller than r_v by equal amounts $\beta = 1/r_v$ per unit charge transfer Δq. For $r_v = 1$ Å the coupling constant is ~ 14 eV elementary charge [3.72, 73]. In solids, a charge Δq on one atom is usually compensated by equal and opposite charges on neighboring atoms. For an ionic crystal this reduces β to $\beta' = (1/r_v - \alpha_M/a)$, where α_M is the Madelung constant and a the lattice constant.

In Si, the proportionality constant β' has been estimated from studies on oxygen-induced shifts [3.74]:

$$\beta'(\text{Si}) \simeq -2.2\,\text{eV/electron} . \qquad (3.41)$$

This value includes the Madelung correction to β.

A rough estimate of the effective charge Δq that enters in (3.41) is readily obtained using the ionicity f_i of a bond [3.75]. The ionicity f_i is expressed in terms of the difference in the electronegativities X_A and X_B of the two atoms forming the bond according to

$$f_i = 1 - \exp[-(X_B - X_A)^2/4] . \qquad (3.42)$$

The effective charge on atom x is $\Delta q_x = \gamma_x \cdot f_i$, where γ_x is the coordination number of atom x ($\gamma_{\text{Si}} = \gamma_{\text{Ge}} = 4$). *Katayama* et al. [3.71] obtained a good correlation between the charges Δq so calculated and the chemical shifts measured for a number of Si compounds, provided they used the electronegativity values of *Phillips* [3.76]. The proportionality factor so obtained ($\Delta E_B/\Delta q = 2.7$) is slightly larger than that of (3.41). The ionicity approach to Δq implies that chemical shifts are additive: two F atoms attached to Si induce twice the chemical shift of one F atom. This additivity is seen to hold even for the most electronegative fluorine (see Sect. 3.3.15 a). In the screening potential approximation, the chemical shifts are the same for all core levels – a result that is confirmed experimentally for the deeper lying core levels [3.77].

Despite the fact that chemical shifts usually amount to less than 1% of the binding energy, they can be determined with great accuracy and yield valuable information about the bonding configurations in fluorine or hydrogen containing silicon alloys (see Sects. 3.3.14, 15) and the formation of Si–O bonds during the oxidation of a-Si (Sect. 3.3.16 b).

3.3.6 Some Experimental Aspects of Photoemission

A photoemission experiment involves a source of monochromatic light, a sample, an electron energy analyzer and a detection system, the latter three elements being housed in a vacuum enclosure. Detailed descriptions of these elements have been given elsewhere [3.11, 12, 78]. We shall, therefore, discuss only those aspects that are of importance in the present context.

a) Photon Sources

A list of photon sources employed in photoemission and the spectral resolution generally obtained with them is given in Table 3.3. Windowless rare gas discharges provide line spectra with useful intensities in the range from $15\,\text{eV} \lesssim \hbar\omega \lesssim 40\,\text{eV}$. They cover the low energy or UPS (ultraviolet photo-

Table 3.3. Photon sources used in photoelectron spectroscopy and resolutions typically obtained with them [3.11]

Source	$\hbar\omega$ [eV]	Typical intensity at the sample [photons s^{-1}]	Typical resolution of photoemission spectrum, FWHM [eV]
Ar I	11.83 (11.62)[a]	6×10^{11}	
Ar II	13.48 (13.30)[a]	8×10^{10}	
Ne I	16.85 (16.67)[a]	8×10^{11}	0.1–0.3
He I	21.22 (23.09)[a]	1×10^{12}	
Ne II	26.9 (27.8)[a]	2×10^{11}	
He II	40.82 (48.38)[a]	2×10^{11}	
Synchrotron	20–500	$\sim 10^{10}$	0.1–1.5
Mg K$_{\alpha_{1,2}}$	1253.6 (1262.1, 1263.7)[a]	1×10^{12}	1.0
Al K$_{\alpha_{1,2}}$	1486.6 (1496.3, 1498.3)[a]	1×10^{12}	1.2 / 0.4–0.7 with monochromator

[a] satellites

emission spectra) range. On the high energy end, referred to as x-ray photoemission spectroscopy or XPS, Al K$_\alpha$ and to a lesser extent Mg K$_\alpha$ characteristic x-rays are used; the former often in conjunction with a quartz crystal monochromator to improve resolution and to suppress unwanted satellites and bremsstrahlung. Spectra obtained with these line sources are labelled "He I", "Al K$_\alpha$", etc.

The continuum of radiation emitted by electron accelerators or storage rings known as synchrotron radiation [3.79] covers, in conjunction with a suitable monochromator, the intermediate range between ~ 20 and ~ 500 eV photon energy. The choice of photon energies is dictated by three considerations: (i) spectral range, (ii) resolution, and (iii) cross sections.

Point (i) is trivial; the binding energies of occupied states accessible with PES is no larger than the photon energy used. That means for Si that UPS is limited to valence band studies. A minimum of $\hbar\omega \simeq 105$ eV is required to excite the least bound Si core level, Si 2p, with a binding energy of $\simeq 99$ eV relative to E_F. In XPS, core and valence levels of all elements are accessible. A list of binding energies for all elements is given in [3.11, 12]. The advantages of UPS over XPS in resolution are evident from Table 3.3.

The single most important consideration in the present context is, however, the cross section. As shown in Fig. 3.6, the hydrogen induced states (A and B in Fig. 3.6) are only discernible for photon energies between ~ 10 and ~ 140 eV [3.80]. Beyond $\hbar\omega \sim 120$ eV σ(H 1s), the partial hydrogen 1s cross section becomes negligible compared to σ(Si 3s) and σ(Si 3p). As a result,

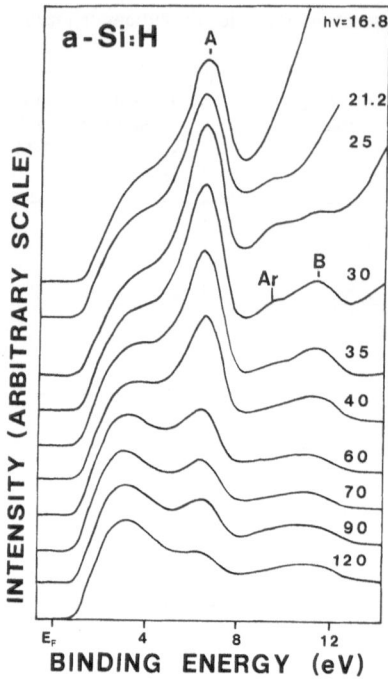

Fig. 3.6. Valence band spectra of a-Si:H as a function of photon energy. A and B are the hydrogen induced states [3.80]. The peak labelled Ar is due to Ar 3p states, Ar being imbedded during the sputter-deposition

XPS valence band spectra yield the pure Si partial density of states independent of hydrogen content (Sect. 3.3.8). The same situation holds for Ge and hydrogen. The cross sections for the valence electrons of fluorine, by contrast, remain comparable to those of silicon even for $\hbar\omega = 1500$ eV (Sect. 3.3.15 b).

b) Sample Preparation for Photoemission Measurements

The short sampling depth of photoemission requires that the specimen surface be free of contaminants on the sub-monolayer level. A few percent of a monolayer of oxygen, for example, change the valence band structure of a-Si completely as will be discussed in Sect. 3.3.16 b, and a fraction of a monolayer of water adsorbed on the surface induces measurable shifts of E_F due to band bending [3.80 a]. It is therefore not possible to obtain reproducible and meaningful photoemission spectra of a sample that has been exposed to atmosphere and thereby collected a variety of surface contaminants in the form of adsorbed H_2O, O_2, CO, etc. without cleaning its surface inside the spectrometer. This is done by bombarding the surface with high energy (0.5–5 keV) rare gas ions (usually argon) to remove successive surface layers. Such a bombardment, however, will not succeed in removing the contamination without drastic changes in the structure and composition of the freshly exposed clean sample surface [3.81].

A procedure that has been followed with few exceptions [3.82] is to prepare the sample inside the photoemission spectrometer or in a separate chamber from which the sample can be transferred into the spectrometer vessel under ultrahigh vacuum conditions. The latter method is preferred because it allows an optimization of the deposition conditions.

Some consideration should also be given to the substrate. The specimens used in photoemission can be rather thin ($\gtrsim 500$ Å). The danger of noticeable interdiffusion between substrate and film, particularly at elevated temperatures, is therefore high. Molybdenum as a substrate material appears to give the least problems in that respect [3.83] and it does not induce premature crystallization of a-Si due to the formation of an eutectic [3.83 a]. It further provides an ohmic contact between spectrometer and sample.

3.3.7 Valence Band Spectra of Amorphous Silicon and Germanium

The valence band spectra of amorphous silicon and germanium have been obtained by low energy photoemission [3.83, 84–87], by XPS [3.88] and by x-ray emission spectra involving transitions from the valence bands of a-Si [3.89].

In Fig. 3.7 a the results of *Ley* et al. [3.88] are presented which were obtained using monochromatized Al K_α x-rays. The spectra of amorphous Si and Ge prepared by room temperature evaporation in the lower frame are compared with their crystalline counterparts (cleaved single crystals) in the upper frames. Also shown are the DOS's calculated using the pseudopotential formalism and a broadened version thereof to facilitate comparison with the experimental spectra (resolution 0.7 eV FWHM). The labels indicate critical points in the theoretical DOS [3.90]. The area under the peaks in the DOS corresponds very nearly to one (peaks II and III) and 2 (peak I) electrons per atom, respectively. The differences in the relative intensities of peaks II and III compared to peak I in Si and Ge are due to the photoemission cross sections of the valence electrons alluded to above (Sect. 3.3.3). In Ge, $\sigma(4s)/\sigma(4p) \simeq 1$, i.e., the EDC is a fair representation of the DOS whereas $\sigma(3s)/\sigma(3p) \simeq 3.4$ in Si leads to an enhancement of the mainly 3s derived states in peaks II and III in the EDC. A density of valence states of a-Si corrected for photoelectric cross sections is shown in Fig. 3.7 b.

The results for a-Si and a-Ge can be summarized as follows:

(i) The overall appearance of the DOS is similar for the amorphous and crystalline materials. We find, in particular, that the total width of the valence bands remains unchanged as well as the valley between the region of predominantly s-derived states (peaks II and III) and the p-like peak I at the top of the DOS.

(ii) The two lower-lying peaks (corresponding to the L_1 and L'_2 critical points in the crystals) merge into one hump. Since the edges of this hump do

not broaden, the authors of [3.88] concluded that the one-hump structure occurs as a result of states that fill the dip between L_1 and L_2' rather
than through a mere broadening of the two peaks.

(iii) The centroid of the p-like peak shifts towards the top of the valence
bands by 0.4 eV in Si and 0.5 eV in Ge. This results in a distinct steepening of the leading edge in the DOS of the amorphous samples.

The shift in peak I is responsible for the red shift in the maximum of $\varepsilon_2(\omega)$
in a-Si that cannot be accounted for by the loss of k-conservation alone
(Sects. 3.2, 3.4.2 a). The first point emphasizes the fact that the principle
features of the DOS are to a large extent determined by the tetrahedral
environment of a Si or Ge atom, the number of nearest neighbors and the
interatomic distance which remain unchanged in the amorphous modifications [3.91]. The considerable density deficit of up to 20% in amorphous

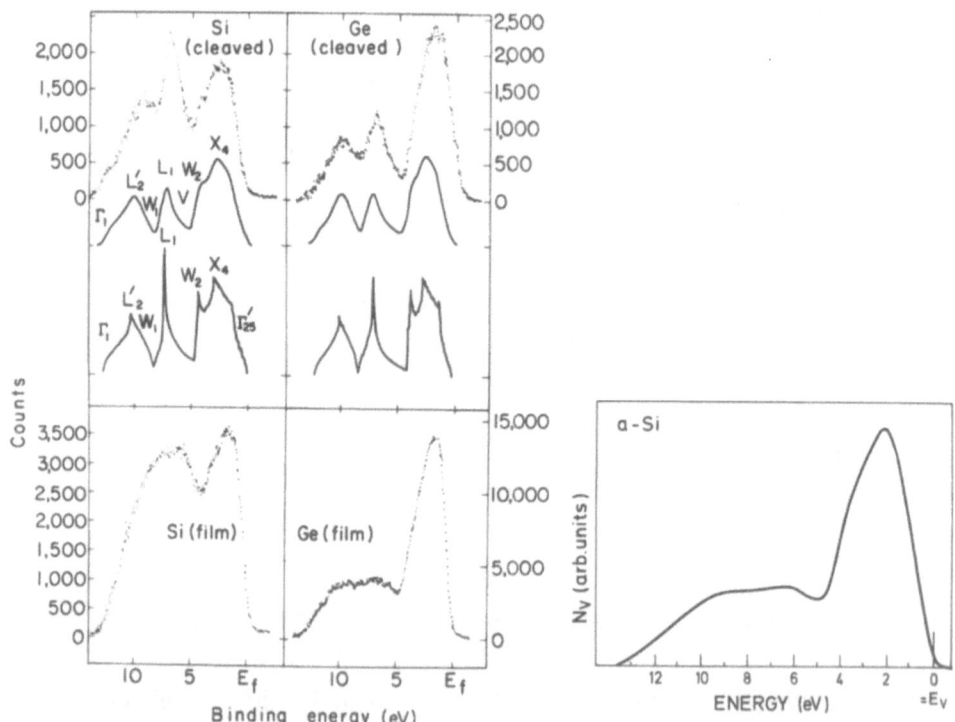

Fig. 3.7. (a) Valence band spectra of crystalline and amorphous silicon and germanium excited
with monochromatized Al K_α x-rays. The spectra have been corrected for a background of
inelastically scattered electrons. Also shown are theoretical densities of states, unbroadened
(lower solid lines) and broadened, in order to facilitate comparison with the experimental
spectra [3.88]. **(b)** Experimental valence density of states of a-Si obtained from spectrum of (a)
after correction for variations in photoelectric cross sections

group IV semiconductors has consequently no influence on the DOS, as first assumed by *Herman* and *van Dyke* [92], since it is due to voids [3.32–35] and not to a dilatation of the lattice as mentioned earlier (Sect. 3.2.1).

The loss of long-range order is obviously of secondary importance for the valence density of states. It introduces some broadening (~ 0.5 eV in peak I) into the valence DOS as shown by the calculations of *Kramer* [3.93] and *Brust* [3.94], hardly sufficient to wipe out fine structure in the density of occupied states completely, let alone to reproduce the characteristic blue shift of peak I and the filling of the valley between peaks II and III.

The extended conduction states are much more susceptible to the absence of long-range order. The structure in the density of unoccupied states is almost completely lost in a-Si according to these calculations [3.93], a result that is in good agreement with the experimental results to be discussed later (Sect. 3.3.10).

Peaks II and III in the DOS's of c-Si and c-Ge are due to the presence of sixfold rings in the diamond structure [3.95, 96]. In models of the amorphous network, a different topological arrangement of the atoms allows for five, seven and perhaps eightfold rings while preserving the basic tetrahedral units. The most common 5-fold ring, for example, can be formed with only a slight adjustment of the tetrahedral bond angle. It has been suggested by *Weaire* and *Thorpe* [3.95] that fivefold rings should introduce states between those of peaks II and III and thereby fill the gap. This conjecture has been confirmed in a series of calculations involving simple models [3.95, 96], complex crystalline polytypes of Si and Ge containing 5, 6 and 7-fold rings [3.97–99] and actual cluster models of the amorphous structure built with as many as 519 atoms [3.100–105]. In summary, we attribute the featureless appearance of the lower portion of the a-Si and a-Ge DOS's to an averaging of a variety of local bonding topologies, i.e., a distribution of rings of bonds.

It should be noted here that Weaire and Thorpe were not the first to point out that fivefold rings appeared to be a necessary and distinguishing ingredient in the structure of a-Si and a-Ge compared to the diamond lattice. *Richter* and *Breitling* [3.106] noticed as early as 1958 that the maxima in the RDF at 4.7 and 7.1 Å which are missing in the RDF of a-Ge correspond to those interatomic distances which change drastically when neighboring tetrahedra are rotated about their common bond. This observation led *Coleman* and *Thomas* [3.107] and *Grigorovici* and *Manaila* [3.108] to explore the possibilities of building small three-dimensional clusters from atoms linked tetrahedrally to each other, but not solely in the staggered configuration that is exclusively present in the diamond lattice. If only eclipsed bonds (known to exist in the wurtzite lattice) are used to link atoms together, fivefold planar rings are formed. The bond angle of 108° differs only slightly from the tetrahedral angle of 109° 28' and this configuration is therefore likely to occur in amorphous networks. The presence of such rings has been detected by *Mader* [3.109] in small Ge particles deposited by electron beam evaporation on NaCl substrates.

As a result of the same calculations mentioned in connection with the s-like states, it appears that the blue shift of peak I is largely due to bond angle variations. The electrons in this part of the valence bands are mainly localized in the bond region between atoms. Deviations from the equilibrium tetrahedral bond angle leads, on the average, to an increase in the energy of these states as a result of increasing Coulomb repulsion between neighboring bond charges [3.96]. Such an increase in the energies of bonding states corresponds, of course, to a weakening of the bonds, in agreement with our earlier arguments regarding the reduction of the Penn gap.

It has been suggested [3.110, 110a] that the states at the top of the valence bands are strongly affected by variations in the dihedral angle. These effects are limited to the top 0.3 eV, however, and are thus less likely to cause the blue shift of peak I as observed.

3.3.8 Valence Band Spectra of Hydrogenated Amorphous Silicon

The addition of hydrogen to amorphous silicon introduces Si–H bonding states between ~ 5 and 11 eV below the top of the valence bands. Figure 3.8 shows a series of EDC's obtained with 21.2 eV (He I) and 40.8 eV (He II) photon energy [3.111]. The spectra are for reactively sputtered films of a-Si prepared at room temperature with increasing amounts of hydrogen in the sputter gas (argon), as indicated in the figure. In the absence of hydrogen, the spectrum is dominated by emission from Si $3p$ states at the top of the valence bands which correspond to peak I in Fig. 3.7 a (notice that the energy scale in the UPS spectra runs in a direction opposite to that of Fig. 3.7 a). The contribution from the Si $3s$ states shows up as the structureless hump between 5 and 14 eV binding energy in the He II spectrum but it is lost in the steeply rising background of secondary electrons in the He I spectrum. The cross section ratio $\sigma(3s)/\sigma(3p)$ is noticeably smaller in the UPS regime than for $\hbar\omega = 1486.6$ eV (see also Fig. 3.6).

With the addition of hydrogen, two new peaks labelled A and B appear with an intensity that is roughly proportional to the hydrogen concentration in the sputter gas. Peak B is hidden in the background of the He I spectrum. The binding energies (relative to E_F) are 5.9 eV (10% H_2) and 6.2 eV (50%) for peak A and 10.2 eV (10% H_2) and 11.2 eV (50% H_2) for peak B. The full width at half maximum (FWHM) of peak A increases from 1.2 eV to 1.9 eV. The width of peak B is 2.2 eV in the spectrum with 50% H_2. The intensity ratio $I(A) : I(B)$ is approximately 2 : 1 for $h\nu = 40.8$ eV. The A–B structure, ascribed to polyhydride configurations (vide infra), is the one always observed for specimens deposited below ~ 300 °C. In Fig. 3.9 we compare the He II EDC's of sputtered a-Si : H with that of a glow discharge sample, both prepared at room temperature [3.83]. Peak A lines up almost perfectly for both specimens. The width of peak A in the glow discharge sample is slightly (~ 0.2 eV) narrower. Peak B, on the other hand, is noticably wider,

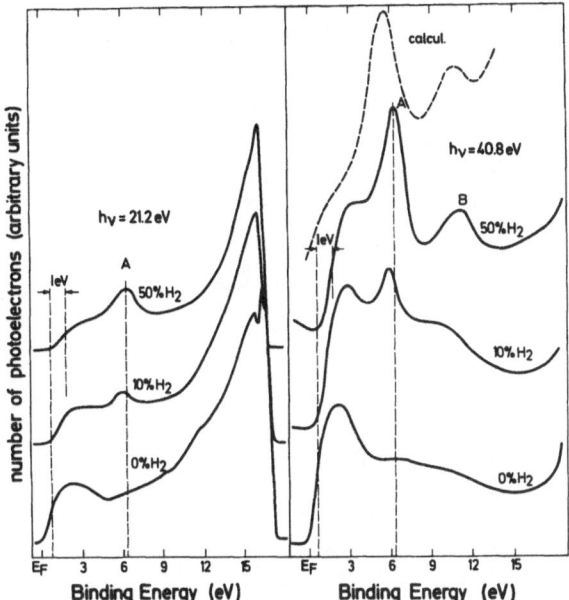

Fig. 3.8. He I (a) and the He II (b) valence band spectra of amorphous silicon films prepared by reactive sputtering in an argon-hydrogen mixture with increasing amounts of hydrogen. The hydrogen concentrations are those of the sputter gas. The dashed curve reproduces the spectrum calculated for hydrogen adsorbed on a Si(111) surface in the form of SiH₃ [111]

somewhat skewed towards the low binding energy side and has a binding energy of 10.8 eV, 0.5 eV lower than in the sputtered film.

Specimens of a-Si:H sputtered onto substrates held at 350 °C exhibit a distinctly different photoemission spectrum (Fig. 3.10). Two small peaks at 5.3 eV (peak C) and 7.3 eV (peak D) take the place of peak A. Instead of peak B, a similar broad peak E at ~ 10.3 eV below E_F is observed. This form of the EDC for $T_D = 350$ °C is independent of the hydrogen concentration in the sputter gas as long as it exceeds 10 vol.%. The spectra for glow discharge a-Si:H films prepared at 350 °C are again virtually identical to the sputtered films with the C–D–E structure somewhat less well defined than in the sputtered sample (see Fig. 3.10). Upon annealing the A–B structure transforms into the C–D–E structure around $T_A = 300°$–350 °C and further annealing to ~ 600 °C removes all hydrogen from the film and the spectrum of unhydrogenated a-Si is recovered [3.83]. It has been argued [3.83] that the transition A–B $\rightarrow$ C–D is not merely due to a loss of hydrogen from polyhydride configurations until only Si–H configurations (structure C–D) remain but that some of the polyhydride configurations actually transform into the monohydrides (see Fig. 3 of [3.111]).

In Table 3.4 the energies of the hydrogen induced peaks in the EDC's of a-Si:H have been collected [3.83, 111]. Also listed in Table 3.4 for compari-

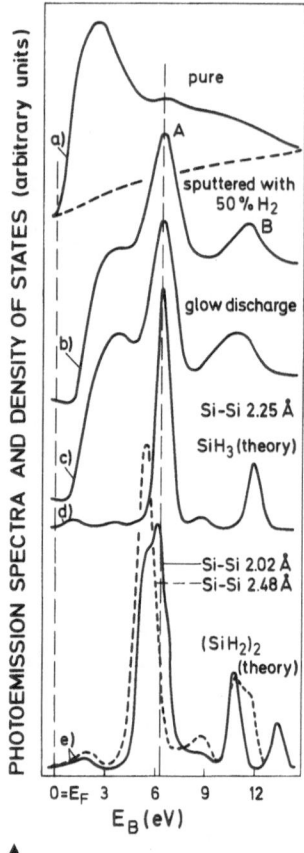

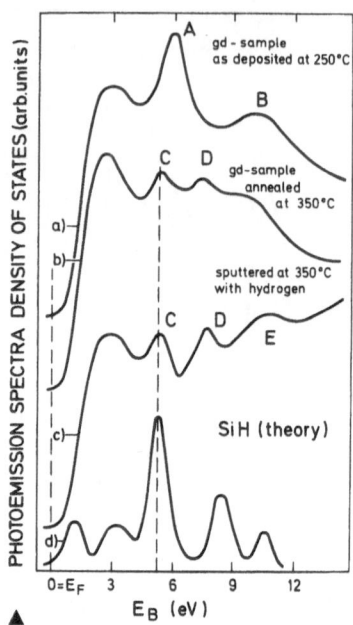

Fig. 3.10. (*Curves a–c*): He II valence band spectra of hydrogenated a-Si annealed or deposited at elevated temperatures. (*Curve d*): calculated local density of states for Si–H; configuration obtained by *Ching* et al. [3.119]. The theoretical spectrum has been shifted by 1.8 eV towards higher energy to line up peaks C [3.83]

Fig. 3.9. (*Curves a–c*): He II valence band spectra of pure and hydrogenated amorphous silicon samples prepared at room temperature by reactive sputtering (*Curves a, b*), and by the glow discharge decomposition of silane (*Curve c*). (*Curves d, e*): calculated local densities of hydrogen derived states in a-Si : H_x for different bonding geometries by *Ching* et al. [3.119]. The calculated spectra have been shifted by 1.8 eV towards lower binding energy so that peaks A line up [3.83]

son are the energies of hydrogenic states in the c-Si–H adsorbate systems [3.112–115], those of silanes and the positions of maxima in the local densities of states calculated for a variety of Si–H_x bonding configurations [3.116–120]. The energies are referenced to E_F with the exception of the silanes. Since the position of E_F in the gap depends, even for undoped samples, somewhat on the preparation conditions, the energy differences A–B and C–D that are also given in Table 3.4 are a more reliable signature of a particular spectrum. The A–B splitting is seen to vary between 4.2 and 5.1 eV depending on preparation condition. Broadly speaking, the separation between A and B decreases with decreasing hydrogen content: specimen 5 contains less hydrogen than specimen 6 and the film prepared at $T_D = 250\,°C$ (specimen 11) less than those at $T_D = $ R.T. (specimens 7, 8, 9 in Table 3.4).

The first identification [3.111] of the hydrogen-induced features in terms of specific Si–H bonding configurations was based on the photoemission spectra of hydrogenated c-Si surfaces [3.112–114] and their interpretation in terms of tight-binding or pseudopotential calculations [3.112, 115, 116]. The spectrum of a c-Si(111) surface saturated with hydrogen in the Si–H$_3$ configuration shows indeed a remarkable similarity to that of the low temperature phase of a-Si : H (Fig. 3.8, dashed line) [3.112, 114]. The energy separation between A–B is, however, reduced at the surface compared to a-Si : H (Table 3.4, entries 15, 16).

Based on these calculations, the origin of peaks A and B as Si 3p–H 1s and Si 3s–H 1s bonding states has been established. *Gruntz* et al. [3.62] obtained from intensity measurements as a function of $\hbar\omega$ and using known photoemission cross sections for σ(H 1s) and σ(Ge 4p) a contribution of 64% H 1s and 36% Ge 4p to peak A. This ratio should not be too different in a-Si : H and it is, in fact, in agreement with the 60% H 1s contribution to peak A in a-Si : H calculated by *Ching* et al. [3.117].

The local (i.e., H 1s derived) densities of states (LDOS) have been calculated by *Ching* et al. [3.118, 119] for several possible types of Si–H bonding configurations, each of them imbedded in an amorphous silicon cluster. These are Si–H$_x$ (x = 1, 2, 3) units, a (SiH$_2$)$_2$ chain fragment and Si–H H–Si – a broken Si–Si bond with two H atoms taking the place of the broken bond. Some of these results are compared in Figs. 3.9, 3.10 with the UPS spectra. The energies of peaks in the LDOS are also listed in Table 3.4.

It is apparent from this comparison that in addition to Si–H$_3$ (Fig. 3.9 d), also Si–H$_2$ (not shown but see Table 3.4) and (SiH$_2$)$_2$ have a similar two-peak structure with a splitting of about 5 eV that could correspond to peaks A and B. Furthermore, the two examples given for the silane fragment (SiH$_2$)$_2$ in Fig. 3.9 emphasize how sensitive the exact splitting and the shape of the LDOS depends on the model parameters. Under these circumstances it appears at present impossible to ascribe a particular polyhydride configuration to a given spectrum. We may at best comment on the tendency of increasing A–B splitting with increasing hydrogen content that we mentioned before. The calculations of *Ching* et al. [3.118, 119] indicate a splitting that is ~ 0.3 eV larger for Si–H$_3$ than for SiH$_2$ or (SiH$_2$)$_2$. This would indicate that the increase in the A–B splitting is associated with a shift from (Si–H$_2$)$_x$ (x = 1, 2) to Si–H$_3$ with increasing hydrogen content.

There is ample evidence both from hydrogen adsorbed on c-Si [3.113–116] and from a number of calculations [3.118–122] that the C–D–E structure signals hydrogen bonded as monohydride Si–H (see Table 3.4). The theoretical result of *Ching* et al. [3.118] obtained for an isolated Si–H unit in a Si cluster is shown in Fig. 3.10. The agreement with the measured spectra is seen to be good if we assume that the two leading peaks in the LDOS are masked by the Si 3p emission between 0 and 5 eV. The densities of states calculated for the broken bond model (Si–H H–Si) can similarly be made to agree well with experiment, although the peak positions depend on the par-

Table 3.4. Binding energies (relative to E_F) of hydrogen-induced states in the valence band spectra of silanes, a-Si:H and of hydrogenated c-Si surfaces. Also given are the energies of peaks in the local densities of states calculated for different Si–H$_x$ configurations in a-Si and on c-Si surfaces

	Substance	Entry No.	Binding energy of A [eV]	of B [eV]	Energy differ- ence B – A [eV]	Technique or method of calcu- lation	Ref.
Polyhydrides	**Silanes**[a]						
	SiH$_4$	1	12.7	18.0	5.3	XPS	[3.67]
	SiH$_4$	2	12.5	18.1	5.6	UPS	c
	Si$_2$H$_6$	3	12.4	16.7	4.3	UPS	c
	Si$_5$H$_{12}$	4	12.4	15.7	3.3	UPS	c
	a-Si:H						
	a) T_D = RT						
	sputtered						
	10 vol.% H$_2$	5	5.9	10.5	4.6	UPS	[3.83]
	50 vol.% H$_2$	6	6.3	11.3	5.0	UPS	[3.83]
	gd cathodic						
	[SiH$_4$]/[Ar] = 0.1	7	6.3	10.8	4.5	UPS	[3.83]
		8	6.4	11.3	4.9	UPS	d
	gd anodic						
	pure SiH$_4$	9	6.6	11.7	5.1	UPS	d
	b) T_D = 200°C						
	gd cathode						
	[SiH$_4$]/[He] = 1	10	6.4	11.1	4.7	UPS	d
	c) T_D = 250°C						
	gd cathode						
	[SiH$_4$]/[Ar] = 0.1	11	6.0	10.2	4.2	UPS	[3.83]
	d) calculations						
	Si–H$_3$	12	8.1	13.5	5.4	a-Si cluster	[3.119]
	Si–H$_2$	13	7.2	12.3	5.1	a-Si cluster	[3.119]
	(SiH$_2$)$_2$	14	7.1	12.5	5.4	a-Si cluster d(Si–Si = 2.48 Å	[3.119]
	c-Si surface						
	a) experiment						
	Si–H$_3$ on						
	Si(111)	15	6.0	10.0	4.0	UPS	[3.112]
	Si(111)	16	6.7	10.1	3.4	UPS	[3.114]
	b) calculation						
	Si–H$_3$ on						
	Si(111)	17	6.0	10.5	4.5	Tight binding	[3.112]
	Si(111)	18	6.0	9.8	3.8	Self-consistent pseudopotential	[3.112]

Table 3.4 (continued)

Substance	Entry No.	Binding energy			Energy difference D − C [eV]	Technique or method of calculation	Ref.
		of C [eV]	of D [eV]	of E [eV]			
a-Si–H							
a) experiment							
sputtered $T_D = 350\,°C$	15	5.2	7.4	10.5	2.2	UPS	[3.83]
sputtered $T_D = RT$; $T_A = 400\,°C$	16	5.5	7.4	n.o.[b]	1.9	UPS	[3.111]
gd $[SiH_4]/[Ar] = 0.1$ $T_D = 250\,°C$ $T_A = 350\,°C$	17	5.2	7.2	10.3	2.0	UPS	[3.83]
b) calculation							
Si–H	18	7.5	10.0	–	2.5	Bethe lattice, Tight binding	[3.122]
Si–H	19	7.0	10.0	–	3.0	Cluster, Tight binding	[3.119]
Si–H H–Si broken bond	20	6.0⎤ 7.0⎦ two ~ 9⎤ ~ 12⎦ peaks			3–4	Cluster, energies depend on Si–H and H–H distance	[3.119]
hydrogenated vacancy	21	5.2	7.6	10.6	2.4	Coherent potential approximation, tight binding	[3.120]
c-Si surface							
a) experiment							
Si–H on Si(100)	22	5.0	6.8	n.o.[b]	1.8	UPS	[3.113]
Si–H on Si(111)	23	5.4	7.4	10.1	2.0	UPS	[3.114]
b) calculation							
Si–H on Si(111)	24	4.5	7.2	–	2.7	Self consistent pseudopotential	[3.116]
Si–H on Si(111)	25	4.8	7.0	–	2.2	tight binding	[3.115]

(left margin label: Monohydrides)

[a] Reference energy is the vacuum level
[b] not observed
[c] H. Bock, W. Ensslin, F. Fehér, R. Freund: J. Am. Chem. Soc. **98**, 668 (1976)
[d] R. Kärcher, L. Ley: unpublished

ameters used for the bond lengths [3.119]. For this configuration a strong peak in the density of states appears at the bottom of the valence band at ~ 14 eV. Such a resonance appears to be characteristic for strongly interacting Si–H units as they are also present in LDOS calculated for hydrogenated vacancies [3.120]. The corresponding peak is not observed in the photoemission spectra.

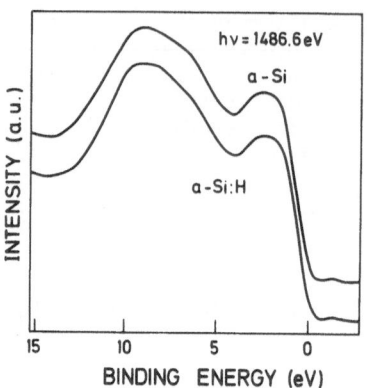

INTENSITY (a.u.)

hν = 1486.6 eV

a - Si

a - Si : H

BINDING ENERGY (eV)

15   10   5   0

Fig. 3.11. XPS valence band spectra of unhydrogenated (*upper curve*) and hydrogenated amorphous silicon [H. Richter, L. Ley: private communication]

Allan et al. [3.123] have investigated the conditions under which Si–H configurations yield the C–D–E structure using the cluster Bethe lattice approach. They concluded that the signature of an isolated monohydride is a single peak C whereas structure D and E appear only under two circumstances: (i) the monohydrides are present as strongly interacting clusters either in the form of a hydrogenated vacancy or on the surfaces of inner voids. In the latter case the situation would be similar to the hydrogenated surface of c-Si. (ii) The second possibility is that an isolated monohydride is part of a network with mainly sixfold rings in the immediate neighborhood as in crystalline silicon. Peaks D and E reflect then the DOS of the remainder of the network corresponding to peaks II and III in the DOS of c-Si (Fig. 3.7a). The idea is [3.124] that hydrogenation of a-Si leads *locally* to the formation of crystalline regions with hydrogenated boundaries. Should this picture apply to a-Si : H, XPS measurements on a-Si : H samples, which are sensitive to the Si LDOS only, would resemble those of c-Si and exhibit peaks II and III. This is, however, not the case as illustrated in Fig. 3.11 and this last possibility does not appear to be the main reason for structures C, D, and E. A number of other calculations on monohydride configurations give two peaks in reasonable agreement with peaks C and D but fail to reproduce peak E [3.120, 122, 125].

It seems therefore premature to try to understand the subtle differences observed in the monohydride spectrum (Table 3.4) that depend on preparation conditions. The possibility of hydrogen clusters is, of course, a very real one. Their presence has been independently postulated by *Shanks* et al. [3.126] based on infrared spectroscopy and by *Reimer* et al. based on NMR studies [3.127].

An estimate of the hydrogen concentration within the sampling depth of photoemission is possible based on a comparison of the intensities of peak A for a-Si : H and for Si–H$_3$ units formed upon chemisorption of H on c-Si (111) surfaces (Fig. 3.8) [3.111]. A Si (111) surface saturated with SiH$_3$ has a surface density of hydrogen of 8×10^{14} cm^{-2}. This corresponds to an effective bulk concentration [H]/[Si] of ~ 35 to 50 at.%, assuming an average escape depth of (10 ± 2)Å. Inspection of Fig. 3.8 indicates a comparable hydrogen concentration in a-Si:H films prepared at room temperature. We shall show in Sect. 3.3.16a, however, that most of the hydrogen is concentrated in one or two surface layers. The high concentration of hydrogen and the Si–H$_x$ bonding configurations determined from photoemission spectra are thus not always representative of the bulk of a-Si:H.

3.3.9 Valence Band Spectra of Hydrogenated Amorphous Germanium

The He I and He II valence band spectra of hydrogenated amorphous Ge(a-Ge:H) have been measured by *Gruntz* et al. [3.62]. The samples were prepared at room temperature from the glow discharge of 10% GeH$_4$ diluted in argon. The spectra (Fig. 3.12) of the as-prepared film exhibit a dominant hydrogen induced peak A at 5.6 eV below E_F and a shoulder A' at 6.8 eV. A deep lying doublet B$_1$ and B$_2$ at 11.1 eV and 12.1 eV, respectively, shows up in the He II spectrum of Fig. 3.12. The small peak at 9.35 eV is due to the 3p levels of argon embedded into the specimen during deposition [3.128]. Their photoemission cross sections are much smaller at $\hbar\omega = 40.8$ eV and they are barely seen in the He II spectra (see also Fig. 3.6). When neon is implanted in a-Ge its 2p levels show up at a binding energy of 14.4 eV relative to E_F [3.129].

Upon annealing hydrogen is driven out, initially from the configurations corresponding to peaks A and A'. This is accompanied by a reduction in B$_1$

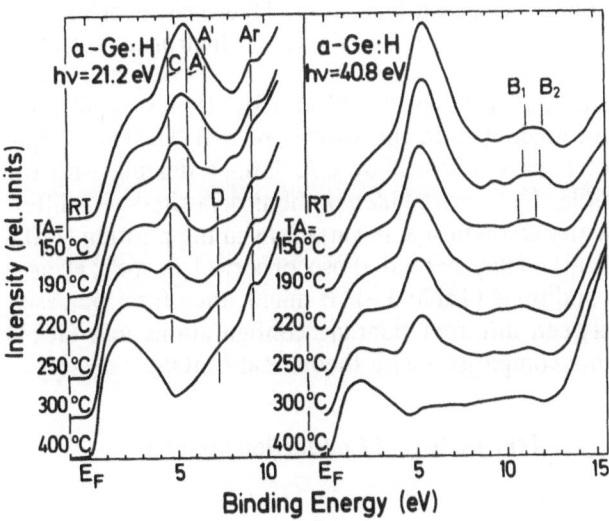

Fig. 3.12. He I and He II valence band spectra of amorphous hydrogenated germanium for a number of annealing steps [3.62]

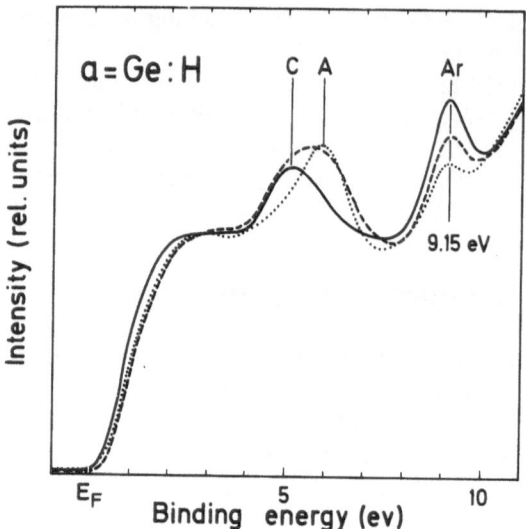

Fig. 3.13. Valence band specta (He I) of gd a-Ge:H films prepared with different flow rates of the GeH$_4$/Ar gas mixture: 0.02 l/h (—), 0.04 l/h (---), and 0.1 l/h (.....) [3.62]

and B$_2$ which shift at the same time to lower binding energies. At an annealing temperature T_A = 220 °C, the first depletion step is completed and hydrogen remains only at sites associated with peak C at 5 eV and a now visible peak D at 7.3 eV. The separation C–D is 2.3–2.9 eV in a-Ge:H, only slightly more than the corresponding 2.0–2.2 eV in a-Si:H.

Hydrogen chemisorbes on c-Ge (110) as monohydride and the photoemission spectrum shows two prominent hydrogen induced levels at 5.2 and 7.8 eV [3.130]. Based on these similarities, *Gruntz* et al. [3.62] identified peaks C and D with monohydride configurations. Hydrogen evolves from this configuration starting at $T_A \simeq$ 220 °C and the film is hydrogen-free at T_A = 400 °C.

Gruntz et al. conjecture that peaks A, A′ and B$_1$, B$_2$ represent polyhydride configurations in analogy with the spectra of a-Si:H. This assignment has, however, not been confirmed by calculations dealing with the electronic structure of a-Ge:H. Spectra taken on films deposited with different flow rates exhibit clearly distinguishable variations in the position (∼ 0.8 eV) of peak A (Fig. 3.13). They exceed those observed in a-Si:H as a function of deposition conditions (Table 3.4). It might, therefore, be easier to try to distinguish between different bonding configurations in a-Ge:H than in a-Si:H based on a comparison with theoretical LDOS.

3.3.10 Yield Spectroscopy and the Density of Conduction States in Amorphous Silicon

As we mentioned in the introduction, optical absorption from sharp core levels into the unoccupied conduction states gives the most direct information about their energy distributions. Since the Si 2*p* core levels have a binding energy of ~ 99 eV, such experiments require synchrotron radiation. The absorption coefficient is, however, not determined from conventional transmission measurements but indirectly via the electron yield that follows the decay of the core hole left behind. This has the advantage that the yield can be measured with the same experimental set-up that is used for photo-emission measurements [3.9, 10]. The essence of yield spectroscopy is as follows (compare Fig. 3.14). The absorption process Si 2*p* $\hbar\omega$ conduction band (CB) leads to a hole in the Si 2*p* core level. The hole state decays via an Auger transition, i.e., a valence electron drops into the Si 2*p* level while the excess energy of about 97 eV is simultaneously transferred to another valence electron that is excited high up into the continuum. This Auger electron escapes with the very small probability $T(E) \lesssim 0.01$ (3.24) from the silicon sample without being inelastically scattered. The elastic Auger electron current constitutes the LVV spectrum where L refers to the Si 2*p* initial hole state and VV to the final state with two holes in the valence bands. The LVV Auger spectrum is essentially a replica of the self-convolution of the Si LDOS and it has been investigated for a-Si : H by *Allie* et al. [3.131].

In yield spectroscopy one selects instead the much more intense $(\sim 1 - T)$ *inelastic* Auger current at its maximum intensity around $E^* \simeq 4$ eV kinetic energy as a measure for the Auger decay rate. The intensity of this current $Y(\omega, E^*)$ as a function of photon energy constitutes the yield spectrum. It has been shown experimentally that $Y(\omega)$ is indeed proportional to $\alpha(\omega)$ [3.10]. Since the initial state is a localized core level, *k*-selection rules play no role in this kind of spectroscopy even for crystals. Instead, the dipole selection rules apply. They allow only transitions between states that differ by one in their

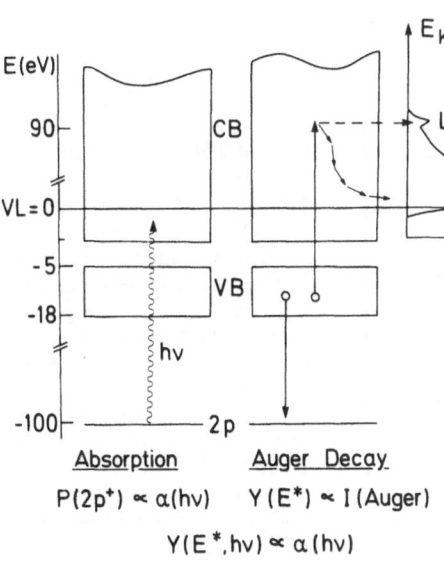

Fig. 3.14. Schematic energy level diagram for yield spectroscopy. The left-hand side indicates the excitation process and the right-hand side the decay of the core hole via the emission of an Auger electron (LVV). The electron yield is measured at the energy E^* of the secondary electron spectrum $Y(E)$. VL, VB and CB refer to the vacuum level, the valence and conduction bands, respectively

orbital angular momentum. For the case at hand, Si $2p \rightarrow$ Si εs, Si εd transitions will dominate the yield spectrum, where ε indicates a continuum state.

The Si $2p \rightarrow$ CB *optical* absorption spectra for c-Si and a-Si have been measured by *Brown* and *Rustgi* [3.132] and the corresponding yield spectra by *Gudat* and *Kunz* [3.133]. In Fig. 3.15 we present the L_{III} (Si $2p_{3/2} \rightarrow$ CB) yield spectra of c-Si, a-Si and a-Si:H (50 vol.% H_2 in sputter gas, $T_D =$ R.T.), all three taken under identical conditions [3.80]. The spectra in Fig. 3.15 have been corrected for contributions from the overlapping L_{II} (Si $2p_{1/2} \rightarrow$ CB) transitions. The silicon $2p_{3/2}$–$2p_{1/2}$ spin-orbit splitting is (0.60 ± 0.05) eV.

The threshold of the L_{III} ($2p_{3/2}$) absorption (point of maximum slope) is 99.90 ± 0.05 eV for c-Si and 99.85 ± 0.05 eV for unhydrogenated sputtered silicon. In hydrogenated a-Si it is shifted by 0.2 eV to 1001.0 eV. The values obtained by *Brown* and *Rustgi* are (99.84 ± 0.06) eV for crystalline and amorphous silicon. The structures between 100 and 103 eV in the spectrum of c-Si are in reasonable agreement with those obtained by *Brown* and *Rustgi* who identified them with transition to maxima in the density of conduction states [3.134] (dotted line in Fig. 3.15). These critical points are, of course, absent in the spectra of the noncrystalline modifications. The complete loss of structure in these spectra – aside from the 1.3 eV wide hump at the onset – is in keeping with the complex band structure results of *Kramer* [3.93]. The delocalized states are less susceptible to the remnants of local order than the more localized valence states.

A remarkable result is the complete lack of discernible hydrogen derived antibonding states in the yield spectrum of a-Si:H. The antibonding states have predominantly Si sp^3 character since the bonding states are mainly H $1s$

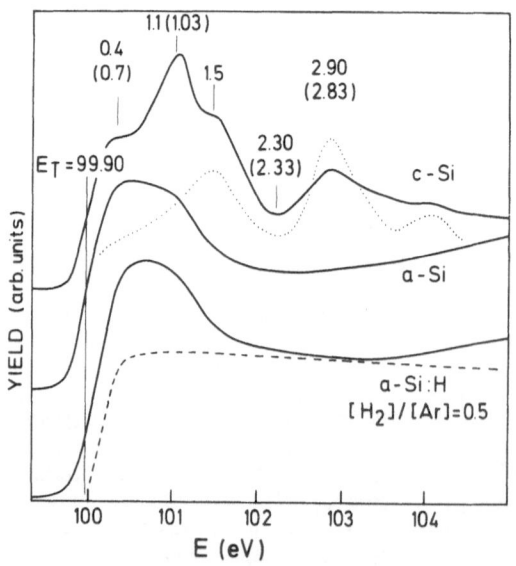

Fig. 3.15. The L_{III} (Si $2p_{3/2} \rightarrow$ conduction band) yield spectra of c-Si, a-Si and a-Si:H. The amorphous films were prepared by reactive sputtering with and without 50 vol.% H_2, respectively. Energies above threshold are given for characteristic features in the spectrum of c-Si and the corresponding numbers obtained by *Brown* and *Rustgi* [3.132] are added in parenthesis. The dotted line is the density of conduction states calculated for c-Si by *Kane* [3.134] and the dashed line indicates the one-electron density of conduction states appropriate for amorphous silicon [3.80].

derived. They are expected to lie at the bottom of conduction band [3.119, 120]. According to the calculation of *Ching* et al. [119], the hybridization of the antibonding states is such that the Si $3p$ partial densities of states exceed the Si $3s$ ones by about a factor of 4 to 5. Si–H antibonding states are therefore expected to be weak in the $2p$ yield spectrum according to the dipole selection rule. It is, therefore, not unlikely that the weak Si–H anti-bonding states are hidden under the initial hump in the yield spectrum of a-Si : H. An altogether satisfying explanation of the yield spectrum of a-Si : H is, nevertheless, still lacking. This is partly due to the fact that calculations of the electronic structure of a-Si : H have been performed using tight-binding methods which are not well suited for conduction states. It has been pointed out by *Brown* and *Rustgi* [3.132] that the initial rise and the region up to about 1 eV above threshold is greatly enhanced over the density of states in the $L_{II, III}$ spectrum of c-Si. They assign this enhancement to the strong Coulomb interaction between core-hole and conduction electron – a view that was subsequently confirmed by the calculations of *Altarelli* and *Dexter* [3.135]. The same Coulomb interaction leads also to the formation of bound excitons in semiconductors and insulators. Individual exciton lines were, however, not resolved in the $L_{II, III}$ absorption spectra but the exciton binding energy E_{xc} has been estimated by comparing the threshold $E_T(L_{III})$ for L_{III} absorption with the sum of the Si $2p_{3/2}$ binding energy relative to the valence band edge $[E_B^v(\text{Si } 2p_{3/2})]$ and the optical gap E_g:

$$E_{xc} = E_B^v(\text{Si } 2p_{3/2}) + E_g - E_T(L_{III}) . \tag{3.43}$$

In this way binding energies of (300 ± 200) [3.136] and (180 ± 20) [3.137] meV have been determined. Our own measurements yield (150 ± 40) meV, for crystalline silicon [3.80]. These values exceed considerably those of the valence band excitons in Si (14 meV) for reasons that are not fully under-stood [3.138]. There is, however, little doubt that the electron-hole interac-tion is also responsible for the sharp onset and the near threshold enhance-ment of the continuum transitions in the $L_{II, III}$ absorption spectra of a-Si and a-Si : H. The spectra of Fig. 3.15 thus do not represent the one-electron density of conduction states. For a-Si, a more realistic shape would be a simple step function as indicated by the dashed line in Fig. 3.15 [3.139] or a structureless average of the crystalline DOS of Fig. 3.15.

3.3.11 The Influence of Hydrogen on Valence and Conduction Band Edges in Hydrogenated Amorphous Silicon

The spectra of Fig. 3.8 reveal an increasing separation of the valence band edge from the Fermi level with the addition of hydrogen to a-Si. When measured at the point of maximum slope, this shift amounts to 1 eV as indicated in Fig. 3.8 for the sample with ∼ 50 at.% hydrogen. *Von Roedern* et

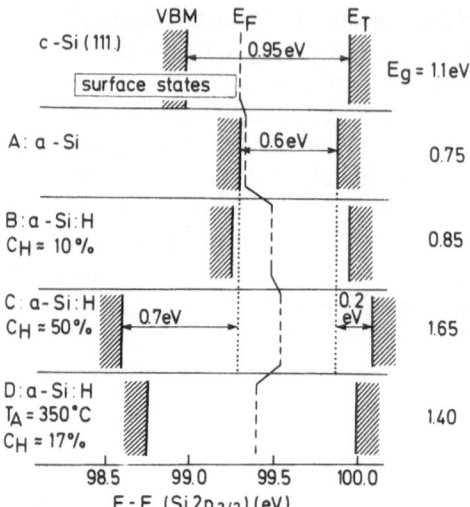

Fig. 3.16. The energies of the valence band maximum (VBM), the Fermi level (E_F) and the photoemission threshold (E_T) are plotted for four samples relative to the binding energy of the Si $2p_{3/2}$ core level. The value of E_g was obtained from E_T − VBM + E_{xc}. The hydrogen concentrations quoted are deduced from the intensities of the chemically shifted Si $2p$ lines as explained in Sect. 3.3.15 a

al. [3.111] proved that the shift is the result of a recession of the valence band edge and not a shift of E_F within the gap because no comparable change in the binding energy of the S $2p$ core levels was observed. They also showed that the correlation between optical gap and hydrogen concentration could be explained by the recession of the valence band alone [3.140]. Recently, the position of the valence band maximum (VBM) of E_F and the L_{III} threshold E_T have been simultaneously measured using synchrotron radiation [3.80]. The results of these measurements are summarized in the form of energy diagrams in Fig. 3.16. All energies are given relative to the unshifted component (Sect. 3.3.14) of the Si $2p_{3/2}$ core level. In this way, mere Fermi level shifts within the gap can be distinguished from changes in the valence band maximum (VBM) and E_T brought about by a redistribution of states near the band edges. The VBM is defined in the usual way through the extrapolation of the steepest descent of the leading edge of the valence band spectrum [3.140], and the definition of E_T has been explained in Sect. 3.3.10. Also given in Fig. 3.16 is the value of the gap E_g defined as

$$E_g = E_T - VBM + E_{xc} , \qquad (3.44)$$

where we used a value of 150 meV for the core level exciton binding energy.

The energy levels so defined are plotted in Fig. 3.16 for three amorphous silicon films and for the (111) surface of c-Si. The three amorphous samples are a hydrogen free film sputtered at room temperature (sample A) and two films, also sputtered at room temperature with different amounts of H₂ in the sputter gas (samples B, C). Film D, finally, is sample C after it has been annealed at 350 °C for ∼ 20 min. The hydrogen concentrations C_H in Fig. 3.16 were obtained from the intensity of peaks A or C–D as explained in Sect.

3.3.8. Sample D has H only in monohydride configurations. The results of Fig. 3.16 may be summarized as follows:

i) The top of the valence bands recedes by as much as 0.7 eV with increasing hydrogen content. This is 0.3 eV less than the recession measured at a point halfway up the leading edge (Fig. 3.8) because the slope of the leading edge also decreases with hydrogenation. In a-Ge : H [3.62, 70] and a-Si : F (Sect. 3.3.15), a similar narrowing of the valence bands by 0.3 and 0.8–0.9 eV, respectively, and a concomitant decrease in the slope of the valence band edge has been observed.

ii) The conduction band edge (E_T) is, by comparison, little affected. The maximum recession is 0.2 eV compared to its position in unhydrogenated a-Si. In fact, the position of E_T is the same in both amorphous and crystalline silicon to within 0.2 eV.

iii) The Fermi level is pinned near VBM in unhydrogenated amorphous Si films. The distance between E_F and VBM may vary between 0 and 0.2 eV for undoped specimens. The smallest amount of hydrogen added to the film frees E_F to move towards the middle of the gap. Further addition of hydrogen results only in minor movements of E_F in such a way that the distance E_T–E_F remains virtually constant at (0.52 ± 0.07) eV, corresponding to an energy of 0.67 eV below the conduction band edge after correction for the exciton binding energy.

The recession of the VBM with hydrogenation is in agreement with the results of most calculations [3.118–123, 125]. It is ascribed to the replacement of the Si–Si bond with the stronger Si–H or Si–F bond which moves states from the top of the valence bands to a position deep inside the valence bands, where they are observed as peaks A, B, C, D, E. The depletion of states is not limited to the Si atoms bonded directly to H or F. According to *DiVincenzo* et al. [3.125], the presence of hydrogen on a dangling bond reduces the bonding charge associated with valence states in the top 0.25 eV out to the third-nearest neighbor. That is the reason why we observe an actual *recession* of the VBM even for small hydrogen concentrations and not just an attenuation in the emission near the top of the valence due to the few atoms bonded directly to H or F. The bottom of the conduction bands is little affected by the addition of H in agreement with the calculations [3.120, 125] and the shift in VBM accounts for almost all the change in the optical gap E_g with hydrogen content (Sect. 3.4.3 b). This result suggests that fluctuations in hydrogen content as they occur apparently in most samples [3.34, 141, 142] will lead to spatial fluctuations of the valence band edge and as a consequence, to localized states near the valence band edge. The localized states exist in regions with low hydrogen concentration and it has been suggested [3.143] that they are responsible for the E_y peak in the density of gap states derived from field effect measurements [3.144]. Moreover, these states could act as deep traps for holes and thus provide a natural explanation for the large disparity between hole and electron mobilities [3.145, 146]. *Brodsky* has taken these considerations further and discusses tail states localized in quantum wells that are formed by regions with low H content surrounded by those with high c_H [3.147].

The pinning of E_F near the VBM in hydrogen-free a-Si is in agreement with the observation of *Beyer* et al. [3.148] that these samples are p-type. The activation energy of 0.77 eV measured by *Beyer* et al. at high temperatures requires, however, that the mobility edge E_v lies about 0.5 to 0.6 eV below the VBM and that the states between E_v and the VBM are localized. A region of localized states below the VBM that is at least 0.3 eV wide is necessary in sample B to ensure the n-type conductivity that is generally observed for hydrogenated amorphous silicon. It is not possible to place E_v in the spectra of samples C and D.

The position of E_F in sample A could be due to the high bulk defect density in the lower half of the gap; it may, however, equally well reflect the surface position of E_F. In c-Si the Fermi level is pinned (0.35 ± 0.15) eV above the valence band edge by a band of surface states [3.149], as indicated in Fig. 3.16. It is possible that a similar band of surface states pins E_F in unhydrogenated a-Si and that E_F moves towards mid-gap as these surface states are passivated by hydrogen. The surface charge Q/e necessary to induce a band bending V_s is [3.150]

$$Q/e = - V_s \sqrt{\varepsilon \varepsilon_0 g} , \tag{3.45}$$

where g is the density of gap states assumed to be constant. With $g = 10^{20}$ eV^{-1} cm^{-3} and $V_s = 0.19$ eV (the shift in E_F between samples A and B), we have $Q/e = 6 \times 10^{12}$ cm^{-2}. This is a reasonable number corresponding to 1 electron per 100 surface atoms and it would reduce the width of localized states below the VBM in the bulk of sample A to 0.3–0.4 eV, a value comparable to that in sample B. Alternatively, one could argue that the states at the VBM in sample A correspond to inner surface states associated with voids. These states are then primarily saturated when a-Si is hydrogenated. We shall return to the question of surface states in a-Si in Sect. 3.3.16 c.

3.3.12 Gap State Spectroscopy

The energy distribution of gap states has been obtained from field effect [3.144, 151, 152], C-V measurements [3.153] and more recently using deep level transient spectroscopy [3.154]. The gap state distribution $g(E)$ has to be deduced from these measurements in an iterative procedure in which many of the details of $g(E)$ are lost.

Optical spectroscopy is, as we shall see (Sect. 3.4), a sensitive spectroscopic method to detect gap states. It is, however, hampered by cross-section effects and the fact that it measures the convolution of occupied and empty states.

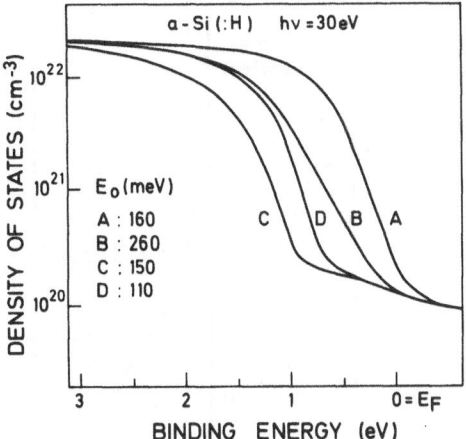

Fig. 3.17. The gap region of the EDC's of samples A to D of Fig. 3.16. The logarithmic slopes E_0 are given in meV

Photoemission offers some advantages in this situation. As pointed out earlier, the photoemission matrix elements are determined by the core parts of the wave functions and they are therefore the same for localized and delocalized states. Also, gap states observed in photoemission are obviously occupied and they can be placed unambiguously in energy relative to the VBM or E_F.

Consequently, a number of attempts have been made to observe band tailing and gap states directly in photoemission from a-Ge [3.155, 156], a-Si [3.157–160] and a-Te [3.161] and for a-Si:H [3.83]. The possibilities offered by this technique and the problems encountered may be illustrated with the help of Fig. 3.17. Here, on a logarithmic scale, the valence band edges of the four amorphous silicon samples of Fig. 3.16 are plotted which were obtained using $\hbar\omega = 30$ eV. The count rates have been transformed into a density-of-states scale using the following argument. The p-like valence states (peak I in Fig. 3.7a) are ~ 4 eV wide and contain 2 electrons per atom. We have, therefore, for the average density of states in this band,

$$\overline{N}_v(\text{peak I}) \simeq \frac{2}{4} \times N_A = \frac{2}{4} \times 5 \times 10^{22}$$

$$= 2.5 \times 10^{22} \, \text{eV}^{-1} \, \text{cm}^{-3} \,, \tag{3.46}$$

where $N_A = 5 \times 10^{22}$ cm^{-3} is the atom density of a-Si. The scale in Fig. 3.17 is obtained by equating $\overline{N}_v$ with the intensity of the photoemission spectrum ~ 3 eV below E_F.

The valence band edges exhibit distinct tails with an exponential energy dependence $N(E) \propto \exp(-E/E_0)$ over one order of magnitude in $N(E)$. The steepness parameter E_0 varies, as indicated in the figure. It is largest for the lightly hydrogenated sample (sample B) and smallest for sample D, the sam-

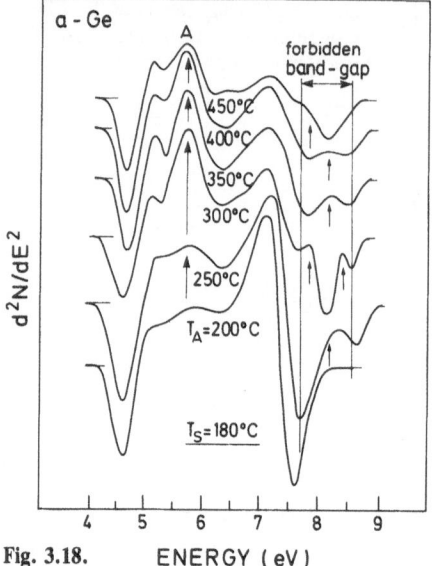

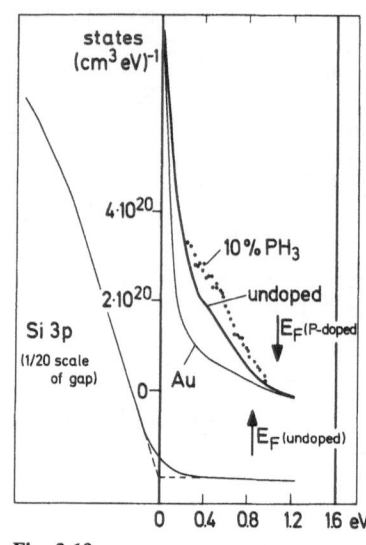

Fig. 3.18.   ENERGY (eV)   **Fig. 3.19.**

Fig. 3.18. Second derivative of the EDC's of a-Ge for a series of annealing temperatures T_A. The arrows indicate features which have been attributed to gap states [3.156]

Fig. 3.19. EDC of the gap region of an undoped and heavily P-doped a-Si : H sample, measured as deposited at 250 °C. The Fermi edge of Au is also shown as a measure of the spectrometer resolution. It is shifted to coincide with the valence band edge of the a-Si : H samples. The zero of energy is the valence band maximum (VBM) [3.83]

ple that has been annealed at 350 °C. The absolute values of E_0 should be taken with caution, however, since they include the finite resolution with which the data of Fig. 3.17 are taken. The *changes* in E_0 reflect, nevertheless, actual changes in the tailing of the valence bands brought about by hydrogenation and annealing.

It is also apparent from Fig. 3.17 that the range in $N(E)$ accessible to photoemission is limited to about two orders of magnitude, in agreement with earlier assessments [3.158], due to a non-negligible background. Electrons originating from states below the VBM contribute to this background in the gap region as a result of the finite experimental resolution. Two orders of magnitude in $g(E)$ are insufficient to explore the density of gap states at a level of 10^{20} eV^{-1} cm^{-3} or less. There are exceptions to the rule, however.

Laude et al. [3.156] were able to identify two peaks in the density of gap states in a-Ge which appeared at the onset of crystallization (Fig. 3.18). They enhanced the sensitivity of their spectra to *structure* in the density of states by measuring the second derivative of the electron current. In this way the two peaks in the gap that are marked by arrows in Fig. 3.18 could be identified. They correspond most likely to bonds that break during the rearrangement of the amorphous network in the course of crystallization. Notice that an estimate of the state density is not possible since the signal strength in Fig. 3.18 depends on the *change* in $g(E)$ with E rather than on its absolute value.

Another example of structure in $g(E)$ that has been identified using photoemission is the spectrum of heavily phosphorus doped a-Si : H shown in Fig. 3.19 [3.83]. These P induced defect states ~ 0.4 eV above the VBM have subsequently been identified [3.162] as the initial states giving rise to the extrinsic absorption band at ~ 1.2 eV in the photoconductivity and optical absorption spectra (Sect. 3.4.5).

In order to exploit the potential of photoemission for gap state spectroscopy further, it seems mandatory to reduce the background. The part of the background that is due to the finite resolution of the experiment may be considerably supressed by using photon energies very close to the photoemission threshold (i.e., $\hbar\omega \gtrsim E_F - $ VBM). This has been done by *Fischer* and *Erbudak* for a-Si [3.159, 160]. They were thus able to suppress most of the valence band emission and probe the gap states with an improved signal-to-background ratio. This method has the further advantage that the electron mean free path is considerably increased (Fig. 3.3) and the spectra are therefore less surface sensitive than spectra taken with $h\nu \gtrsim 20$ eV.

3.3.13 The Position of E_F

The position of E_F within the gap varies with doping and with temperature. Changes in the activation energy E_σ obtained from temperature-dependent dark conductivities are interpreted in terms of changes in the position of E_F relative to a mobility edge. An implicit assumption usually made is that E_F shifts linearly with T:

$$E_F(T) = E_F(0) - \beta T \ . \tag{3.47}$$

In that case the temperature dependence of E_F drops out of the expression for E_σ and $E_\sigma = (E_F - E_v)_0$ or $(E_c - E_F)_0$, where the subscript indicates that the energies are appropriate for zero temperature (Chap. 2).

The shift of E_F with doping and with temperature has been independently measured with photoemission. The basis of this method has been explained in Sect. 3.3.4 in connection with Fig. 3.5.

a) The Shift of E_F with Doping

The position of E_F as a function of doping has been investigated by *Williams* et al. [3.82] and *von Roedern* et al. [3.163]. The former determined the change in the binding energy relative to E_F of the Si $2p$ core levels of three samples: one undoped, one heavily P doped and one heavily B doped. The samples were prepared by glow discharge outside the spectrometer and their surfaces were oxidized when measured. The binding energy difference was

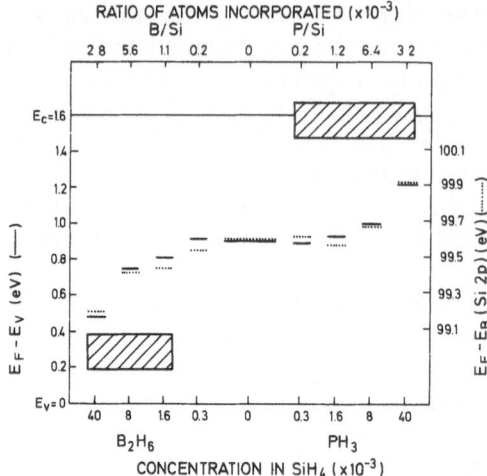

Fig. 3.20. Position of the Fermi level E_F in the gap of a-Si : H as determined spectroscopically for a range of P and B-doped samples. The hatched regions indicate the energies at which charge transport takes place as deduced by combining the activation energies with E_F. The conduction band edge E_c is derived by adding an optical gap of 1.6 eV to the valence band maximum E_v. The number of B or P atoms incorporated (*upper abscissa*) has been derived as described in the text [3.163]

1.2 eV between the two doped samples in good agreement with the 1.2 eV difference in E_σ obtained from conductivity measurements performed on the same samples. They concluded, therefore, that changes in E_σ were indeed due to corresponding shifts in E_F.

Von Roedern et al. came essentially to the same conclusion from measurements on a series of gd a-Si : H ($T_D = 250\,°C$, $T_A = 350\,°C$) films prepared in situ and therefore unoxidized. This enabled them to follow the separation between E_F and the top of the valence bands (E_v in Fig. 3.20) in addition to $E_F - E_B$ (Si 2p). The two quantities track each other well as a function of doping and the total variation in E_F of 0.75 eV agrees with the change in E_σ observed for the same samples. By adding (or subtracting for p-type samples) the activation energies to (or from) E_F, *von Roedern* et al. obtained the average energies of the conduction paths as indicated by the hatched regions in Fig. 3.20. For n-type samples this region coincides with E_c, the bottom of the conduction bands. For p-type samples a region 0.3 eV above the VBM appears to carry the hole current. This is a direct spectroscopic confirmation of the analysis of transport data by *Persans* [3.164] using the formalism proposed by *Döhler* [3.165]. The average energy $\bar\varepsilon$ of the conductance path defined through the Kubo-Greenwood formula for $\sigma(T)$ is [3.164]

$$\bar\varepsilon = \frac{\int \varepsilon\, e^{(-\varepsilon/kT)}\, g(\varepsilon)\mu(\varepsilon)d\varepsilon}{\int e^{(-\varepsilon/kT)}\, g(\varepsilon)\,\mu(\varepsilon)d\varepsilon}\ , \tag{3.48}$$

where all energies are measured from E_F and $\mu(\varepsilon)$ is the energy-dependent carrier mobility. A large Boltzmann factor (small ε) in conjunction with a large $g(\varepsilon)$ may thus shift $\bar\varepsilon$ into a region of low mobility, i.e., into the tails. A high density of gap states $g(\varepsilon)$ near the VBM was already observed in the undoped samples of *von Roedern* et al. [3.83] (Fig. 3.19) and there is evidence that B doping increases this density further [3.166]. The exact character of the tail states and the transport mechanisms in these states is still a matter of controversy [3.145, 164]. There is no comparable tailing of the conduction band edge (Sect. 3.4.4).

Another aspect of Fig. 3.20 worth noting are the very small shifts of E_F obtained for low dopant concentrations. Significant movement sets in only for $[PH_3]/[SiH_4] \gtrsim 8 \times 10^{-3}$ and $B_2H_6/SiH_4 \gtrsim 1.6 \times 10^{-3}$. This is in contrast to the variation in E_σ with doping: rapid changes for small concentrations and saturation for concentrations approaching 1×10^{-3} in the gas phase [3.167]. It is possible that this difference is due to a near surface density of gap states $g_s(E)$ which differs from that of the bulk. The photoemission measurements are, of course, sensitive to the position of E_F within the first 20 Å or so, whereas E_σ measures the shift of E_F in the bulk unless we are dealing with an accumulation layer. The maximum difference of 0.2 eV (the width of the hatched regions) allowed between ΔE_F and ΔE_σ, according to Fig. 3.20, corresponds to a difference of 2×10^{19} eV^{-1} cm^{-3} between $g_s(E)$ and $g(E)$. For a Debye length of 100 Å, this difference may be accounted for by one true surface state for every fifty surface atoms. It seems worthwhile to pursue this question further using more accurate data in view of the known fact that the surfaces of a-Si:H films exhibit electrical properties that differ from those of the bulk [3.168]. *Williams* et al. concluded from a similar analysis for their films that the density of states at the a-Si:H $-$ SiO$_2$ interface was less than 10^{11} cm^{-2} [3.82].

That doping shifts the Fermi level in a-Ge:H much in the same way as in a-Si:H has been demonstrated by *Gruntz* et al. using the Ge $3d_{5/2}$ and the Ar $3p$ core levels [3.62]. Their results indicate a maximum shift of 0.5 eV between heavily B and P doped samples. This is much smaller than the band gap of a-Ge:H of about 1.0 eV.

The actual concentration of dopant atoms has also been determined by *von Roedern* et al. [3.163] utilizing the relative intensities of the B 1s, P 2p, and Si 2p core levels of the most heavily doped films in conjunction with (3.39). The result that 80% of the P atoms in PH$_3$ and 70% of the B atoms in B$_2$H$_6$ are incorporated into the films is in good agreement with the investigations of *Thomas* [3.169]. The incorporation ratios for a-Ge:H are virtually identical: 80% for B and 70% for P [3.62].

b) The Statistical Shift of E_F

The Fermi level shifts as a function of T when E_F falls in a range of the DOS where $dg(E)/dE \neq 0$. Consider Fig. 3.21. The number of electrons is given by

$$n(E, T) = g(E) f(E, T) = g(E) \frac{1}{1 + \exp[(E - E_F)/(kT)]} . \tag{3.49}$$

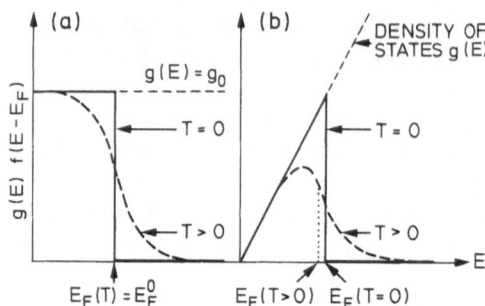

Fig. 3.21 a, b. Model densities of gap states $g(E)$ and the statistical shift of E_F.

The Fermi function $f(E, T)$ is a step function for $T = 0$ with $E_F^0 = E_F(T = 0)$ positioned at the discontinuity of $f(E, 0)$. The Fermi function broadens with increasing temperature. For the case shown in Fig. 3.21 b, more states above E_F^0 are populated than states below E_F^0 are emptied. The conservation of the number of electrons requires, therefore, that $E_F(T > 0)$ moves in the direction of a decreasing DOS. This is referred to as the statistical shift. For a DOS that varies linearly in the neighborhood of E_F^0

$$g(E) = g(E_F^0) + m(E-E_F^0) , \qquad (3.50)$$

the shift in E_F is approximately given by

$$E_F(T) - E_F^0 = -\frac{1}{6} \frac{m(2kT)^2}{N(E_F^0) - mE_F^0 + \frac{1}{2} m(E_F^0 + E_F)} . \qquad (3.51)$$

For a density of states that increases by a factor of 100 over 100 meV ($4\,kT$), the shift in E_F between 0° and room temperature is only about 40 meV. To obtain the necessary precision, *Gruntz* [3.70] measured the energy difference between E_F and three well-defined points in the spectrum of a-Si:H. The position of E_F was remeasured for each temperature using a clean molybdenum film. In this way a reproducibility of 20 meV has been obtained in the shift of the three fiducial points with respect to E_F.

The statistical shifts of four glow discharge a-Si:H films between -200 and $+200$°C are shown in Fig. 3.22. From these measurements, *average* statistical shifts δ between -150°C and $+150$°C have been obtained which are compared in Fig. 3.23 with those calculated by *Jones* et al. [3.170] from the field effect DOS of *Madan* et al. [3.144]. δ is plotted in Fig. 3.23 as a function of $E_F(RT) - E_v$, where $E_v \equiv$ VBM for the photoemission data. For the heavily doped films the values of δ are in reasonable agreement. The statistical shift in this range of E_F is determined by the rapidly rising DOS towards the conduction band. Whereas the field effect DOS levels out towards the middle of the band gap ($\delta \to 0$), the measurements of *Gruntz* [3.70] require a minimum in $N(E)$ at $E_F - E_v \simeq 1.2$ eV and a new rise below

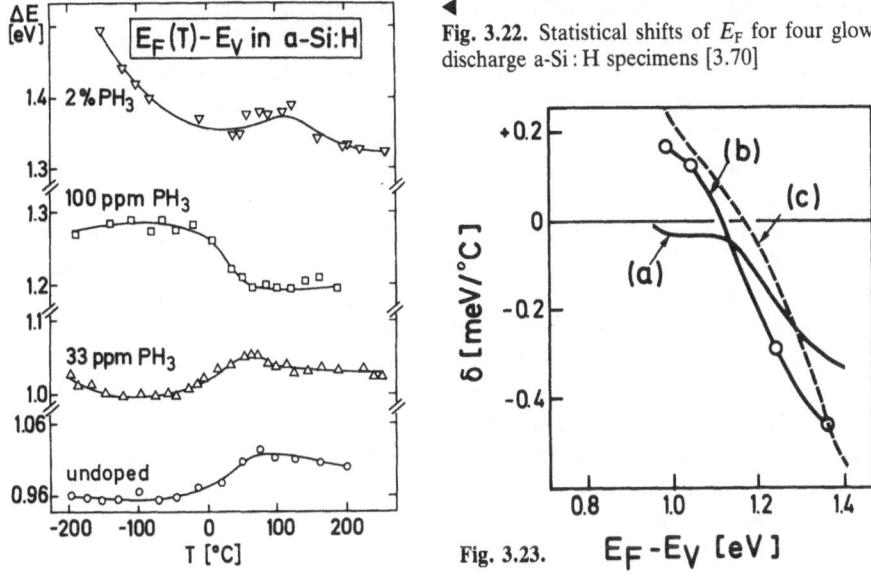

◄
Fig. 3.22. Statistical shifts of E_F for four glow discharge a-Si:H specimens [3.70]

Fig. 3.23. The temperature coefficient of E_F, δ, as a function of the Fermi level position; (*Curve a*) calculated from the field-effect density of states, taken from [3.144], by *Jones* et al. [3.170]; (*Curve b*) obtained between $-150\,°C$ and $+150\,°C$ from the data of Fig. 3.22, and (*Curve c*) derived from $\sigma(T)$ and $S(T)$ measurements on Li-doped samples by *Overhof* and *Beyer* [3.166]. The data of *Jones* et al. and *Overhof* and *Beyer* have been replotted relative to E_v assuming a mobility gap of 1.6 and 1.7 eV, respectively

that value to explain the positive δ observed for undoped and lightly P doped samples. Such densities of states have recently been obtained using field effect [3.152] and DLTS [3.154] measurements.

The results of *Gruntz* for $E_F(T)$ are very similar to those obtained by *Overhof* and *Beyer* [3.166] from an analysis of dark conductivity and thermopower measurements. Their results are included as dashed lines in Fig. 3.23 for comparison and the density of states required to reproduce the positive δ is shown in Fig. 3 of [3.166]. The origin of the structure in $E_F(T)$ that is more pronounced in the photoemission data than in the data of *Overhof* and *Beyer* is not understood at present.

3.3.14 Charge Fluctuations in Amorphous Silicon

There are three possible sources of electric charges which could give rise to a chemical shift of the Si core levels according to the mechanism described in Sect. 3.3.5: charged defects, charge transfer between Si and H upon formation of a Si–H bond and charge fluctuations on the Si atoms as a result of bond-length and bond-angle variations. According to the model calculations of *Guttman* et al. [3.171], the latter mechanism leads to a Gaussian distribution of charges on the Si atoms with an rms width of 0.2 elementary charges. A detailed analysis reveals that bond-angle variations and the concomitant changes in the second nearest-neighbor distance are mostly responsible for these charges. Nearest-neighbor distances are virtually rigid in a-Si and the orbital overlap between neighbors more than one removed is too small to induce a sizable rehybridization upon distance changes.

The most likely charged defect in a-Si is a dangling bond occupied with either no or two electrons. In the terminology of *Adler* [3.172], this corresponds to states T_3^+ and T_3^-, respectively. The analog of these defects is the dangling-bond states on a reconstructed c-Si (111) surface. The reconstruction is driven by a Jahn-Teller distortion in which the originally singly occupied dangling bonds acquire alternatively positive and negative charges via a charge transfer between neighboring rows of dangling-bonds [3.173]. This corresponds to the reaction $T_3^0 \rightarrow T_3^+ + T_3^-$ and it is accompanied by the loss of the spin on the T_3^0 dangling-bond state. A similar situation is expected to hold for the inner surfaces of voids in a-Si. It would explain the disparity between the spin density in unhydrogenated a-Si and the hydrogen concentration necessary to quench the spin signal [3.174].

The charge transfer connected with the reconstruction of the Si(111) surface has been measured through the chemical shift of the Si $2p$ core level of the surface atoms [3.175, 176]. Figure 3.24 shows the spectrum of the Si $2p$ lines obtained on a clean Si(111) surface using synchrotron radiation [3.177]. The photon energy ($h\nu = 140$ eV) is chosen such that the photoelectrons have a kinetic energy (~ 40 eV) near the minimum of the electron mean free path ($\lambda_e \simeq 4$ Å) to obtain maximum surface sensitivity. The spectrum has been fitted by three spin-orbit split doublets ($2p_{3/2}$, $2p_{1/2}$, $\Delta_{so} = 0.60$ eV). They represent emission from bulk Si atoms and two doublets (S1 and S2 in Fig. 3.24) which correspond to Si atoms in the reconstructed top surface layer. The energy separation of S1 and S2 from the bulk lines is -0.39 and $+0.33$ eV, respectively. This corresponds to an average charge transfer between the surface atoms $\Delta q_{surf} \simeq 0.16$ electrons using the calibration factor of 2.2 eV/e mentioned earlier (Sect. 3.3.5). The charge transfer is less than one due to the formation of bands of surface states. Nevertheless, this example gives an indication of the magnitude of chemical shifts expected for reconstructed inner void surfaces in a-Si.

Upon hydrogenation the reconstruction is removed and a uniform chemical shift towards *higher* binding energy is expected as a result of the charge transfer to the more electronegative hydrogen atom. Measurements on a-Si and a-Si:H [3.177] indicate, however, that the charged dangling-bond states are of secondary importance in determining the charges on Si atoms in a-Si. The other two mechanisms dominate. In Fig. 3.25 we compare the $2p$ spectra of c-Si and a-Si obtained with $h\nu = 107.5$ and 110 eV, respectively. The spectra represent bulk Si atoms because of the large (~ 40 Å) electron mean free path of electrons with only ~ 5 eV kinetic energy. The Si $2p$ doublet of the Si(111) surface (Fig. 3.25a) is well described by the convolution of a

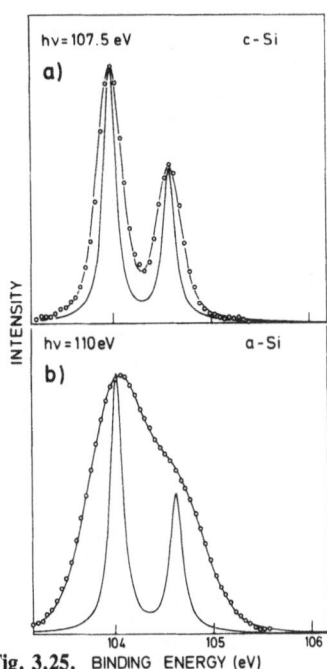

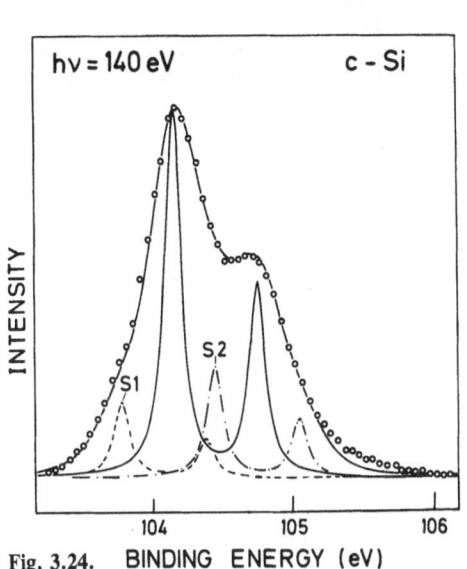

Fig. 3.24.   BINDING ENERGY (eV)

Fig. 3.25.   BINDING ENERGY (eV)

Fig. 3.24. Si $2p$ core level spectra of a c-Si(111) surface. The data points have been fitted with three doublets: one corresponding to bulk Si atoms (——) and two (S1 and S2) corresponding to Si atoms in the topmost reconstructed layer. Only the Lorentz contributions of the component lines are shown. Binding energies are referenced to the vacuum level [3.177]

Fig. 3.25. Si $2p$ core level spectra of a-Si and c-Si. The photon energy chosen ensures that mainly bulk atoms contribute to the spectra. Binding energies are referenced to the vacuum level [3.177]

Table 3.5. Parameters obtained in the least squares fits of Si $2p$ core levels of Figs. 3.25, 3.26. Errors in the last digits are given in parentheses

Sample		Photon energy [eV]	Binding energy relative to vacuum level Si $2p_{3/2}$ [eV]	2Γ [meV]	σ_{exp} [meV]	σ_{amorph} [meV]	Δq (rms) [|e|]
c-Si		107.5	103.99(05)	210(30)	60(5)	–	–
a-Si		110.0	104.01(05)	106(30)	80(5)	256(10)	0.11
a-Si:H	Si_0	110.0	104.21(05)	106(30)	60(5)	192(10)	0.09
	Si_1		104.55(05)			128(10)	0.06
	Si_2		104.89(05)			64(10)	0.03
	Si_3		105.23(05)			0	0

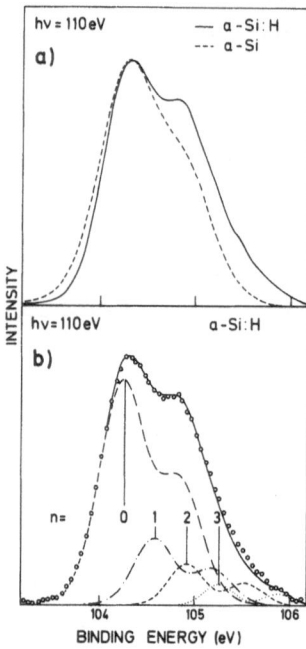

Fig. 3.26. (a) Comparison of the Si 2p spectra of unhydrogenated (a-Si) and hydrogenated (a-Si:H) amorphous silicon. The a-Si spectrum is shifted by 0.2 eV to higher-binding energies. Binding energies are referenced to the vacuum level. (b) Least squares fit of the a-Si:H spectra to four doublets corresponding to Si atoms bonded to $n = 0, 1, 2,$ or 3 hydrogen atoms, respectively [3.177]

Lorentzian, $[(\omega - \omega_0)^2 + \Gamma^2]^{-1}$, which takes into account the lifetime of the core hole, and a Gaussian, $\exp[-(\omega - \omega_0)^2/2\sigma^2]$, representing the combined resolution of monochromator and electron analyzer. The parameters obtained from a least squares fit to the data points are listed in Table 3.5.

The corresponding spectrum (Fig. 3.25 b) taken with $h\nu = 110$ eV from an amorphous silicon sample prepared in situ by sputter deposition is considerably broader so that the spin-orbit components are no longer resolved. The data points are again fitted by the convolution of a Lorentzian and a Gaussian with parameters displayed in Table 3.5. The width of the Gaussian is partitioned into a contribution σ_{exp} due to the experimental resolution obtained from a spectrum of c-Si measured under identical conditions and a contribution σ_{amorph} which takes the homogenous broadening into account such that $\sigma_{exp}^2 + \sigma_{amorph}^2 = \sigma^2$.

The Si 2p core level spectrum ($h\nu = 110$ eV) of hydrogenated a-Si (prepared by adding 50 vol.% H_2 to the sputter gas) is compared with that of unhydrogenated a-Si in Fig. 3.26a [3.177]. The leading (low-binding energy) edge of the spectrum of a-Si:H is clearly steeper than that of a-Si and extra intensity is found towards higher-binding energy. The data can be fitted with a minimum of four doublets: an unshifted doublet corresponding to emission from Si atoms not bonded directly to hydrogen (Si_0) and three components representing Si_{4-n}–Si–H_n configurations (referred to as Si_n hereafter). The authors assumed a fixed chemical shift ΔE_H between Si_n and Si_{n+1} configurations in accordance with the additivity of chemical shifts (Sect. 3.3.5). The fit assumes that σ decreases with increasing n towards $\sigma_{\exp}$. Two equally satisfactory fits were obtained assuming a linear or an exponential decrease from $\sigma(Si_0)$ to $\sigma(Si_3) \equiv \sigma_{\exp}$. The former is shown in Fig. 3.26 b and the fit parameters are displayed in Table 3.5. The hydrogen induced shift $\Delta E_H = 0.34 \pm 0.10$ eV is independent of the assumptions made for σ. It corresponds to a charge transfer of $\Delta q = 0.15$ electrons from Si to hydrogen, noticeably larger than expected on the basis of *Pauling's* electronegativity [3.75] ($\Delta q = 0.02$) and that obtained by *Usami* et al. [3.178]. *Kramer* et al. [3.179] obtained a value of $\Delta q = 0.27$ for the charge transfer from calculations using a hydrogenated silicon cluster model.

The addition of hydrogen reduces the charges in proportion to the number of hydrogen atoms attached to each Si atom. The charges on silicon atoms not bonded directly to hydrogen is only slightly reduced (Table 3.5). This excludes the assumption that all broadening in a-Si is due to surface chemical shifts of atoms surrounding voids. It must rather be attributed to random charge fluctuations as the result of bond-length variations in the amorphous network. The average charge is estimated to be ± 0.07 electrons in a-Si, only about half as much as obtained by *Guttman* et al. [3.171]. In a-Si : H the incorporation of hydrogen leads, apparently, to an overall reduction in the bond-length fluctuations as manifested in the reduced σ_{amorph}, corresponding to charge fluctuations (rms) of only ± 0.09 electrons. This is in agreement with the 20% attenuation of the TO bands in the *ir* spectrum of a-Si upon hydrogenation [3.180].

The diminishing values of σ_{amorph} for Si_n ($n \geq 1$) also point more towards bulk fluctuations in q as a result of constraints rather than to inner surface chemical shifts resulting from dangling bonds. In the latter case the saturation of the dangling bonds would immediately reduce σ_{amorph} to zero, whereas we observe a gradual reduction in accordance with a gradual relaxation of the constraints put on a Si atom as one, two, or three bonds are released from the network. Dangling bonds are present in a-Si and their number is greatly reduced by the addition of hydrogen. However, their concentration ($\lesssim 10^{20}$ cm^{-3}) is approximately two orders of magnitude lower than the hydrogen concentration normally found in a-Si : H [3.174]. Dangling bonds which are saturated with hydrogen represent, therefore, only a small fraction of the total number of Si–H bonds observed.

Ley et al. [3.177] interpreted the Si 2p line broadening in terms of the spatially fluctuating charges Δq on the Si atoms. Connected with these charges are, of course, potential fluctuations which can be determined once the local distribution of the charges are known in addition to their magnitude. Remembering what we said about the origin of the chemical shift in Sect. 3.3.5, it is clear, however, that the binding-energy shifts themselves are equal to the variations in the potential at the sites of the Si atoms. The potential fluctuations thus have a Gaussian distribution with an rms magnitude of 256 meV in a-Si and 192 meV in nonannealed a-Si : H. As we shall see in Sect. 3.4.4, the electric fields connected with these potential fluctuations may contribute to the exponential absorption tail in amorphous silicon. Chemical shifts induced by C in Si have been measured by *Katayama* et al. [3.181] and those for N in a-Si by *Kärcher* and *Ley* [3.182]. The charge transfer due to the Si–F bond will be discussed in the following section.

The areas under the components Si_n in Fig. 3.26 are a measure of the absolute hydrogen content of the films within the sampling depth of photoemission. This aspect has not yet been exploited in a-Si : H but it has been used successfully in a-Si : F (Sect. 3.3.15).

3.3.15 Amorphous Fluorinated Silicon

Hydrogen may be replaced by fluorine in a-Si with very similar beneficial effects for the electrical properties of the material [3.183, 184]. The photoemission spectra of a-Si : F have been investigated by *Gruntz* et al. [3.185, 186]. The films used in their investigation were prepared by sputtering silicon at room temperature in an argon-SiF$_4$ atmosphere (0–50 mol.% SiF$_4$).

a) The Fluorine Distribution in Amorphous Fluorinated Silicon

The highly electronegative F shifts the Si $2p$ line by $\Delta E(F) = (1.15 \pm 0.02)$ eV for each attached F atom. The shift is additive, i.e., Si–F$_n$ units have a chemical shift of $n \times \Delta E(F)$. A minimum of five lines corresponding to Si–F$_n$ ($n = 0\ldots4$) units is necessary to fit the spectra, as illustrated by the two examples of Fig. 3.27. The Si–F$_4$ units correspond to trapped SiF$_4$ molecules.

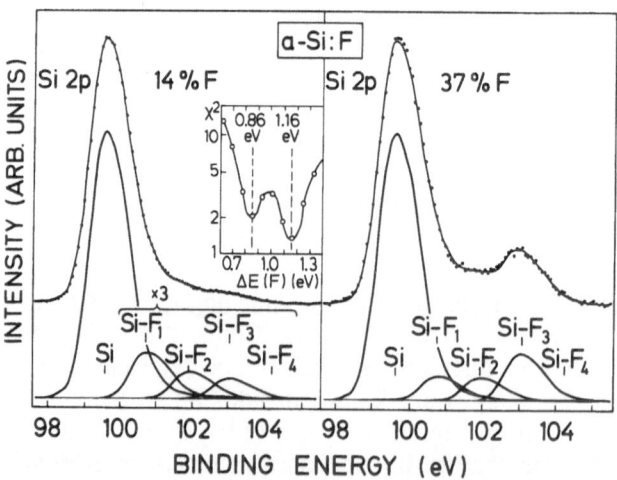

Fig. 3.27. Si $2p$ core level spectra of two a-Si : F specimens. The spectra ($\cdots$) were obtained with monochromatized Al K$_\alpha$ radiation. The solid line is the result of a fit to five components corresponding to Si atoms bonded to $0\ldots4$ fluorine atoms. The inset represents the χ^2 test for a particular fit with the chemical shift per fluorine atom $\Delta E(F)$ as the parameter [3.186]

They are only observed for the highest F concentrations ($C_F \gtrsim 35$ at.%) and their existence has been confirmed by infrared absorption measurements [3.187].

The absolute F concentration and the distribution of the F atoms among the different configurations is simply related to the areas $I(\text{Si–F}_n)$ under the component lines.

$$N_{F_m} = m\, I(\text{Si–F}_m) \quad \text{and} \tag{3.52}$$

$$N_{Si} = \sum_{n=0}^{4} I(\text{Si–F}_n) \tag{3.53}$$

are the number of F atoms in configuration Si–F$_m$ and the number of Si atoms within the sampling depth of photoelectrons, respectively. The total, $C_F(\text{Si}\,2p)$ and partial, $C_F(\text{Si–F}_m)$, fluorine concentrations are given in terms of the N_{F_m} and N_{Si} by

$$C_F(\text{Si}\,2p) = \frac{\displaystyle\sum_{n=1}^{4} N_{F_n}}{\displaystyle\sum_{n=1}^{4} N_{F_n} + N_{Si}} \quad \text{and} \tag{3.54}$$

$$C_F(\text{Si–F}_m) = \frac{N_{F_m}}{\displaystyle\sum_{n=1}^{4} N_{F_n} + N_{Si}}. \tag{3.55}$$

In panel (a) of Fig. 3.28 the values of $C_F(\text{Si–F}_m)$ are plotted versus $C_F(\text{Si}\,2p)$ for samples with F concentrations between 0 and 45 at.%. The same data are replotted in Fig. 3.28b to indicate the relative contribution each configuration makes to the total fluorine content. The distribution of the Si–F$_n$ units derived from a simple statistical model for the incorporation of fluorine in a-Si is also shown for comparison in panel (c) of Fig. 3.28. The model considers a SiX$_4$ cluster and asks for the probability of finding a Si–F$_n$ configuration under the assumption that the four places X are statistically occupied by F or Si atoms. The boundary condition is that the F/Si ratio corresponds to a given fluorine concentration C_F.

For low C_F the Si–F$_1$ units dominate as expected. With increasing fluorine concentration the relative contribution of Si–F$_1$ decreases and Si–F$_2$ and Si–F$_3$ configurations become more important. SiF$_4$ clusters are only formed according to the model for fluorine concentrations higher than ~ 40 at.%. The concentration dependent distribution of the F atoms among the possible configurations found experimentally is in basic agreement with the model. A notable exception is the excess in Si–F$_3$ configurations compared to the statistical probability. Si–F$_3$ takes its intensity initially from Si–F$_1$ units and for C_F

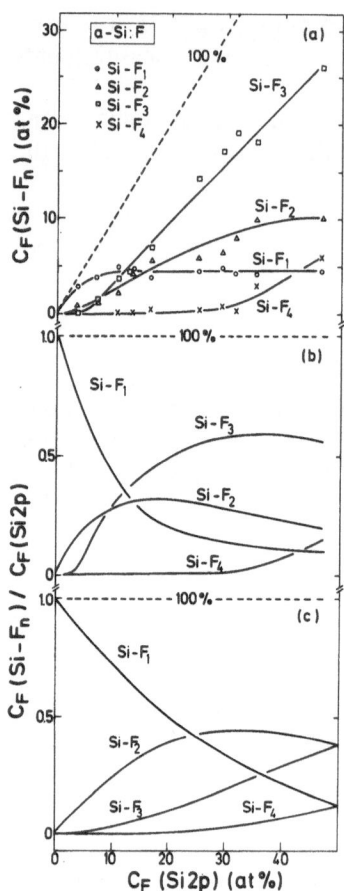

Fig. 3.28 a–c. The absolute (**a**) and relative (**b**) fluorine concentrations in each Si-F_n configuration as a function of the total fluorine content C_F. (**c**) The relative contribution of each Si-F_n unit to C_F calculated for a statistical distribution of F atoms in a-Si : F [3.185]

above about 20 at.% from the then statistically dominant Si–F_2 units. Below ~12% an enrichment of Si–F_2 is also evident.

The overemphasis of Si–F_3 (and Si–F_2) configurations has been attributed to a surface enrichment in fluorine by *Gruntz* et al. [3.185, 186]. They could show that an excess fluorine concentration of ~9 at.% was concentrated in a surface layer no thicker than 9 Å. This corresponds to an excess surface density of F atoms of 3.4×10^{15} cm^{-3} or 2.5 fluorine atoms per surface Si atom. This concentration is reached, for example, if ~80% of the available surface sites are saturated with Si–F_3 units. The excess fluorine concentration builds up between $0 \lesssim C_F \text{(Si }2p) \lesssim 12$ at.% and remains constant thereafter.

These results impose certain requirements on any growth model of a-Si : F films. (i) The growth has to take place in such a way as to ensure a surface rich in Si–F_3 and possibly Si–F_2 units (for $C_F \lesssim 12$ at.%). (ii) The incorporation of F into the growing network is governed primarily by the laws of statistics and not by chemical effects. Chemical effects would, for instance,

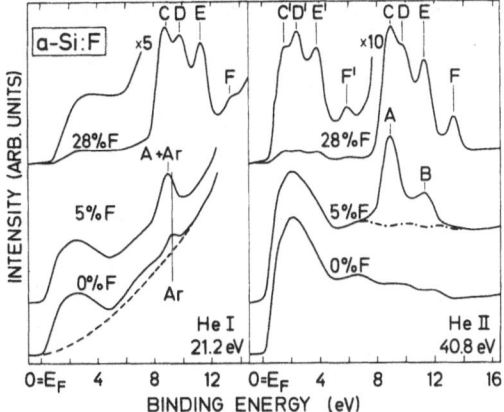

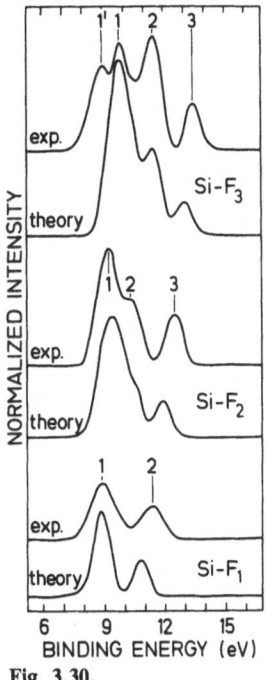

▲
Fig. 3.29. He I and He II valence band spectra of fluorinated a-Si:F. The fluorine concentrations are calculated from the Si $2p$ spectra as explained in the text. The features C′, D′, E′, are due to He II* satellite excitation (hv = 48.38 eV). Emission from Ar $3p$ levels implanted during sputter deposition is labelled Ar [3.185]

Fig. 3.30.

Fig. 3.30. Partial F $3p$ densities of states derived for different Si-F$_n$ configurations from the He II valence band spectra of a-Si:F [3.185]. The theoretical curves are due to *Ching* [3.189]

favour the bonding of a F⁻ ion with a Si⁺–F⁻ unit because of the Coulomb interaction between Si⁺ and F⁻.

A growth model in agreement with these requirements has been suggested in connection with hydrogenated amorphous silicon. It will be discussed in Sect. 3.3.16 a. It is also expected that the saturation of the a-Si:F film surface with fluorine makes these films particularly resistant to oxidation.

b) The Fluorine Derived Partial Densities of States in Amorphous Fluorinated Silicon

The valence band spectra (Fig. 3.29) of a-Si:F reveal Si–F bonding states between 7 and 14 eV binding energy. These states are of almost pure F $2p$ character as a result of the strong electronegativity of the F atoms [3.188]. A narrow (2.8 eV FWHM) F $2s$ band is situated at 31.3 eV binding energy (not seen in Fig. 3.29). Unlike the hydrogen derived states, the cross section of the F $2p$ states remains comparable with that of the Si $3p$, $3s$ states up to 1500 eV.

The structure of the F $2p$ derived states depends on the fluorine content as seen in Fig. 3.29. The energies of characteristic features (A–F in Fig. 3.29)

Table 3.6. Binding energies E_B of core levels and structure in the valence band spectra of a-Si : F (all energies are given in eV relative to E_F)

	E_B [eV]	Excited with	Remarks
Si 2p	99.6 ± 0.1	Al K$_\alpha$	–
F 1s	686.2 ± 0.2	Al K$_\alpha$	–
F 2s	31.3 ± 0.1	Al K$_\alpha$	–
Valence bands			
A	8.9	He II	5 at.% F
B	11.3	He II	5 at.% F
C	9.0	He II	28 at.% F
D	10.8	He II	28 at.% F
E	11.3	He II	28 at.% F
F	13.3	He II	28 at.% F
Ar 3p	9.3	He I	very weak
volume plasmon	17.0[a]	Al K$_\alpha$	Si 2p, F 1s
F related loss	7.5[a]	Al K$_\alpha$	F 1s Si 2p (weak)

[a] relative to core level

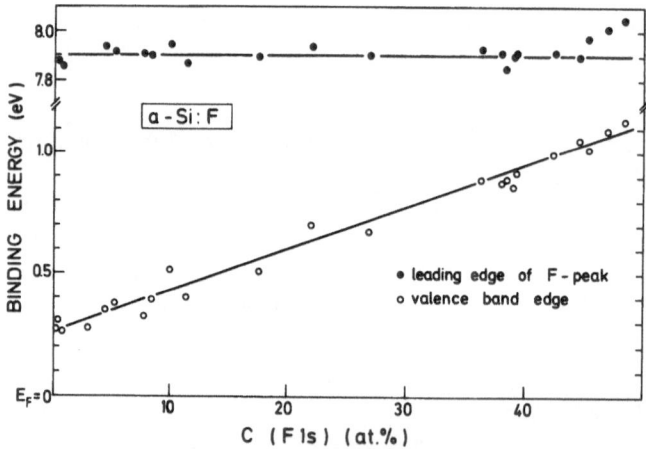

Fig. 3.31. Shift of the valence band edge (○) relative to E_F with fluorine concentration in a-Si : F. Also shown is the onset of F 2p emission (●). The F concentration C(F 1s) was obtained from the relative intensities of F 1s and S 2p core lines [3.185]

are listed in Table 3.6, together with core level binding energies and plasmon loss energies. *Gruntz* et al. [3.185] succeeded in extracting partial DOS's (PDOS) for each Si–F$_n$ (n = 1, 2, 3) configuration from the He II spectra with the aid of the Si 2p level analysis described in Sect. 3.3.15 a.

The result shown in Fig. 3.30 is in remarkable agreement with the theoretical F $2p$ partial densities of states obtained by *Ching* [3.189] after they have been rigidly shifted by -3.0 eV. The calculated spectra are consistently too narrow by $\sim 10\%$. The only true discrepancy occurs for the Si-F$_3$ PDOS. The experimental spectrum shows a splitting of peak 1 into two components that is not reproduced by the calculation. A PDOS in good agreement with the spectrum is, however, obtained for a (Si–F$_2$)$_2$ chain fragment [3.190].

Figure 3.31 finally indicates that the incorporation of fluorine leads to a similar albeit larger recession of the valence band edge as it was observed for a-Si:H. The top of the valence bands recedes linearly by 0.9 eV for F concentrations between 0 and 48 at.%. For comparison the onset of the F $2p$ states is also plotted as the full dots in Fig. 3.31. There is hardly a shift in this quantity with C_F, indicating that the increase in E_F − VBM is truly a recession of the valence band edge.

3.3.16 Surface Properties of Amorphous Silicon

We conclude the section on photoemission with a few remarks about the surface properties of amorphous silicon which have been deduced from photoemission experiments.

When a-Si:H films are deposited in situ for photoemission measurements, the deposition process may be interrupted at will and the as-grown surface of the specimen can be investigated immediately. In this way use is made of the surface sensitive, analytic capabilities of the photoelectron spectrum as regards hydrogen and impurity concentrations, hydrogen bonding and the position of E_F. This procedure comes as close to the investigation of the growing surface as one may hope to get. The method has been extended to study the reaction of the a-Si:H surfaces with oxygen (Sect. 3.3.16 b).

a) Hydrogen Enrichment of the Hydrogenated Amorphous Silicon Surface

The photoelectron spectrum of clean a-Si:H surfaces distinguishes, broadly speaking, only between a polyhydride phase characterized by peaks A–B and a monohydride phase with structures C, D and E (Sect. 3.3.8). The polyhydride phase is always observed for specimens prepared at temperatures below $\sim 350°C$, independent of the deposition method used (glow discharge, dc or rf sputtering). It has therefore been conjectured that the polyhydride phase is indigenous to the growing a-Si:H surface, independent of the concentration and prevailing bonding configurations of H in the bulk of the specimen [3.191]. The conjecture has been substantiated by a series of experiments in which photoemission provided the surface sensitive method and infrared spectroscopy was the means of determining the concentration and bonding geometries of hydrogen in the bulk of the specimens [3.192]. Figure 3.32 shows the infrared (ir) and He II valence band spectra of three

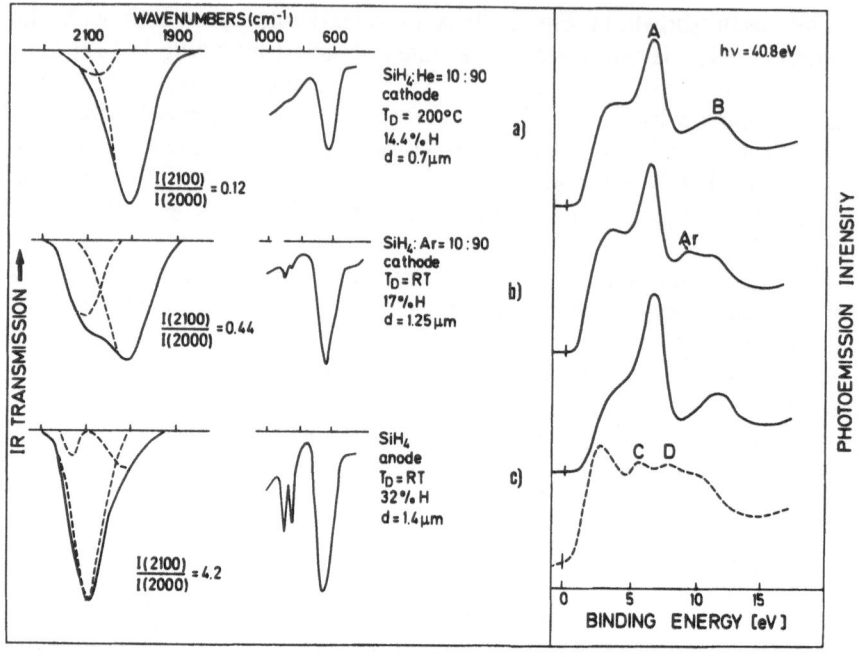

Fig. 3.32 a–c. Infrared and valence band spectra of three a-Si : H films prepared in such a way that the bulk Si-H bonding configurations are dominated by monohydrides (**a**), polyhydrides (**c**), and a mixture of both (**b**). The photoelectron spectrum containing only monohydrides at the surface ($T_D = 350\,°C$) is shown for comparison by the dashed curve [3.192]

a-Si : H specimens. The photoemission spectra were taken during the growth of the specimens and the ir spectra after the films had grown to a thickness d between 0.7 and 1.4 μm. All films were prepared by rf glow discharge in situ. The deposition conditions are also specified in Fig. 3.32. They are varied in such a way that for sample (a) the hydrogen is almost exclusively bonded in the form of monohydride, as witnessed by the weak absorption at 2100 cm^{-1} and the absence of bending modes around 850 cm^{-1} [3.193]. In sample (c) the polyhydride configurations Si–H$_2$, Si–H$_3$ and possibly (SiH$_2$)$_n$ are dominant and sample (b) finally contains a mixture of poly- and monohydrides as indicated by the comparable strength of the 2100 and 2000 cm^{-1} absorption bands. The hydrogen content of these films was determined from the strength of the 640 cm^{-1} wagging band according to the description of *Shanks* et al. [3.126].

Despite the considerable variations in hydrogen bonding and hydrogen concentration in the bulk of these samples, the valence band spectra of all three samples are dominated by the signature of the polyhydride configurations. The hydrogen concentration, furthermore, lies between ~ 40 and 50 at.% within the sampling depth of ~ 10 Å using the estimate described in Sect. 3.3.8. This exceeds the bulk concentration by factors of 2.5 and 3 for samples (b) and (a), respectively. It cannot be excluded from the data that the monohydrides are also present at the surface but they are clearly of secondary importance. Sputter etching experiments ensured that the hydrogen enrichment is indeed limited to the first 2 or 3 atomic layers [3.192].

These results are not in contradiction to the hydrogen profiling data of *Müller* et al. [3.194]. These authors found a subsurface hydrogen depletion layer of ~ 1000 Å thickness but dismissed a two to threefold increase in hydrogen concentration right at the surface (measured with a 30 Å resolution) as being due to adsorbed water.

The observations just described support the growth model for a-Si:H proposed by *Kampas* and *Griffith* [3.195, 196] and *Scott* et al. [3.197]. In this model (Fig. 3.33), the active species in the gas phase is the SiH_2 radical. It "inserts" itself into a Si–H bond at the surface and forms an Si–H_3 surface species. The cross linking of the Si–$H_{2,3}$ units at the surface occurs through the hydrogen elimination reactions as indicated in Fig. 3.33. In this model a surface rich in Si–H_2 and Si–H_3 units is thus pushed ahead of the growing film. The fluorine enrichment of the a-Si:F films discussed in the previous section suggest that a similar growth mechanism with SiF_2 taking the place of SiH_2 might apply in this case.

At deposition temperatures above ~ 300°C, the a-Si:H surfaces exhibit only monohydride species. Whether this is due to an annealing process that occurs faster at the surface than it takes to transfer the sample from the deposition chamber to the measuring chamber (~ 1 min) or due to a different growth mechanism remains to be seen [3.196].

b) The Oxydation of Amorphous Silicon

The oxydation of amorphous silicon is studied by exposing the surface of a freshly prepared sample in the photoelectron spectrometer to controlled amounts of dry oxygen. The exposure is measured in terms of the product of

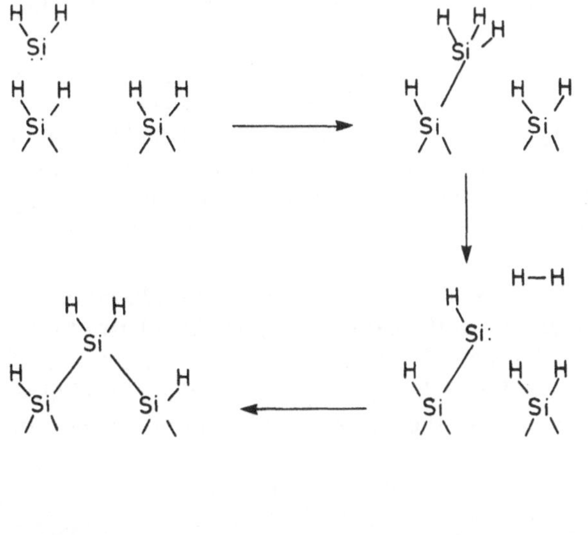

Fig. 3.33. The a-Si:H growth model due to *Kampas* and *Griffith* [3.196]

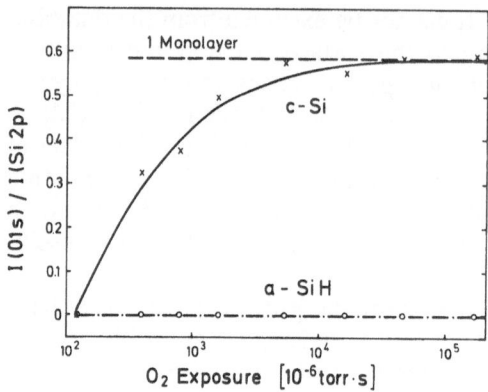

Fig. 3.34. O $1s$ to Si $2p$ core level intensity ratio as a function of O_2 exposure in c-Si and a-Si : H [3.186]

exposure time and gas pressure in units of Langmuir ($1\,\mathrm{L} \triangleq 1 \times 10^{-6}$ torr · s). One Langmuir is the exposure necessary to form a monolayer of adsorbed gas on a surface provided each molecule or atom that hits the surface sticks to it. A measure for the oxygen coverage is the relative intensity of the O $1s$ and Si $2p$ core levels.

Oxygen is readily adsorbed on a c-Si surface as shown in Fig. 3.34 [3.191]. After an exposure of about 5×10^4 L the surface is saturated with one monolayer of oxygen atoms [3.198]. Any further adsorption proceeds with a sticking probability orders of magnitude lower. The monolayer coverage corresponds to an intensity ratio $I(O/s)/I$ (Si $2p$) = 0.58 using Al K_α x-rays and an electron take-off angle of about 30° from the surface normal. Molecular oxygen does not adsorb to any measurable extent on hydrogenated a-Si for exposures up to 10^6 L (Fig. 3.34). The reactivity of the c-Si surface is due to unsaturated surface states (dangling bonds) which are available to form Si–O bonds. In a-Si : H these states are already saturated with hydrogen. The oxygen uptake as a function of exposure is virtually the same for unhydrogenated a-Si and for c-Si [3.199]. This proves that disorder alone does not reduce the reactivity of the Si surface by an autocompensation of dangling bonds, for instance.

The surface of a-Si : H is attacked by "activated" oxygen i.e., oxygen admitted in the presence of a hot ($T \simeq 1600\,^\circ\mathrm{C}$) filament which cracks the O_2 molecules. The uptake of oxygen (Fig. 3.35 a) follows approximately that of c-Si subject to the same treatment. Notice, however, that no plateau at the monolayer coverage is observed in the a-Si : H films. This almost certainly has something to do with the absence of well-defined crystal planes in a-Si : H.

As soon as oxygen bonds to a Si atom, a chemical shift is induced on the $2p$ level. The intensity $I(\mathrm{Si}\ 2p\ O_x)$ of the shifted components relative to the unshifted Si $2p$ level [$I(\mathrm{Si}\ 2p)$] is plotted in Fig. 3.35 b and the distribution of the shifted Si $2p$ components in Fig. 3.36. The oxidation of the a-Si : H films proceeds from the very beginning of the oxygen uptake; an initial stage of

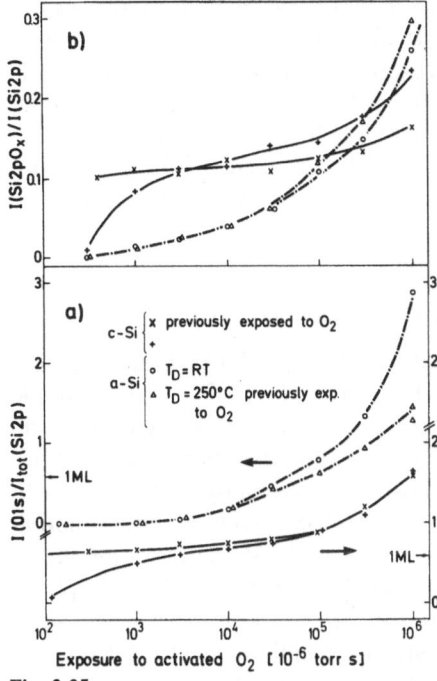

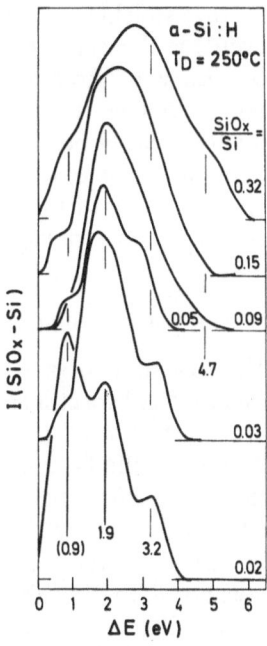

Fig. 3.35.   Fig. 3.36.

Fig. 3.35 a, b. Oxygen coverage (**a**) and relative amount of oxidized silicon (**b**) as a function of oxygen exposure for c-Si and a-Si : H [3.191]

Fig. 3.36. Si $2p$ difference spectra between clean and oxidized a-Si : H films. The amplitudes are not to scale. The relative intensities of the shifted (SiO$_x$) and unshifted (Si) components are given as a parameter next to each curve [3.191]

solely physisorbed oxygen (i.e., one that does not involve any chemical shift) is not observed. At least four chemically shifted components can be distinguished in the difference spectra of Fig. 3.36. With increasing oxidation the components with the larger chemical shift gain in relative intensity much like it was observed in the Si $2p$ spectra of a-Si : F. However, an interpretation of the individual components in terms of simple Si–O units is hampered by two facts:

(i) oxygen is expected to occur as terminal as well as bridging (like in SiO$_2$) oxygen [3.200], and

(ii) the charge transfer between O and Si depends on the Si–O–Si bond angle [3.201].

It is nevertheless possible to define an *average* stoichiometry x of the SiO$_x$ species present at the surface from the number of oxygen atoms adsorbed [$\propto I(\text{O } 1s)$] and the number of Si atoms bonded to oxygen [$\propto I(\text{Si } 2p \, \text{O}_x)$] using bulk SiO$_2$ as a standard. The result (Fig. 9 in [3.191]) indicates that the average stoichiometry of a-Si:H films deviates considerably from that of c-Si even for the same oxygen coverage. It also depends on the preparation conditions of the a-Si:H specimens presumably through the differences in polyhydride vs. monohydride configurations. In particular, the stoichiometry for RT a-Si:H has been taken as evidence for Si–H–O bonding configurations in which oxygen is not bonded directly to silicon [3.191]. A fuller understanding of the oxydation of a-Si:H must obviously await a detailed analysis of spectra like those of Fig. 3.36.

c) The Valence Band Spectra of Oxidized Amorphous Silicon

The oxidation of a-Si induces predominantly O $2p$ derived states in the valence band spectra as seen in the He II spectra of Fig. 3.37 [3.202]. Sample (d) is a sputtered a-Si film which was exposed to activated oxygen. The strength of the O $1s$ line at 532 eV binding energy (also shown in Fig. 3.37) indicates a coverage of 0.03 monolayers (ML) of oxygen by comparison with Fig. 3.34. The O $2p$ derived states show up at 6.7 eV (O_I in Fig. 3.37) and at 11.5 eV (O_{II}). Increasing the oxygen coverage to 0.1 ML increases the intensity and width of peak O_I and shifts O_{II} to 11.8 eV. *Miller* et al. [3.199] have extended the oxidation studies of a-Si to O_2 exposures up to 10^5 L and have found essentially the same features: O_I at 7.1 eV and O_{II} at 11.8 eV relative to the top of the valence bands. Energies of oxygen derived features observed on the c-Si(111) surface are found in [3.203]. For an exposure of 10^8 L O_2 the spectrum of a-Si changes. Instead of two oxygen derived peaks one finds three peaks at 7.0, 11.0 and 13.5 eV below E_F [3.199]. The 7 eV peak dominates and the Si $3p$ derived states at the top of the valence bands are only barely visible. The spectrum resembles those of thick SiO_2 layers [3.204, 205] with the exception of the lowest peak which is displaced by ~ 1 eV. Also the two lower peaks are much more distinct in the SiO_2 spectra.

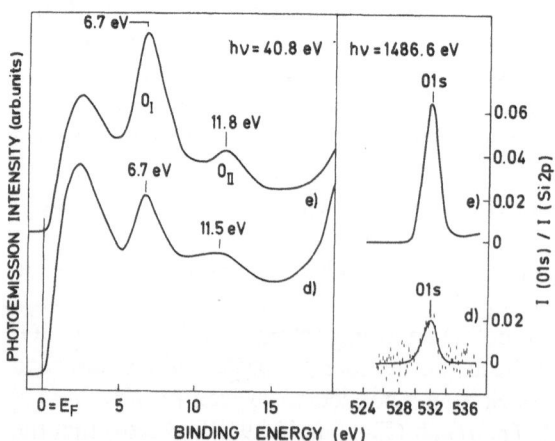

Fig. 3.37. Valence and O $1s$ core level spectra of unhydrogenated a-Si films covered with ~ 0.03 (d) and ~ 0.1 (e) monolayers of oxygen [3.202]

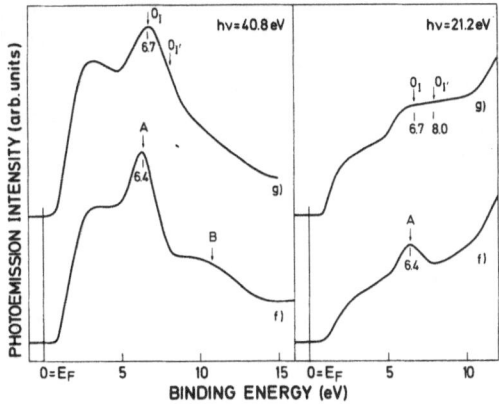

Fig. 3.38. He I and He II valence band spectra of an a-Si : H film before (*f*) and after (*g*) exposure to oxygen leading to an oxygen coverage of 0.06 monolayers [3.202]

This points towards different bonding configurations in a heavily oxidized a-Si surface compared to bulk SiO_2, a result that ties in with the differences in the average stoichiometry during the formation of the SiO_2 layer discussed in the previous section.

When the surface of *hydrogenated* a-Si : H is covered with 0.06 ML of oxygen, a shoulder appears at 8.0 eV below E_F in addition to O_I ($O_{I'}$ in Fig. 3.38). This shoulder has been tentatively assigned by *Kärcher* and *Ley* [3.202] to Si-H bonds. For higher oxygen coverages the spectra are indistinguishable from those of unhydrogenated a-Si.

There is a remarkable resemblance in the spectra of hydrogenated and oxidized a-Si as far as the hydrogen or oxygen induced structures (A, B and O_I, O_{II}) are concerned which has not been explained. There is, however, also a significant difference: the photoemission cross section for peaks O_I and O_{II} is about a factor of 10 to 20 larger than that of the hydrogen induced features A, B [3.202]. Consider spectrum (f) of Fig. 3.38. The hydrogen content of the film is about 16 at.% using the estimate of Sect. 3.3.8. The comparable spectrum of Fig. 3.37 d is brought about by a mere 0.7 at.% oxygen assuming a sampling depth of 5 atomic layers. This corresponds to a twentyfold enhancement of the photo-yield in peak O_I (Si–O bonds) compared to peak A (Si–H bonds). This explains why the top of the valence bands is virtually unaffected in oxidized a-Si samples compared to the recession in the VBM for similar spectra of a-Si : H samples. Spectrum (f) of Fig. 3.38 represents the spectrum of a silicon-hydrogen *alloy* $SiH_{0.16}$. The spectra of Fig. 3.37 constitute the *superposition* of the state density of a very thin surface layer of SiO_x on an otherwise undisturbed valence band of a-Si.

d) Are There Surface States in Amorphous Silicon?

The c-Si(111) exhibits a 1 eV wide band of filled surface states at the top of the valence bands which extends into the gap and pins the Fermi level (0.35 ± 0.15) eV above E_v (Fig. 3.39) [3.149, 206, 207]. This band is selectively

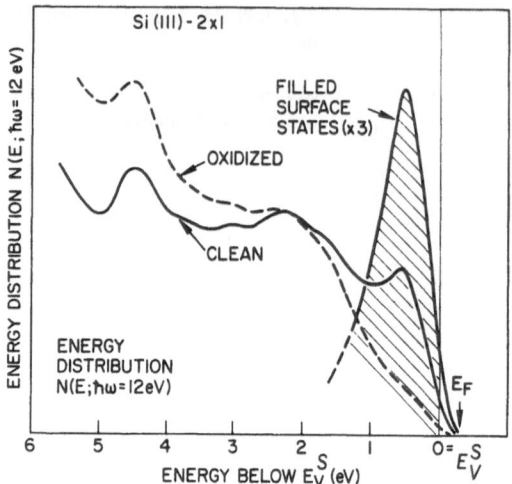

Fig. 3.39. He I valence band spectra of clean and oxidized $S(111)$ 2×1 surfaces. The difference curve is a measure of the density of occupied intrinsic surface states [3.206]

attenuated by at least 90% when one monolayer of oxide is grown on the surface [3.207]. The difference curve as shown in Fig. 3.39 is a measure of the filled band of surface states. The band is due to dangling bands and its density is $\sim 8 \times 10^{14}$ cm^{-2} eV^{-1}, i.e., one state per surface atom. Since no comparable attenuation occurs upon oxidation of a-Si samples, *Miller* et al. concluded that the density of surface states in a-Si cannot be larger than 4×10^{13} states cm^{-2} [3.199].

We argue against this conclusion. The level scheme of Fig. 3.16 indicates that the valence band edge of unhydrogenated amorphous silicon overlaps the band of surface states in c-Si. The smallest amount of hydrogen removes the top 0.3 eV of the valence bands in the spectra of a-Si. It is therefore not unreasonable to assume that the corresponding states are defect states. The commonly assumed density of 10^{21} cm^{-3} eV^{-1} for the bulk defect states represents an equivalent density of surface states of 2×10^{14} cm^{-2} eV^{-1} in the valence band spectrum assuming the standard 5 atomic layer sampling depth. Thus, the removal of a fraction of the 8×10^{14} cm^{-2} eV^{-1} surface states upon oxidation could go unnoticed even under more favourable conditions than the ones assumed here (lower bulk defect density, for example).

The similarity in the oxidation rate of c-Si and a-Si and the dramatic reduction of that rate in a-Si:H strongly suggest that there are surface states in a-Si. Furthermore, *Miller* et al. [3.199] observed a de-pinning of E_F in n-type a-Si upon oxidation which is similar to that found in n-type c-Si [3.207].

The hydrogenated and the oxidized surfaces have a density of surface states that is low compared to the bulk density times the Debye length, as illustrated by the Fermi level shifts discussed in Sect. 3.3.13.

3.4 Optical Absorption Spectra

3.4.1 Determination of Optical Constants

The optical constants n and κ of amorphous Si and Ge are usually obtained from transmission and reflectance measurements of thin films on transparent substrates such as glass or quartz. The measured quantities are the transmittance T and the reflectance R, where

$$T = \frac{\text{Transmitted intensity}}{\text{Incident intensity}} \tag{3.56}$$

and

$$R = \frac{\text{Reflected intensity}}{\text{Incident intensity}} . \tag{3.57}$$

Conservation of energy defines the absorptance A through

$$A + R + T = 1 . \tag{3.58}$$

Multiple reflections within the substrate and the film make the evaluation of R and T in terms of the optical constants nontrivial. A number of simplifying assumptions are usually made in relating measured values for R and T to n and κ. The assumptions differ from author to author and are often not clearly stated. We shall, therefore, summarize the salient results here (see also *Cody* et al. [3.208]). General formulae for R and T have been given by *Wolter*

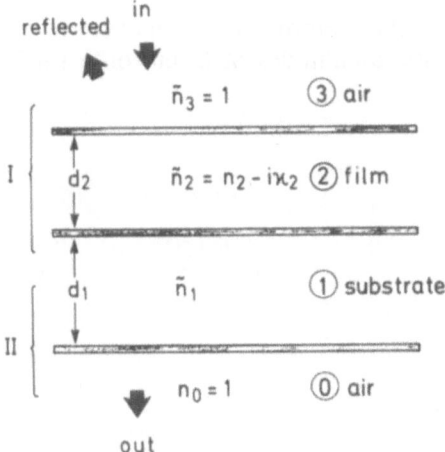

Fig. 3.40. Optical constants used in the calculation of transmittance and reflectance of a thin film on a transparent substrate

[3.209] and in the book of *Heavens* [3.210] and we shall make use of them assuming that light of frequency ω impinges perpendicular onto the surface of the specimen, as shown in Fig. 3.40.

R and T are then expressed in terms of complex transmission ($\tilde{t}$) and reflectance ($\tilde{r}$) amplitudes according to

$$T = \frac{n_0}{n_3} |\tilde{t}|^2 \quad \text{and} \quad R = |\tilde{r}|^2 , \tag{3.59}$$

where

$$\tilde{t} = \frac{\tilde{t}_{32} \tilde{t}_{21} \tilde{t}_{10} \exp\left[-i(\beta_1 + \beta_2)\right]}{(1 + \tilde{r}_{32}\tilde{r}_{21} e^{-i2\beta_2}) + (\tilde{r}_{21} + \tilde{r}_{32} e^{-i2\beta_2})\tilde{r}_{10} e^{-i2\beta_1}} \tag{3.60}$$

and

$$\tilde{r} = \frac{(\tilde{r}_{32} + \tilde{r}_{21} e^{-i2\beta_2}) + (\tilde{r}_{32} \tilde{r}_{21} + e^{-i2\beta_2})\tilde{r}_{10} e^{-i2\beta_1}}{(1 + \tilde{r}_{32}\tilde{r}_{21} e^{-i2\beta_2}) + (\tilde{r}_{21} + \tilde{r}_{32} e^{-i2\beta_2})\tilde{r}_{10} e^{-i2\beta_1}} . \tag{3.61}$$

The $\tilde{r}_{ij}$ and $\tilde{t}_{ij}$ are the Fresnel coefficients for a light beam going from medium i with complex refractive index $\tilde{n}_i = (n_i - i\kappa_i)$ to medium j with refractive index $\tilde{n}_j$, and $\tilde{\beta}_l$ is the complex phase factor for a traversal of medium l with thickness d_l:

$$\tilde{r}_{ij} = \frac{\tilde{n}_i - \tilde{n}_j}{\tilde{n}_i + \tilde{n}_j} ; \quad \tilde{t}_{ij} = \frac{2\tilde{n}_i}{\tilde{n}_i + \tilde{n}_j} \tag{3.62}$$

$$\beta_l = \frac{2\pi}{\lambda} \tilde{n}_l d_l = \frac{\omega}{c} \tilde{n}_l d_l , \tag{3.63}$$

where λ is the wavelength in vacuo.

For the sake of further discussion we shall reformulate $\tilde{r}$ and $\tilde{t}$ in terms of $\tilde{r}_I$ and $\tilde{t}_I$, the reflection and transmission amplitudes of a film on a semi-infinite substrate (see Fig. 3.40):

$$\tilde{t} = \tilde{t}_I \frac{\tilde{t}_{10} e^{-i\beta_1}}{1 - \tilde{r}'_I \tilde{r}_{10} e^{-i2\beta_1}}$$

$$= \tilde{t}_I \tilde{t}_{II} \tag{3.64}$$

and $\tilde{r} = \tilde{r}_I + \dfrac{\tilde{t}_I \tilde{t}'_I \tilde{r}_{10} e^{-i2\beta_1}}{1 - \tilde{r}'_I \tilde{r}_{10} e^{-i2\beta_1}}$

$$= \tilde{r}_I + \tilde{r}_{II} \tag{3.65}$$

with $\tilde{t}_{\mathrm{I}} = \tilde{t}_{32}\,e^{i\beta_2}\,t_{21}\,\underbrace{\dfrac{1}{1 - \tilde{r}_{23}\,\tilde{r}_{21}\,e^{-i2\beta_2}}}_{A}$ (3.66)

and $\tilde{r}_{\mathrm{I}} = \dfrac{\tilde{r}_{32} + \tilde{r}_{21}\,e^{-i2\beta_2}}{1 - \tilde{r}_{23}\,\tilde{r}_{21}\,e^{-i2\beta_2}}$ (3.67 a)

$= \tilde{r}_{32} + \underbrace{\dfrac{\tilde{r}_{21}\,\tilde{t}_{32}\,\tilde{t}_{23}\,e^{-i2\beta_2}}{1 - \tilde{r}_{23}\,\tilde{r}_{21}\,e^{-i2\beta_2}}}_{B}$ (3.67 b)

using $\tilde{t}_{32}\,\tilde{t}_{23} = 1 - \tilde{r}_{32}^2$.

The factors A and B in (3.66, 67 b) and the corresponding factors in $\tilde{r}_{\mathrm{II}}$ and $\tilde{t}_{\mathrm{II}}$ allow for multiple coherent reflections in the film and substrate, respectively.

$r_{\mathrm{I}}' = \dfrac{\tilde{r}_{12} + \tilde{r}_{23}\,e^{-i2\beta_2}}{1 - \tilde{r}_{23}\,\tilde{r}_{21}\,e^{-i2\beta_2}}$ (3.68 a)

$= \tilde{r}_{12} + \dfrac{\tilde{t}_{12}\,\tilde{t}_{23}\,\tilde{t}_{23}\,e^{-i2\beta_2}}{1 - \tilde{r}_{23}\,\tilde{r}_{21}\,e^{-i2\beta_2}}$ (3.68 b)

and $\tilde{t}_{\mathrm{I}}' = \tilde{t}_{\mathrm{I}}$

refer to the reflection and transmission amplitudes for light impinging from the substrate on the film.

The coherence condition

$$\Delta\lambda \cdot 2\pi \cdot n_1 \cdot d_1/\lambda^2 \ll 1$$ (3.69)

is usually not met for the substrates used ($d_1 \simeq 1$ mm). The multiple reflections in the substrate add therefore incoherently and instead of $T = n_0/n_3\,|\tilde{t}_{\mathrm{I}}\tilde{t}_{\mathrm{II}}|^2$ and $R = |\tilde{r}_{\mathrm{I}} + \tilde{r}_{\mathrm{II}}|^2$, we have

$T = \dfrac{n_1}{n_3}\,|\tilde{t}_{\mathrm{I}}|^2\,\dfrac{n_0}{n_1}\,|\tilde{t}_{\mathrm{II}}|^2$ (3.70 a)

$= \dfrac{n_1}{n_3}\,|\tilde{t}_{\mathrm{I}}|^2\,\dfrac{\dfrac{n_0}{n_1}\,|\tilde{t}_{10}|^2\,e^{-\alpha_1 d_1}}{1 - |r_{\mathrm{I}}'|^2\,|\tilde{r}_{10}|^2\,e^{-2\alpha_1 d_1}}$ (3.70 b)

$\equiv T_{\mathrm{f}}\,T_{\mathrm{s}}$

and

$$R = |\tilde{r}_{\mathrm{I}}|^2 + |\tilde{r}_{\mathrm{II}}|^2 \qquad (3.71\,\mathrm{a})$$

$$= |r_{\mathrm{I}}|^2 + \frac{|\tilde{t}_{\mathrm{I}}|^2|\tilde{t}'_{\mathrm{I}}|^2|\tilde{r}_{10}|^2\,e^{-\alpha_1 d_1}}{1 - |\tilde{r}'_{\mathrm{I}}|^2|\tilde{r}_{10}|^2\,e^{-2\alpha_1 d_1}} \qquad (3.71\,\mathrm{b})$$

$$= R_{\mathrm{f}} + R_{\mathrm{s}}$$

with $\alpha_1 = 4\pi\kappa_1/\lambda$ the absorption coefficient of the substrate.[1] The subscripts f and s refer to the contributions of film and substrate, respectively. The optical constants n_2 and κ_2 of an amorphous film of known thickness are obtained rigorously through an iteration procedure such that $\tilde{n}_2$ minimizes $|R - R_{\mathrm{meas}}|$ and $|T - T_{\mathrm{meas}}|$ [3.211, 212].

It is not necessary, however, to deal with the exact expressions for R and T under all circumstances. Equations 3.70, 71 can be simplified considerably by a judicious choice of approximations.

Nonabsorbing Substrate. In this case ($\alpha_1 \equiv 0$) $\tilde{t}_{10}$ and $\tilde{r}_{10}$ are real,

$$1 - r_{10}^2 = 1 - R_{10} = \frac{n_0}{n_1}\, t_{10}^2 \qquad (3.72)$$

and we have for the transmittance of the substrate

$$T_{\mathrm{s}} = \frac{1 - R_{10}}{1 - |\tilde{r}'_{\mathrm{I}}|^2 R_{10}} \cdot \qquad (3.73)$$

When quartz is used as a substrate, $n_1 = 1.5$, $n_0 = 1$ and $R_{10} = 0.04$. Since $|\tilde{r}_{\mathrm{I}}|^2$ cannot exceed one, we may replace the denominator in (3.73) by an average value of 0.98 ± 0.02 resulting in $T_{\mathrm{s}} = 0.98$ with an error of $\pm 2\%$. Similarly, the contribution of the substrate to the reflectance (3.71)

$$R_{\mathrm{s}} = |\tilde{r}_{\mathrm{II}}|^2 = \frac{|t_{\mathrm{I}}|^4 R_{10}}{1 - |\tilde{r}'_{\mathrm{I}}|^2 R_{10}}, \qquad (3.74)$$

---

1 The definitions of t_{ij} and T used here (3.59, 60) differ from those of *Cody* et al. [3.208]. Cody sets

$$\tilde{t}'_{ij} = \frac{2\tilde{n}_j}{\tilde{n}_i + \tilde{n}_j} \quad \text{and} \quad T'_{ij} = \frac{n_i}{n_j}\,|\tilde{t}_{ij}|^2 .$$

The two definitions are, in general, not equal: $T_{ij} \neq T'_{ij}$. Nevertheless, the quantity T for the transmission of the film is the same in both cases because the film is sandwiched between two media with real refractive indices. We have adopted here the definition of t_{ij} and T that is generally used throughout the literature.

yields with the same approximation

$$R_s \simeq |t_1|^4 \, 0.0408 \pm 2\% \tag{3.75}$$

for a quartz substrate.

Low Absorption Limit. When $\kappa_2 \ll |n_2-n_1|$, $|n_2-n_3|$, all Fresnel coefficients are approximated by their real parts:

$$\frac{n_j}{n_i} |\tilde{t}_{ij}|^2 = \frac{n_j}{n_i} t_{ij}^2 = 1 - r_{ij}^2 = 1 - R_{ij} \; . \tag{3.76}$$

The transmittance of the film (3.60, 70)

$$T_f = \frac{(1 - R_{32})(1 - R_{21}) \, e^{-\alpha_2 d_2}}{1 + R_{23} R_{21} \, e^{-2\alpha_2 d_2} - 2 R_{23}^{1/2} R_{21}^{1/2} \, e^{-\alpha d} \cos\left(4\pi n_2 d_2/\lambda\right)} \tag{3.77}$$

is an oscillatory function of λ due to interference in the film with maxima and minima given by

$$T_f{}^{\max}_{\min} = \frac{(1 - R_{32})(1 - R_{21}) \, e^{-\alpha_2 d_2}}{(1 \mp R_{23}^{1/2} R_{21}^{1/2} \, e^{-\alpha_2 d_2})^2} \; . \tag{3.78}$$

Similarly, we have for the reflectance of the film

$$R_f{}^{\max}_{\min} = |r_1|^2{}^{\max}_{\min} = \frac{(r_{32} \mp r_{21} e^{-\alpha_2 d_2})^2}{(1 \mp r_{23} r_{21} e^{-\alpha_2 d_2})^2} \; . \tag{3.79}$$

The maxima (minima) in T_f and R_f occur when $4\pi n_2 d_2/\lambda$ is an even (odd) multiple of π. The interference condition is used to derive values for n_2 in the low absorption regime from the separation of the transmission maxima when d_2 is known. It is also possible to get n_2 without knowledge of d_2 from the value of T_f at its minimum or maximum value using (3.78). A graphical solution for three values of n_1, the substrate refractive index, is due to *Cody* et al. [3.208] and is shown in Fig. 3.41.

Averaging over the interference fringes in a plot of $\log T$ or $\log R$ versus λ amounts to taking the logarithmic averages

$$(T_f^{\min} T_f^{\max})^{1/2} = \frac{(1 - R_{32})(1 - R_{21}) \, e^{-\alpha_2 d_2}}{1 - R_{23} R_{21} \, e^{-2\alpha_2 d_2}} \tag{3.80}$$

and

$$(R_f^{\min} R_f^{\max})^{1/2} = \frac{R_{32} - R_{11} \, e^{-2\alpha_2 d_2}}{1 - R_{23} R_{21} \, e^{-2\alpha_2 d_2}} \; . \tag{3.81}$$

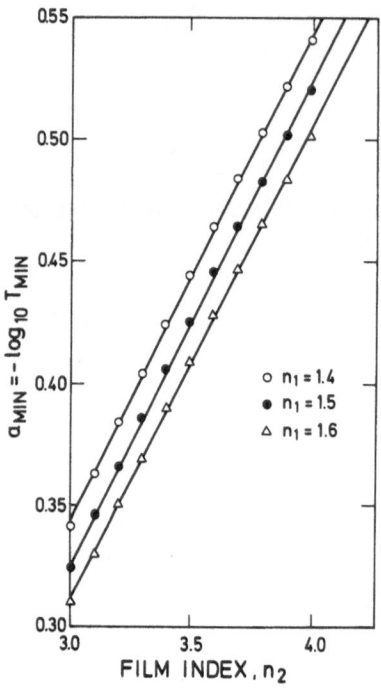

Fig. 3.41. The negative logarithm of the transmission minimum in the fringe regime as a function of the specimen index n_2 and the substrate index $n_s \equiv n_1$ corresponding to (3.78) in the text [3.208]

In the plot legend:

○ $n_1 = 1.4$
● $n_1 = 1.5$
△ $n_1 = 1.6$

y-axis: $a_{MIN} = -\log_{10} T_{MIN}$

x-axis: FILM INDEX, n_2

The logarithmic average of T_f is equal to T_f in the incoherent limit (i.e., $\Delta\lambda \cdot n_2 \cdot \pi d_2/\lambda^2 \gtrsim 1$), but for R_f the incoherent result

$$R_{f,\,incoh} = R_{32} + \frac{(1 - R_{32})^2 R_{21}\, e^{-2a_2 d_2}}{1 - R_{32} R_{21}\, e^{-2a_2 d_2}} \tag{3.82}$$

is different from the logarithmic average.

The High Absorption Regime; Reflectivity Measurements. When $n_2 \simeq \kappa_2$ and $a_2 d_2 \gtrsim 1$, (3.70 a)

$$T = \frac{n_1}{n_3}\, |\tilde{t}_I|^2\, \frac{n_0}{n_1}\, |\tilde{t}_{II}|^2 \tag{3.70 a}$$

simplifies to

$$T = \frac{n_0}{n_3}\, |\tilde{t}_{32}|^2\, |\tilde{t}_{21}|^2\, e^{-a_2 d_2} \,. \tag{3.83}$$

Similarly, the reflectivity is just that of the first interface

$$R = |\tilde{r}_{32}|^2 \tag{3.84}$$

and multiple reflections in the film and substrate can be neglected.

In the fundamental absorption region (i.e., $\alpha \gtrsim 10^5$ cm^{-1}), transmission measurements become impractical and the optical constants have to be determined from reflectivity measurements alone. This is possible, in principle, when the amplitude r and phase angle θ of the complex reflectivity

$$\tilde{r}_{32}(\omega) = r(\omega)\,e^{i\theta(\omega)} \tag{3.85}$$

are both determined as a function of $\hbar\omega$. It is achieved in ellipsometry where the polarization state of a light beam impinging at non-normal incidence onto the specimen surface is analyzed [3.213, 214]. Since ellipsometry is very surface sensitive, it has so far been applied to study the growth [3.215] and the oxidation [3.216] of plasma deposited amorphous silicon. An ε_2 spectrum of HF etched a-Si:H is given in [3.216]. It is also possible to derive both $r(\omega)$ and $\theta(\omega)$ from measurements of $|r_2|^2$ at normal incidence alone using the Kramers-Kronig dispersion relationship between r and θ [3.1]:

$$\theta(\omega_0) = -\frac{\omega_0}{\pi} \int\limits_0^\infty d\omega \frac{\ln r(\omega)}{\omega^2 - \omega_0^2} . \tag{3.86}$$

The Kramers-Kronig relationship links the value of θ at a particular frequency ω_0 to an integral over reflectivities extending over all energies $\hbar\omega$. The latter is, however, only known over a limited region in ω and methods have therefore been developed to extrapolate $r(\omega)$ to very low or very high frequencies in a physically reasonable way [3.1]. The sum rules mentioned earlier (see Sect. 3.2.1) are of considerable help in this procedure. The most useful of these in the present context is (3.12) which relates ε_∞, the long wavelength dielectric constant to an integral over $\varepsilon_2(\omega)$.

a) Techniques for the Very Low Absorption Regime: Bolometric Methods

By this we mean absorption constants $\alpha \lesssim 10^2$ cm^{-1}. This region is particularly interesting in amorphous semiconductors since such low absorption constants involve transition from or to states deep in the pseudogap. *Connell* and *Lewis* [3.217] have discussed comprehensively the various pitfalls encountered when applying approximate formulas for R and T in this regime. But even when the exact formulae are used, the accuracy of the measurements (at best a few tenths of a percent) limits the obtained accuracy of α to

$$\Delta\alpha = \frac{(\Delta R/R + \Delta T/T)}{d} . \tag{3.87}$$

This amounts to $\Delta\alpha = 100$ cm^{-1} for a 1 μm thick film when R and T are measured with an accuracy of half a percent. The only way out is to use thicker samples which are, however, often difficult to obtain in high enough quality.

The problem is further aggravated by light scattering from density fluctuations or imperfections in the specimen which reduces the apparent transmittance without actually absorbing intensity from the beam [3.218]. There are, of course, ways to correct for these losses by measuring the intensity of the scattered light separately in an integrating sphere [3.219].

It is, nevertheless, desirable to measure the absorptance which is the energy deposited in the sample directly. The absorptance A, as we recall, is given by $A = 1 - R - T$ and can thus be obtained in terms of the formulae for R and T. In the weak absorption limit (i.e., $\alpha_2 d_2 \ll 1$), we have

$$A_{f\ \max}^{\ \min} = \alpha_2 d_2 \frac{(1 - R_{32})(1 + R_{21})}{(1 \mp \tilde{r}_{32}\tilde{r}_{21})^2} \tag{3.88}$$

which reduces to

$$A_{f,\ \text{inc}} \simeq \alpha_2 d_2 \frac{(1 - R_{32})(1 + R_{21})}{1 - R_{32}R_{21}} \tag{3.89}$$

in the incoherent limit or after averaging over the interference fringes. To this the absorptance in the substrate has to be added. A is seen to be linear in α_2 and it needs to be measured no more accurately than we wish to obtain α_2, in contrast to the considerable accuracy required in measuring T and R for small values of $\alpha_2 d_2$. The methods used to measure A fall into two categories: the bolometric and the photoconductive techniques. In the former the temperature rise of the sample due to the absorbed light is measured when the specimen is illuminated with a beam of light of known intensity.

The temperature rise is measured directly by a thermometer for samples cooled to ~ 1.5 K in the experiment by *Bubenzer* et al. [3.220, 221]. The low temperatures used improve the accuracy with which small temperature differences can be measured, and the temperature rise itself is higher due to the reduction in specific heat of the specimen. *Yamasaki* et al. [3.222] employed the photoaccoustic technique whereby the periodic heating of a small gas volume through the sample which is illuminated by a chopped light beam is detected as sound in an attached microphone. *Jackson* et al. [3.223, 224], finally, measured the temperature rise in the sample through the temperature gradient that it produces in a gas or a liquid immediately in front of the sample surface. The temperature gradient induces a gradient in the refractive index of the gas (liquid) which in turn is detected by the deflection of a laser beam passing just in front of the specimen. The technique is referred to as photothermal deflection spectroscopy.

All three methods have been applied to a-Si:H and they allow, after appropriate calibration, measurements of α down to ~ 1 cm^{-1} on samples no thicker than 2 μm.

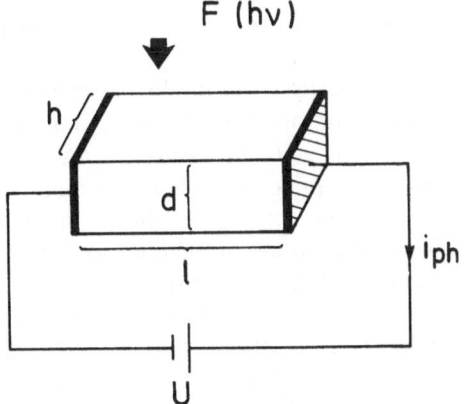

F (hν)

Fig. 3.42. Principle arrangement for photoconductivity measurements

b) Photoconductivity

The high photoresponse of glow discharge a-Si and a-Ge first noticed by *Chittick* et al. [3.225, 226] is being used as a very sensitive method to measure low ($\alpha < 10^3$ cm^{-1}) absorption coefficients in these materials. Consider the setup of Fig. 3.42 where a photon flux F [photons cm^{-2} s^{-1}] impinges on a film of thickness d. $A \cdot F$ of these photons are absorbed and create carriers at a rate [3.227]

$$\frac{\partial n}{\partial t} = \eta \cdot A \cdot F \cdot l \cdot h \tag{3.90}$$

in the region between the contacts where η is the generation efficiency. With an average lifetime τ of these carriers, an excess carrier concentration

$$\Delta n = \eta \cdot A \cdot F \cdot l \cdot h \cdot \tau \tag{3.91}$$

will persist which leads to a photocurrent

$$i_{ph} = e \cdot \Delta n \cdot \frac{1}{t_{tr}} = e \cdot \Delta n \frac{\mu \mathscr{E}}{l} . \tag{3.92}$$

The inverse of the transit time t_{tr} required to cross the distance l between the contacts is proportional to the product of carrier mobility μ and applied electric field $\mathscr{E} = U/l$. The proportionality between current density $j_{ph} = i_{ph}/(hd)$ (Fig. 3.42) and electric field $\mathscr{E}$ in (3.92) defines a photoconductivity σ_{ph}:

$$j_{ph} = \sigma_{ph} \mathscr{E} \tag{3.93}$$

with

$$\sigma_{ph} = \frac{e \cdot \Delta n \cdot \mu}{h \cdot d \cdot l} = \frac{e \cdot A \cdot F \cdot \tau \cdot \mu \cdot \eta}{d}. \tag{3.94}$$

The usefulness of photoconductivity to measure A in the low absorption limit as expressed in (3.94) depends on the assumption that $\eta \cdot \mu \cdot \tau$ remains essentially independent of photon energy. This assumption has been tested by *Loveland* et al. [3.228] in a range of photon energies where A could be measured independently by optical means. The results are shown in Fig. 3.43. There is an enormous variation in the magnitude of the photoresponse from different samples covering almost eight orders of magnitude. It is therefore mandatory to match photoconductivity data with direct optical measurements over as wide a range as possible. The variation of $\eta \cdot \mu \cdot \tau$ for a particular sample appears bearable in view of the fact that α varies much more rapidly in the absorption edge region. A simple extrapolation of the results of Fig. 3.43 to even lower energy, that is, into the range where photoconductivity is often the only means to determine α, is dangerous, however. The $\eta \cdot \mu \cdot \tau$ product is likely to drop rapidly whenever the photon energy is insufficient to excite carriers into extended states [3.228].

A further complication arises because the recombination rate $1/\tau$ of the carriers depends on the number n of excess carriers and therefore on $A \cdot F$ [3.227]. In a-Si:H one finds

$$i_{ph} \propto (A \cdot F)^{\gamma} \tag{3.95}$$

with $0.5 \leq \gamma \leq 1$ [3.228, 229, 229a].

The exponent γ is independent of F over several orders of magnitude in the photon flux but it depends critically on temperature [3.230]. The exponent γ should thus be determined unless measurements are made with a constant $A \cdot F$ product [3.229b].

The above discussion applies to ohmic contacts, i.e., the photocurrent increases in proportion to the applied voltage U. Holes and electrons contribute under these circumstances to i_{ph} in any proportion, depending on their

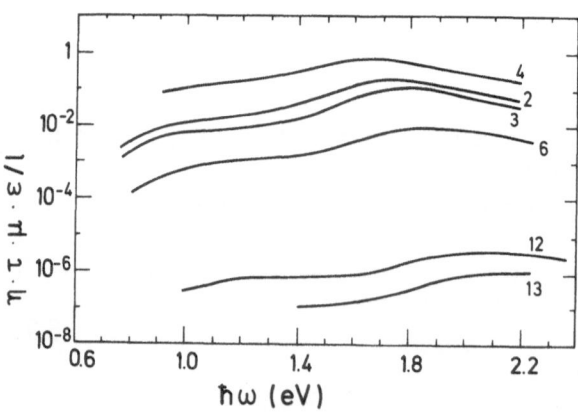

Fig. 3.43. The $\eta \cdot \mu \cdot \tau$ product as a function of photon energy for a number of glow discharge a-Si:H samples. ε is the electric field applied over a distance l [3.228]

respective $\mu \cdot \tau$ product. When blocking contacts are used instead, the photo-current saturates at a value equal to the creation rate $(\eta \cdot A \cdot F \cdot l \cdot h)$. A determination of A without knowledge of $\mu \cdot \tau$ is thus possible. *Crandall* [3.230 a] has realized this situation using an a-Si:H Schottky cell in reverse bias to measure α and also to determine $\mu \cdot \tau$ by comparing the saturation or primary photocurrent with the secondary photocurrent of (3.31) that flows in the forward direction.

Abeles et al. [3.229] compared primary and secondary photoconductivity with each other and with direct optical absorption measurements where possible. They found excellent agreement in the values of α obtained from all three methods in glow discharge a-Si:H. The exponent γ was constant at $\gamma = 0.7$ between 1.23 and 1.76 eV and for photon fluxes varying by 3 orders of magnitude. They concluded that the $\mu \cdot \tau$ product did not vary appreciably between 1.4 and 2.2 eV and that the photogeneration efficiency η is close to unity at room temperature for their samples (see Fig. 3.49). A critical assessment of the two photoconductivity modes is given in [3.230 b].

3.4.2 The Fundamental Absorption Band

a) Amorphous Silicon

The ε_2 spectrum obtained by *Pierce* and *Spicer* [3.86] on evaporated a-Si has already been compared with that of c-Si in Fig. 3.1. Similar spectra have been measured by *Beaglehole* and *Zavetova* for a-Si, also prepared by evaporation [3.231].

Using the valence and conduction densities of states of a-Si as described in Sects. 3.3.7, 3.3.10, one may calculate $\varepsilon_2(\omega)$ in the framework of the

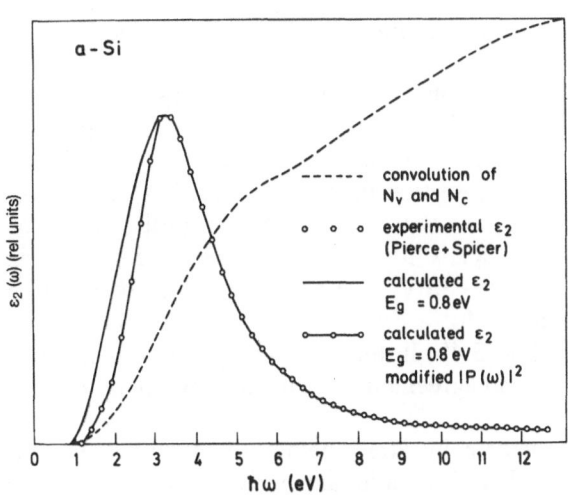

Fig. 3.44. Experimental and calculated $\varepsilon_2(\omega)$ spectrum of amorphous silicon. Also shown is the convolution of the experimentally determined valence (N_v) and conduction band (N_c) densities of states

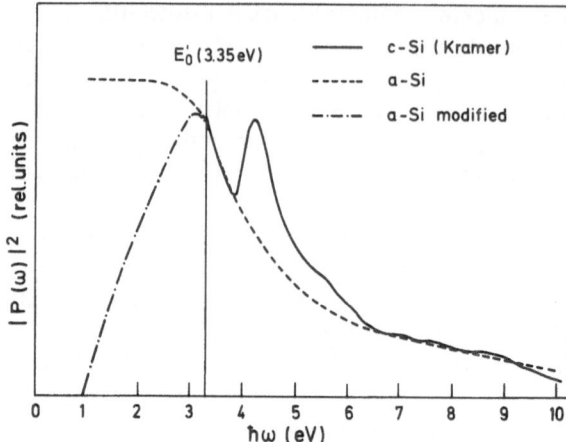

Fig. 3.45. The dipole matrix element $|P(\omega)|^2$ for c-Si as calculated by B. Kramer, M. Maschke, P. Thomas: Phys. Stat. Sol. (b) **49**, 525 (1972). Also shown are two approximations for $|P(\omega)|^2$ used in the calculation of $\varepsilon_2(\omega)$ of a-Si in Fig. 3.44

indirect transition model (3.11). The result shown in Fig. 3.44 is indeed in remarkably good agreement with the measured spectrum for energies above ~3.5 eV. The red shift of the maximum in ε_2 in a-Si compared to c-Si is brought about by the corresponding blue shift in the leading peak of $N_v(E)$. Calculating ε_2 with the crystalline $N_v(E)$ gives an ε_2 spectrum displaced to higher energies by ~0.8 eV. The one-peak structure is the result of the monotonically increasing convolution of N_v and N_c (dashed line in Fig. 3.44) and the factors $1/\omega^2$ and $|P(\omega)|^2$ which drop rapidly with $\hbar\omega$. For $|P(\omega)|^2$ we used, following *Maschke* and *Thomas* [3.232], essentially the crystalline matrix element except for the Umklapp enhancement of $|P(\omega)|^2$ at 4.3 eV as shown in Fig. 3.45. The Umklapp enhancement of $P(\omega)$ is a band-structure effect and it has no counterpart in amorphous solids [3.3]. According to Fig. 3.16 the gap between the VBM and conduction band minimum is 0.6 eV in a-Si. Only a small correction to 0.8 eV was necessary to bring the position of the calculated and measured ε_2 spectra into agreement. This value for E_g is noticeably smaller than the 1.2 to 1.5 eV quoted for E_G from optical measurements on unhydrogenated amorphous silicon [3.233, 246].

In c-Si the lowest direct gap is at 3.35 eV (E_0') [3.76] and transitions below E_0' are forbidden as a result of the electron momentum selection rules. In the strict application of the indirect transition model, the selection rule breaks down completely in a-Si and we have therefore extended $|P(\omega)|^2$ below 3.35 eV as indicated in Fig. 3.45. This leads, however, to noticeable deviations in the calculated ε_2 spectrum below ~3 eV. The discrepancy can be made to disappear completely if we allow for a gradual drop in $|P(\omega)|^2$ in a-Si, as shown by the chain curve in Fig. 3.45. In this way the apparent *optical gap* also increases to (1.3 ± 0.1) eV, in agreement with the measurements quoted above [3.233].

It is tempting to identify the drop in $|P(\omega)|^2$ below ~3 eV with a remnant of the k selection rule in a-Si. There is additional evidence from the pressure coefficients of the optical gap in a-Si and a-Si : H (Sect. 3.4) which suggests that states at the band edges retain a "memory" of the crystalline states they are derived from. The complex-band-structure calculations of *Kramer* [3.234] also indicate that the influence of disorder on the Bloch states in the crystal acts selectively and depends on the magnitude and orientation of the *k*-vector.

b) Amorphous Hydrogenated Silicon

Peak height ($\varepsilon_{2,\,\mathrm{max}}$) and peak position ($\omega_{\mathrm{max}}$) of the ε_2 spectrum depend sensitively on the preparation conditions of the a-Si specimens. Films evaporated at 250 °C reach their peak at $\hbar\omega_{\mathrm{max}} = 3.05$ eV with $\varepsilon_{2,\,\mathrm{max}} = 20$ [3.231]. Very slow evaporation at temperatures close to crystallization shifts $\hbar\omega_{\mathrm{max}}$ to 3.4 eV with $\varepsilon_{2,\,\mathrm{max}} \sim 20$, as shown in Figs. 3.1 and 3.44 [3.86], and amorphous silicon prepared by chemical vapor deposition at $T_{\mathrm{D}} = 600$ °C reaches values of $\varepsilon_{2,\,\mathrm{max}} = 26$ at $\hbar\omega_{\mathrm{max}} = 3.45$ eV [3.235]. Even higher values of $\hbar\omega_{\mathrm{max}}$ between 3.55 and 3.75 eV have been measured by *Weiser* et al. on glow discharge a-Si : H as shown in Fig. 3.46 [3.235, 236]. They interpret the increase in ω_{max} in the framework of the Penn model as a gradual increase in the average bond strength brought about by a progressive relaxation of the strain in the amorphous network. It is evident then that hydrogen is more effective in relieving the strain than gentle deposition conditions in the absence of hydrogen. This observation corroborates our interpretation of the charge fluctuations in a-Si and their reduction upon hydrogenation given in Sect. 3.3.14.

Once hydrogen is incorporated, the Penn gap ($\equiv \omega_{\mathrm{max}}$; see Sect. 3.2.2) depends only little on substrate temperature T_{s} and hydrogen content (see Table 3.7); a minimum of 4 at.% hydrogen is sufficient to place $\hbar\omega_{\mathrm{max}}$ at 3.60 eV (sample 1 in Fig. 3.46). This excludes the possibility that the replacement of Si–Si bonds by the stronger Si–H bonds is responsible for the relatively high $\hbar\omega_{\mathrm{max}}$ in a-Si : H which falls, nevertheless, still short of the Penn gap of c-Si (4.8 eV).

The shapes of the ε_2 spectra in Fig. 3.46 are very similar once they have been normalized to the same $\varepsilon_{2,\,\mathrm{max}}$. Only the room temperature specimen (sample 5 a in Fig. 3.46) which contains 50 at.% hydrogen exhibits a tail of increased absorption beyond ~ 4 eV. *Weiser* et al. attributed the tail to transitions from the hydrogen induced bonding states deep in the valence bands to empty conduction states (Sect. 3.3.8). These transitions appear to be exhausted at ~ 15 eV judging from Fig. 4 in [3.236].

Most conspicuous in the fundamental absorption bands of amorphous silicon are the low values of $\varepsilon_{2,\,\mathrm{max}}$ which vary, moreover, between 20 for rapidly evaporated a-Si and 32 for glow discharge a-Si : H ($T_{\mathrm{s}} = 400$ °C). Applying the sum rule (3.13) to the ε_2 spectra consequently yields values of n_{eff} which lie, with the exception of the high temperature glow discharge a-Si : H film, up to 24% lower than n_{eff} of c-Si at a photon energy of 10 eV

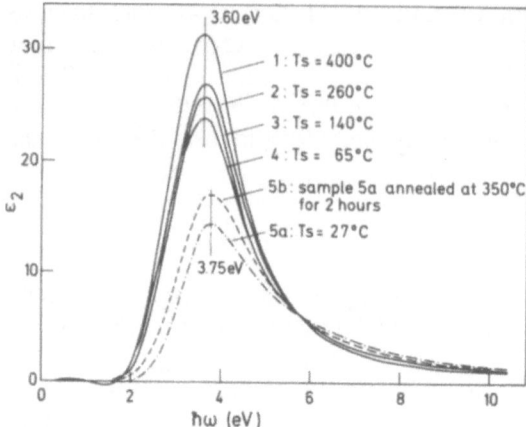

Fig. 3.46. $\varepsilon_2(\omega)$ spectra of a number of glow discharge a-Si:H films obtained from a Kramers-Kronig analysis of normal incidence reflectance spectra [3.236]

Table 3.7. Results of the $\varepsilon_2(\omega)$ analysis of three glow-discharge a-Si:H samples. The experimental data are from [3.236]. The sample numbers refer to Figs. 3.46, 47

Sample	1	3	5a	Units
T_s	400	140	27	°C
$\varrho/\varrho_c{}^a$ ($\varrho_c = 2.33$ g cm^{-3})	0.97	0.95	0.85	
$n_\infty(2\text{–}5\ \mu\text{m})^a$ $n_\infty(\text{c-Si}) = 3.41$	3.68(4)	3.46(4)	2.92(4)	
$C_H = \dfrac{[\text{H}]}{[\text{Si}]}$ [a, b]	4	11	50	at.%
$M_{-1}(\varepsilon_2)$	12.56	9.80	5.67	
$M_{-1}(n_\infty)$	12.54	10.97	7.53	
$\beta = \dfrac{M_{-1}(n_\infty)}{M_{-1}(\varepsilon_2)}$	0.98	1.12	1.32	
$n_{\text{eff}}(10\text{ eV})$ c-Si : 3.15	3.05	2.71	2.40	
$n_{\text{eff}}(\text{corr}) = \beta n_{\text{eff}}(10\text{ eV})$	3.00	3.05	3.17	
$\dfrac{n_{\text{eff}}(\text{corr}) - n_{\text{eff}}(\text{c-Si})}{n_{\text{eff}}(\text{c-Si})}$	− 4.8	− 3.2	± 0	%

[a]   G. Weiser: private communication
[b]   determined by N^{15} nuclear reaction

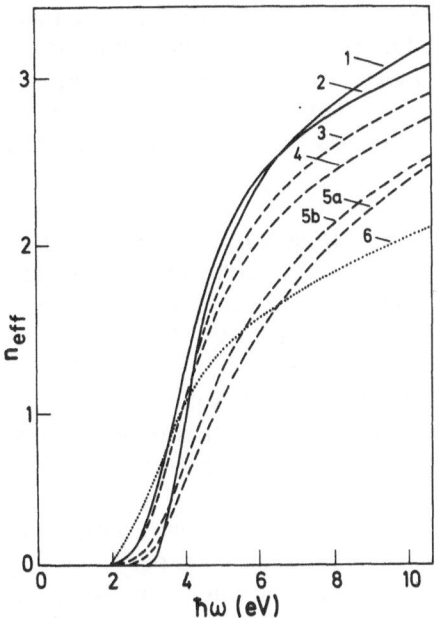

Fig. 3.47. The effective number of electrons per atom (n_{eff}) contributing to the optical absorption. (*1*) c-Si; (*2–5*) glow discharge a-Si : H; (*2*) $T_S = 400\,°C$, (*3*) $T_S = 260\,°C$; (*4*) $T_S = 140\,°C$; (*5 a*) $T_S = 27\,°C$; (*5 b*) $T_S = 27\,°C$, $T_A = 350\,°C$; (*6*) evaporated a-Si [3.235]

(see Fig. 3.47). Similar deviations have been observed in a-Ge where the sum rule could be followed to $\hbar\omega = 18$ eV, a value high enough that n_{eff} (c-Ge) reached the proper value of 4 (neglecting the contribution from d-electrons) [3.237]. One could argue that the wave functions of valence and conduction states change in a-Si and a-Ge in such a way that the sum rule (3.3) is exhausted at higher energies than in c-Si or c-Ge. This means that $\varepsilon_2(\omega)$ for amorphous samples at high energies exceeds that of their crystalline counterparts to make up for the deficit below ~ 10 eV. If one were to measure $\varepsilon_2(\omega)$ beyond 10 eV using synchrotron radiation, for example, this hypothesis could thus be tested. We consider such an explanation highly unlikely, however.

Instead, we propose that a simple *scaling* of the whole ε_2 spectrum is sufficient to account for the deficit in n_{eff} and a redistribution of oscillator strength is not required. To this end we consider the second sum rule (3.12, 15) which relates the moment of order -1 (M_{-1}) to the long wavelength refractive index n_∞ (Sects. 3.2.1, 2). This sum rule is complementary to 3.13 since it weights the *low energy* parts of $\varepsilon_2(\omega)$ through the factor $1/\omega$, whereas (3.13) weights $\varepsilon_2(\omega)$ in favour of the *higher energy* portions through the factor ω.

In Table 3.7 we compare M_{-1} (ε_2) calculated from the ε_2 spectrum with M_{-1} (n_∞) calculated from n_∞ using (3.15) for three of the glow discharge a-Si : H samples of Figs. 3.46, 47. The ratio of the two values for M_{-1} gives a correction factor β with which we multiply ε_2 to calculate n_{eff} (corr). It is apparent that this simple procedure removes the discrepancy in n_{eff} between a-Si : H and c-Si completely, even for sample 5 which had a 24% deficit in n_{eff}

before the correction. In fact, one could even argue that the increase in n_{eff} (corr) from sample 1 to 5 reflects at least in part the extra contribution of the one hydrogen valence electron which increases n_v to $4 + 1 C_H$ without contributing to the weight of the sample appreciably. The 5% increase in n_{eff} (corr) represents about one third of the required 16% increase in n_v due to a C_H of 50 at.%.

What could be the reason for the loss in amplitude of ε_2 that depends so critically on the substrate temperature? The problem of scattering in reflectance measurements on a-Si:H films has been addressed by *Donnadieu* et al. [3.238, 239] and scattering losses could contribute to an apparent reduction in $\varepsilon_2(\omega)$. It is also possible and in our opinion more likely that the density corrections which have to be made in order to evaluate n_{eff} from $\varepsilon_2(\omega)$ introduce an error. The ratios ϱ/ϱ_c in Table 3.6 represent bulk values, whereas the penetration depth of light in the energy range of interest is only ~ 100 Å. A density deficit in the top layer would therefore account for the observed reduction in $\varepsilon_2(\omega)$.

There is ample evidence that the surface properties of a-Si films differ from those of the bulk and it seems not unreasonable to expect a density deficit at the surface which depends on the preparation conditions. *Aspnes* et al. [3.216] took such a density deficit explicitly into account when they tried to fit their $\varepsilon_2(\omega)$ spectrum obtained from ellipsometry measurements. They had to assume a void fraction of 0.32 in a 100 Å thick surface layer compared to a void fraction of 0.16 for the bulk of the material to account for the low $\varepsilon_{2, max}$ of about 11. With this in mind, we may consider the ε_2 spectrum of the high temperature glow discharge a-Si:H film of Fig. 3.46 as representative of the intrinsic optical properties of a-Si:H with low hydrogen concentration because it has the highest $\varepsilon_{2, max}$ and it is the only one to fulfill the sum rule in n_{eff}.

The refractive indices of the three specimens in Table 3.7 cover the range of n_∞, $2.9 \leq n_\infty \leq 3.8$, commonly observed in a-Si and a-Si:H [3.226, 228,

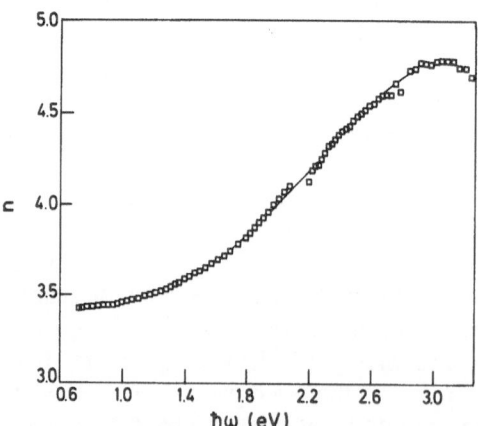

Fig. 3.48. Index of refraction of glow discharge a-Si:H ($T_D = 370\,°C$) [3.252]

240–250]. Generally, samples with high hydrogen content have a large density deficit ($\varrho_{min} \simeq 1.9$ g cm^{-3}) and a low refractive index, and vice versa. A similar decrease in refractive index with hydrogenation was found in a-Ge:H by *Connell* and *Pawlik* [3.251]. Values of n_∞ which *exceed* the refractive index of c-Si ($n_\infty \equiv n = 3.41$) are possible because of the reduced Penn gap [cf. (3.18)] in a-Si(:H). Values of n_∞ in unhydrogenated a-Si range from 3.4 to 4.1 [3.240, 245–250], and the refractive index decreases with annealing presumably because the Penn gap increases with the relaxation of the strained amorphous network [3.245].

The refractive index $n(\omega)$ has been measured as a function of energy between 0.8 and 4.0 eV by *Cody* et al. [3.208] and between 0.6 and 3 eV by *Klazes* et al. [3.252] for hydrogenated a-Si:H (Fig. 3.48). n exhibits a strong dispersion in this energy range due to the onset of interband transitions and reaches values as high as $n = 4.8$ to 4.9 at its maximum at 3.0 eV.

3.4.3 The Absorption Edge

Figure 3.49 shows the absorption edge of glow discharge a-Si:H deposited at $T_D = 240\,°C$ and containing 16 at.% hydrogen [3.208, 229]. The absorption constant was obtained from optical transmission measurements above $\alpha = 10^2$ cm^{-1}, from secondary photoconductivity using a diode and a coplanar configuration of the contacts for α below $\sim 10^3$ cm^{-1}, and from the collection efficiency of a Schottky-type solar cell covering the range $1 \lesssim \alpha \lesssim 3 \times 10^4$ cm^{-1} [3.229]. The latter method is essentially the same as the primary photoconductivity discussed in Sect. 3.4.1 b and used by *Crandall* to obtain α in a-Si:H [3.253, 254]. The agreement between all methods argues strongly in favour of the notion that Fig. 3.49 represents the true absorption spectrum of the specimen.

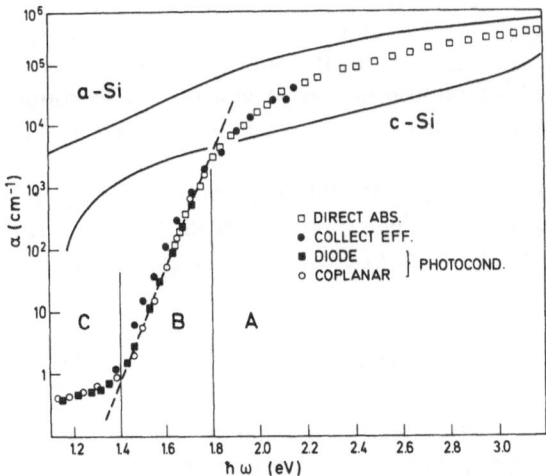

Fig. 3.49. Absorption edge spectrum of glow discharge a-Si:H$_{0.16}$ prepared at $T_D = 240\,°C$ [3.208, 229]. Also shown are the absorption edges of evaporated hydrogen-free a-Si [3.86] and of c-Si [3.255]

We distinguish three regimes: (A) a power-law regime above $\alpha \simeq 10^3$ cm^{-1} where $\alpha \propto (E - E_g)^r$, (B) the so-called Urbach edge between 1.4 and 1.8 eV where the absorption coefficient depends exponentially on $\hbar\omega$, and (C) an absorption tail below 1.4 eV ascribed to optical excitations from defect states deep in the gap.

The spectrum of Fig. 3.49 resembles that of bulk glasses such as the chalcogenide glasses or selenium [3.5], in particular, as far as the clear Urbach edge extending over three orders of magnitude in α is concerned. In most of the earlier absorption measurements on a-Si:H, a large weakly energy-dependent tail sets in at α between 5×10^3 to 10^2 cm^{-1} which extends to energies as low as 0.8 eV and thus masks the Urbach edge [3.228, 242, 244]. *Cody* et al. [3.208] argued that the tail is an artifact due to light scattering but its sensitivity to preparation conditions and on doping suggests that optical absorption from intrinsic defects contributes as well (Sect. 3.4.5). The measurements of *Freeman* and *Paul* [3.240] on sputtered films ($T_D = 200\,°C$) with varying hydrogen partial pressures in the sputter gas indicate a continuous transition from absorption spectra similar to that of Fig. 3.49 for a-SiH$_{0.16}$ to that of unhydrogenated a-Si also shown in Fig. 3.49 (compare Sect. 3.4.3 b). The a-Si spectrum of Fig. 3.49 was obtained on evaporated a-Si [3.86] but it is within half an order of magnitude in α representative for most published a-Si absorption "edges" independent of preparation conditions.

a) The Definition of an Optical Gap

The absorption edge of a-Si:H changes in shape and position with preparation conditions such as hydrogen content and deposition or annealing temperature. The shape is least affected in the power law regime (A of Fig. 3.49), i.e., for absorption coefficients above $\sim 5 \times 10^3$ cm^{-1} (see, e.g., [3.240]). The power law regime is therefore used to characterize the position of the absorption edge by a single energy referred to as the energy gap E_G.

A number of procedures are in use to define the optical gap. The simplest of these is to take E_{03} or E_{04}, the photon energy at which the absorption coefficient reaches 10^3 or 10^4 cm^{-1}, respectively, as a measure of the position of the absorption edge [3.240].

A definition of E_G that attaches more physical significance to the optical gap and the way it is derived was introduced by *Tauc* et al. [3.256]. They assumed that the transitions in region A of Fig. 3.49 take place between delocalized states. We then have, using (3.1 c) and (3.11),

$$\alpha \propto \frac{\omega}{n(\omega)}\, \varepsilon_2(\omega) \propto \frac{1}{\omega\, n(\omega)}\, |P(\omega)|^2 N_{vc}(\omega)\;, \tag{3.96}$$

where $N_{vc}(\omega)$ is the convolution of the valence and conduction densities of states. If N_v and N_c follow a simple power law behaviour as a function of E:

$$N_v = (E_v - E)^{r_1} \tag{3.97a}$$

$$N_c = (E - E_c)^{r_2} , \tag{3.97b}$$

then

$$\alpha \propto \frac{1}{\omega \, n(\omega)} \, |P(\omega)|^2 \, (\hbar\omega - E_G)^{r_1 + r_2 + 1}$$

where $E_G = E_c - E_v$.   (3.98)

For free-electron-like densities of states, $r_1 = r_2 = \frac{1}{2}$, and with a constant matrix element, we obtain the result originally proposed by *Tauc* et al. [3.256]:

$$\sqrt{\alpha \cdot \omega \cdot n(\omega)} = c_2 \cdot (\hbar\omega - E_G) . \tag{3.99}$$

When plotted in this form the linear extrapolation of $\sqrt{\alpha \cdot \omega}$ versus $\hbar\omega$ yields E_G, as shown in Fig. 3.50. This Tauc plot is generally used to define E_G. The omission of the refractive index $n(\omega)$ in (3.99) does not affect the linearity of the plot and the value of E_G changes by no more than 50 meV when n is omitted [3.208, 252].

The range of linearity in the Tauc plot of Fig. 3.50 is typical for a-Si:H samples and rather limited. Over a wide energy range the data deviate markedly from a straight line and the value of E_G obtained depends on the region used for the extrapolation (see, e.g., [3.208, 246]. Also, the slope parameter c_1 in (3.99) varies roughly linearly with E_G in the limits $5.6 \leq c_1' \leq 9.0$ $(\mathrm{eV}^{-1/2}\mu\mathrm{m}^{-1/2})$ for $1.6 \leq E_G \leq 1.85$ eV, where c_1' applies to the form of (3.99)

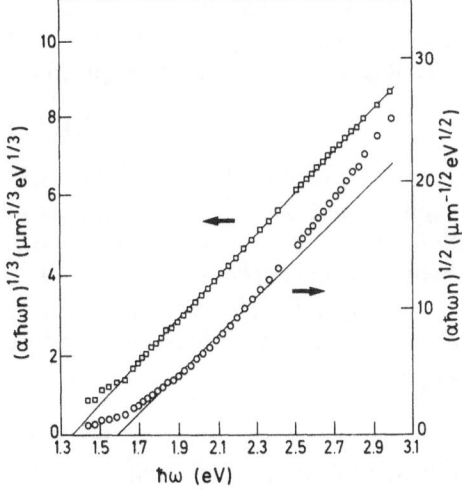

Fig. 3.50. The absorption coefficient plotted in the form suggested by *Tauc* et al. [3.256] for optical transitions between parabolic bands *(right ordinate)* and in the form proposed by *Klazes* et al. [3.252] for transitions between state densities that depend linearly on energy *(left ordinate)*. The sample is glow discharge a-Si:H prepared at $T_D = 370\,°C$

without the refractive index [3.208]. For specimens evaporated or sputtered in the absence of hydrogen, values for E_G and c_1' between 1.2 and 1.5 eV and 5.4 and 6.5 ($eV^{-1/2}\mu m^{-1/2}$), respectively, have been observed [3.231, 245–250].

A linear fit over a wider range in $\hbar\omega$ is obtained when one assumes that

$$\alpha \cdot \omega \cdot n = c_3 (\hbar\omega - E_G)^3 , \tag{3.100}$$

as suggested by *Klazes* et al. [3.252] and illustrated on the left-hand side of Fig. 3.50. Such an energy dependence of $\alpha \cdot \omega \cdot n$ is expected for densities of states that depend linearly ($r_1 = r_2 = 1$) on energy. For the valence band a linear edge is in accordance with the photoemission measurements described in Sect. 3.3.8 (Fig. 3.19). The slope parameter c_3 varies between 125 and 250 ($eV^{-2}\ \mu m^{-1}$). The value of E_G depends, of course, on the choice of the approximation made for $\alpha(E)$. In general, E_G determined from the E^3 plot is 0.2 to 0.3 eV smaller than E_G derived using Tauc's prescription. It is clear, therefore, that not too much significance should be attached to the absolute value of the optical gap determined by any of the methods discussed so far. E_{03}, $E_G(E^2)$ and $E_G(E^3)$ lie close to the transition from the power law regime to the Urbach edge and they are equally well suited to describe changes in the energetic position of the absorption edge as long as the shape of the "edge" itself is not affected.

When not otherwise stated, E_G will refer to the gap derived from the Tauc plot in what follows.

b) The Optical Gap as a Function of Hydrogen Content and Annealing Temperature

The primary source of variations in E_G is the hydrogen content of the specimens. We have already seen that the valence band edge recedes with increasing hydrogen content of the a-Si : H films, whereas the conduction band edge is relatively unaffected. A plot of E_G values versus hydrogen concentration for a-Si : H specimens prepared by rf glow discharge or sputtering in Fig. 3.51 yields a linear relationship between optical gap and hydrogen concentration:

$$E_G = 1.5 + 0.015 \cdot c_H , \tag{3.101 a}$$

where E_G is measured in eV and c_H in at.% [3.208]. The data of Fig. 3.51 cover hydrogen concentrations between 10 and 30 at.%. The films were prepared at temperatures above $\sim 200\,°C$ with the exception of two RT glow-discharge films of [3.244] which had hydrogen concentrations in excess of 20 at.%. A similar linear relationship between c_H and E_G for $2 \lesssim c_H \lesssim 17$ at.% was obtained by *Matsuda* et al. [3.257] for a variety of films prepared by reactive sputtering or glow discharge at $T_D = 300\,°C$:

$$E_G = 1.48 + 0.019\, c_H , \tag{3.101 b}$$

with parameters that agree well with those of *Cody* et al. [3.208].

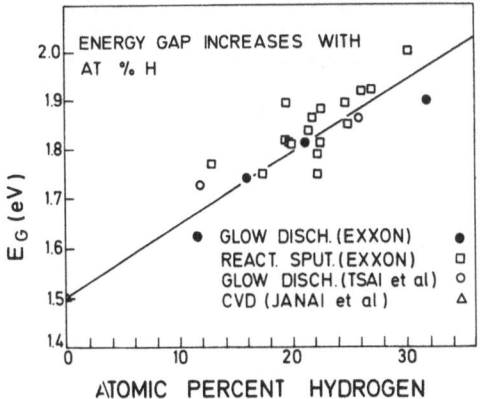

Fig. 3.51 Optical gap E_G versus hydrogen content of a-Si:H samples [3.208]

The two sets of data described by (3.101 a, b) illustrate the fact that the deposition conditions (sputtering, glow discharge, T_D, rf power, bias, etc.) determine E_G (to within ~0.1 eV) primarily in so far as they control the hydrogen content of the specimens. Films prepared at high substrate temperatures generally contain less hydrogen and have, therefore, a lower optical gap. The relationship (3.101 b) covers phosphorus and boron doped a-Si:H samples as well as undoped ones. The decrease in the optical gap with boron doping measured by *Tsai* et al. [3.241] can thus – at least for boron concentrations below a few percent – be traced to a decrease in hydrogen concentration in the specimen. This decrease and a similar increase for P-doped a-Si:H was measured by *Müller* et al. [3.194] and it is believed to be the result of altered plasma conditions in the presence of B_2H_6 or PH_3.

The proportionality between E_G and c_H as expressed in (3.101) does not hold for samples prepared by sputtering with low hydrogen content of the sputter gas and with deposition temperatures $\leq 200\,°C$ [3.221, 240, 258]. When prepared without hydrogen the value of E_G is (1.20 ± 0.05) eV for these films. As hydrogen is added the gap increases again linearly with c_H. The proportionality constant is, however, two to three times that of (3.101 a, b) for films deposited at room temperature [3.221, 258] and even higher for the $T_D = 200\,°C$ specimens [3.240] until the E_G versus c_H curves approach those of (3.101). The reason for this rapid initial increase in E_G is apparent from the absorption spectra of Fig. 3.52. The hydrogen-free amorphous silicon has an absorption spectrum in which the tail due to defects (region C in Fig. 3.50) is so high that it interferes with the power law regime (A in Fig. 3.50) and E_G as derived from the Tauc plot is lowered as a result. The initial hydrogen content passivates these defects as judged by the decrease in the ESR signal [3.259], α drops and E_G increases as a result. This corresponds to the step from sample A to sample B in the valence band spectra of Fig. 3.16. The total variation in E_G for low temperature a-Si:H specimens estimated by combining the results of [3.221, 240, 258] is about 1 eV for $0 \leq c_H \leq 50$ at.%,

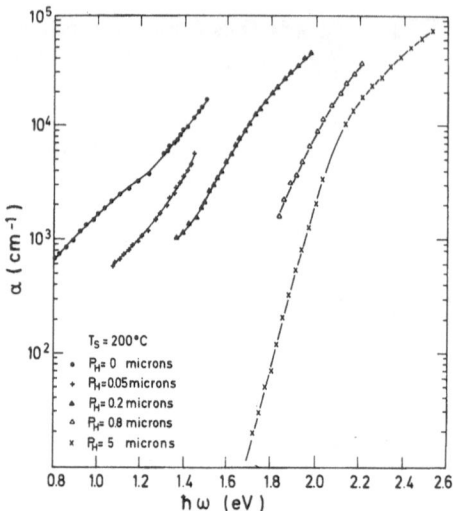

Fig. 3.52. Absorption coefficient of sputtered a-Si films as a function of the hydrogen partial pressure. The argon pressure is 5×10^{-3} torr for all specimens [3.240]

in excellent agreement with the 0.9 eV change in E_g, the separation of the band edges discussed in Sect. 3.3.11.

The variation of E_G with annealing temperature T_A is governed by two competing factors: the loss of hydrogen which tends to lower E_G with increasing T_A and the relaxation of the network which lowers the absorption tails that are due to defects and thus increases E_G, provided the tail begins close enough to E_G. This leads to the following situation. E_G decreases monotonically for samples prepared at high deposition temperatures ($T_D \gtrsim 200\,°C$) when hydrogen begins to evolve above $T_A \simeq 350\,°C$, as illustrated in Fig. 3.53

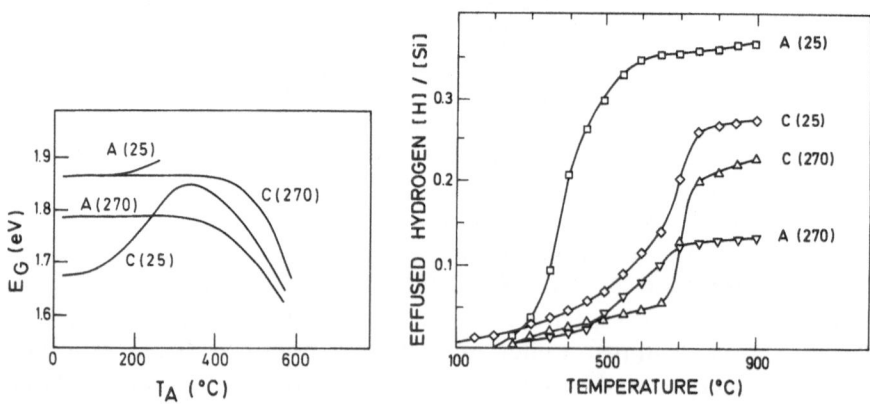

Fig. 3.53. (a) The optical gap E_G as a function of 30 min anneals at various temperatures T_A [3.244]. (b) Hydrogen evolution with annealing temperature increasing at a rate of 15°/min [3.260]. The samples were prepared by the glow discharge decomposition of a SiH₄-Ar mixture (1 : 60) on the cathode (C) or anode (A) of the reactor at substrate temperatures which are given in parentheses

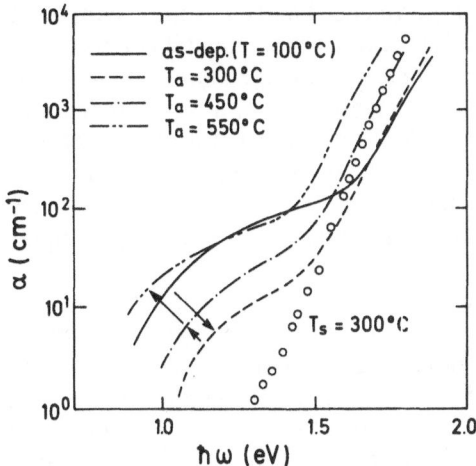

Fig. 3.54. Optical absorption spectra of glow discharge a-Si : H ($T_D = 100\,°C$) for a number of isochronal annealing steps. The spectrum for a film prepared at $T_D = 300\,°C$ is also shown. Note the absence of a defect induced tail in the high T_D specimen [3.222]

for the films prepared at 270 °C [3.244, 260]. Similar observations have been made by *Freeman* and *Paul* [3.240] and the relationship between E_G and c_H follows roughly that of (3.101) [3.222].

The initial *increase* in E_G for samples prepared at $T_D \lesssim 200\,°C$ (see Fig. 3.53) or at high deposition rates [3.244, 261, 262] has been traced by *Yamasaki* et al. [3.222] and *Deneuville* et al. [3.262] to the *shape* of the absorption spectrum. These films exhibit an unusually high tail due to defects (see Fig. 3.54). As the specimens are annealed, the defect density decreases by several orders of magnitude as illustrated by ESR measurements [3.263], the tail is reduced and E_G increases. At $T_A \simeq 300\,°C$ the increase is overtaken by the tendency of E_G to decrease with the loss of hydrogen as illustrated for specimen C(25) in Fig. 3.53. The magnitude of the initial increase in E_G depends on the defect density and is usually no more than a few hundreds of an eV [3.222, 262]. The high defect density of low temperature films despite hydrogen concentrations of 20 at.% and more poses problems that are beyond the scope of this chapter. A correlation of E_G with hydrogen configurations has been attempted by *Deneuville* et al. [3.262].

The examples given so far illustrate that it can be dangerous to describe the "edge" in a-Si : H by a single parameter E_G without due attention to the shape of the optical absorption spectrum. In fact, the sensitivity of E_G which is derived from α values $\gtrsim 10^3$ cm^{-1} to the defect density suggests that some of the defect related optical transitions extend to energies well beyond 1.6 eV. This view is supported by results on ion bombarded c-Si where α is affected deep into the fundamental absorption band [3.264].

It has been argued that the increase in E_G with c_H is mainly due to the reduction of defects [3.262] and not the result of an alloy formation similar to the Si$_{1-x}$Ge$_x$ system where E_G also varies linearly with x [3.265]. *Moddel* et al. [3.266], however, found the same gap (E_{04}) for samples with defect state densities that differ by one order of magnitude, a result that is in contrast to this opinion. The drastic recession of the valence band edge upon hydrogenation also appears to support a model whereby a fundamental change in the electronic structure takes place *in addition* to a stabilization of the amorphous structure. The stabilization finds its manifestation, among other things (Sect. 3.3.14), in the value of $E_G = 1.49$ eV extrapolated for hydrogen-free specimens from (3.101 a, b). A gap of about 1.5 eV is approached when a-Si films are annealed just short of crystallization [3.233, 267] or for specimens prepared at $T_D \gtrsim 550\,°C$ by chemical vapor deposition [3.250, 268] (compare Fig. 3.51).

We close this section with two remarks about the connection between E_G and the mobility edges. In Sect. 3.3.11 we noted that the separation between E_F and the bottom of the conduction bands remained fixed at 0.67 eV, independent of c_H between 0 and 50 at.% for a-Si:H samples prepared by reactive sputtering at room temperature. *Jeffrey* et al. [3.258] found for similarly prepared samples that the activation energy remains constant at 0.66 eV despite variations in E_G between 1.3 and 1.7 eV. Both observations are thus in agreement if we place the mobility edge E_c at the spectroscopically determined conduction band edge.

Solomon et al. [3.268 a], on the other hand, demonstrated that the activation energy E_σ scales with the optical gap such that E_σ is half the optical gap for $E_G = 1.8$ eV and slightly smaller than half gap for $E_G = 1.67$ eV ($E_\sigma = 0.75$ eV). We attribute the differences between the two sets of data to differences in the preparation conditions. In the RT sputtered samples, E_F is pinned by an as yet unspecified defect ~ 0.66 eV below E_c and in the low defect samples of *Solomon* et al. the position of E_F is determined by tails extending from E_v and E_c.

c) Temperature and Pressure Dependence of the Absorption Edge

The absorption edge of amorphous silicon shifts towards lower energies with increasing temperature. The temperature dependence of three sputtered a-Si:H samples shown in Fig. 3.55 is representative of all specimens investigated [3.240, 269, 270]. The shift of the edge is nearly linear with T between about 200 and 400 K and tends to zero for temperatures approaching zero degrees Kelvin. The temperature coefficients γ of the three samples of Fig. 3.54 determined from the linear portion of the data are -3.87 (sample A), -4.37 (B) and -2.90×10^{-4} eV/K (C), respectively. The hydrogen content of the three specimens decreases from A to C and a correlation of γ with hydrogen content is not observed with the possible exception that unhydrogenated a-Si has a slightly smaller γ. Very similar values of γ between 4.2 and 4.8×10^{-4} eV/K have been reported by *Tsang* and *Street* [3.269] and *Chmel* and *Fritzsche* [3.270] for glow-discharge a-Si:H. The addition of fluorine or dopants has no significant influence on γ [3.270] and the temperature coefficient depends little on the energy (E_{03}, E_{04}, E_G) choosen to define the edge.

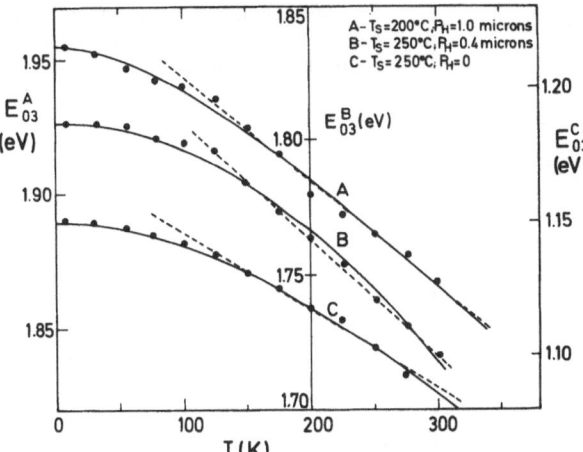

Fig. 3.55. Temperature dependence of the optical gap E_{04} of three a-Si:H specimens sputtered with varying hydrogen partial pressures [3.240]

A parallel shift of the absorption edge above $\alpha \simeq 1000$ cm^{-1} with an average temperature coefficient $\gamma \simeq -(4.4 \pm 0.4) \times 10^{-4}$ eV/K for temperatures above 200 K is thus a common feature of all amorphous hydrogenated silicon samples. The corresponding values of γ for crystalline silicon are -2.8×10^{-4} eV/K for the indirect gap and -2.2×10^{-4} eV/K for the two direct transitions E_1 and E_2 at 3.48 and 4.4 eV, respectively; that is, of the same order albeit somewhat smaller than those in a-Si:H [3.271].

The optical absorption of amorphous silicon as a function of pressure was first measured by *Connell* and *Paul* using sputtered a-Si samples [3.272]. They obtained a small positive pressure coefficient of $+0.25 \times 10^{-6}$ eV/bar. Repeating these experiments on similarly prepared samples, *Welber* and *Brodsky* [3.273] deduced an average pressure coefficient $\Delta E/\Delta P \simeq -(1 \pm 0.5) \times 10^{-6}$ eV/bar for absorption coefficients between 2000 and 7000 cm^{-1} and pressures up to 48 kbar.

Minomura and *Tsuji* [3.274, 275] extended the optical measurements to pressures as high as 180 kbar. For pressures below ~50 kbar the pressure coefficient of a-Si and a-Si:H is 1×10^{-6} eV/bar, in agreement with the data of *Welber* and *Brodsky*. Beyond 50 kbar the optical gap drops more rapidly as seen in Fig. 3.56 until dE_G/dP reaches about -12×10^{-6} eV/bar in glow-discharge a-Si:H, just short of the phase transition to a metallic phase when E_G drops to zero. The semiconductor-metal transition occurs at about 130 kbar in glow-discharge a-Si:H independent of hydrogen content, at 100 kbar in a-Si and at 150 kbar in c-Si. While these phase transitions appear to be of first order, a gradual and ill-defined transition is observed in sputtered a-Si:H and the metallic state is reached at ~180 kbar.

All films of amorphous silicon exhibit considerable hysteresis for pressures below the phase transition [3.273–275]. The structures of some of the high pressure modifications have been determined by x-ray diffraction [3.274, 275]. The high pressure phase of c-Si has the α-Sn structure. X-ray

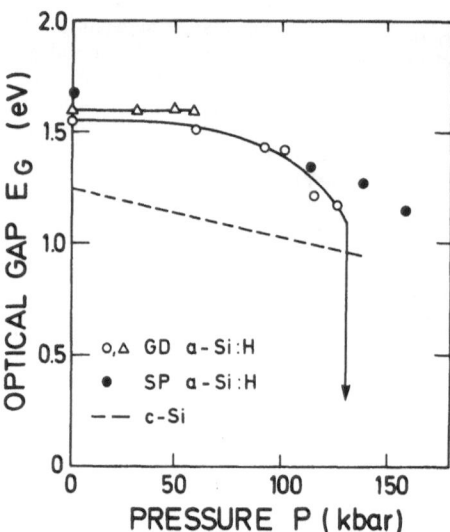

Fig. 3.56. Optical gap of amorphous and crystalline silicon as a function of hydrostatic pressure [3.274]

diffraction patterns of evaporated a-Si reveal the growth and development of α-Sn like modifications in addition to the amorphous background above the transition pressure of 100 kbar. The metallic phase is obviously heterogenous with a mixture of two metastable phases, one amorphous and one crystalline. The pressure induced crystallization is reversible for pressures below 150 kbar. Measurements of the optical edge covering the amorphous to α-Sn transition have not been made.

The pressure coefficient of the luminescence energy in glow-discharge a-Si : H was measured between 0 and 40 kbar to be $-(2 \pm 0.5) \times 10^{-6}$ eV/bar [3.276], i.e., identical to the pressure coefficient of the optical edge. This was taken by *Weinstein* [3.276] as evidence that the luminescence takes place between tail states connected to the valence and conduction band edges. A strong quenching of the luminescence intensity with pressure was explained by an increase in the number of nonradiative recombination centers deep in the gap (by about 10^{17} cm^{-3}). This conclusion is in agreement with an increased tailing observed in the absorption spectrum of pressurized amorphous silicon films. It is almost certain that these defects are broken bonds because the pressure induced crystallization requires a new arrangement of the silicon atoms [3.274, 275].

The pressure coefficients of the band gaps in diamond and zincblende type semiconductors exhibit remarkable regularity [3.277]. We denote the direct gap at Γ by E_0 and the indirect transitions from the top of the valence bands to the conduction band minima at L and X by $E_{\Gamma L}$ and $E_{\Gamma X}$, respectively. Broadly speaking, the data show that dE_0/dP is about two to four times $dE_{\Gamma L}/dP$, both coefficients being positive, with dE_0/dP of the order of 10^{-5} eV/bar. The coefficient $dE_{\Gamma X}/dP$, on the other hand, is about 10^{-6} eV/bar and negative. The gap $E_{\Gamma X}$ is the fundamental gap of c-Si with an energy of 1.1 eV and a pressure coefficient of $-(1.5 \pm 0.1) \times 10^{-6}$ eV/bar [3.271]. The similarity in the pressure coefficients of the gaps in a-Si and c-Si may be taken as additional evidence for the remnance of a "band structure" in amorphous silicon, as pointed out earlier in Sect. 3.4.2 a.

The temperature coefficient of the optical gap has two contributions:

$$\gamma \equiv \left(\frac{\partial E_G}{\partial T}\right)_P = \left(\frac{\partial E_G}{\partial T}\right)_V - \frac{\beta}{k_p}\left(\frac{\partial E_G}{\partial P}\right)_T. \tag{3.102}$$

The second term on the right-hand side of (3.102) takes the change in E_G into account which accompanies the thermal expansion of the specimen. $\beta = (dV/dT)_p/V$ is the volume expansion coefficient and k_p is the compressibility $-(dV/dP)_T/V$. Using the pressure coefficient of E_G just discussed (-1×10^{-6} eV/bar) and values for β and k_p from c-Si: $\beta = 7.7 \times 10^{-6}$ K^{-1} and $k_p^{-1} = 9.8 \times 10^{11}$ (dyn cm^{-2}) [3.271], the thermal expansion term turns out to be 7.5×10^{-6} eV/K. This is small compared to the measured value of γ and the direct contribution to γ [the first term on the rhs of (3.102)] due to the electron-phonon interaction dominates [3.277 a, 278]. The explicit temperature dependence of E_G in crystalline semiconductors has been written as [3.278]

$$E_G(T) = E_G(0) - D(\langle u^2 \rangle_T - \langle u^2 \rangle_0) , \tag{3.103}$$

where D is a second-order deformation potential and $\langle u^2 \rangle_T$ is the ensemble average of the displacements u of the atoms from their equilibrium positions. The average of the zero-point motion $\langle u^2 \rangle_0$ has been added to (3.103) so that $E_G(T)$ equals $E_G(0)$, the zero temperature gap, for $T = 0$. In the Einstein approximation

$$\langle u^2 \rangle_T = \frac{\hbar\omega_0}{\omega_0^2 M}\left(\langle n \rangle + \frac{1}{2}\right) = \frac{\hbar\omega_0}{\omega_0^2 M}\frac{1}{2}\coth\left(\frac{\hbar\omega_0}{2kT}\right) , \tag{3.104}$$

where $\langle n \rangle$, the average number of phonons with energy $\hbar\omega_0$, is given by the Bose-Einstein factor $[\exp(\hbar\omega_0/kT) - 1]^{-1}$ and M is the atomic mass. For $kT \gg \hbar\omega_0$, $\langle u^2 \rangle_T$ is proportional to kT as expected, and for $T \to 0$, $\langle u^2 \rangle_T$ tends towards a constant value equal to $\hbar/2M\omega_0$, the zero-point motion. Thus, for a temperature independent deformation potential, (3.103) yields a temperature dependence of the optical gap, in agreement with $E_G(T)$ found in amorphous silicon (Fig. 3.55). In a realistic calculation (3.104) has to be evaluated by integrating over the phonon spectrum (see, e.g., [3.278]).

3.4.4 The Urbach Edge

An exponential part in the absorption spectrum of a-Si: H corresponding to part B of Fig. 3.49 has been observed by a number of authors using photoconductivity [3.208, 229 a, 253, 265, 279, 280], photoacoustic [3.222], and photothermal deflection techniques [3.223] and also direct optical absorption [3.240, 281] to obtain α below $\sim 10^3$ cm^{-1}. The logarithmic slope

$$E_0 = \partial \hbar \omega / \partial \ln \alpha$$

varies between 60 and 120 meV depending on preparation conditions. *Cody et al.* [3.280] measured the temperature dependence of the absorption spectrum for a number of glow-discharge a-SiH$_{0.13}$ samples and established that the exponential part can be expressed as

$$\alpha(E, T) = \alpha_0 \exp\left[E - E_1\right)/E_0(T, X)] , \qquad (3.105)$$

where $E_0(T, X)$ is the logarithmic slope or "width" of the exponential tail, $\alpha_0 = 1.5 \times 10^6$ cm^{-1} and $E_1 = 2.2$ eV. E_1 is the "focal" point towards which all exponential edges converge, independent of temperature and sample. The parameter X differentiates between different specimens and its meaning will become clear in a moment.

The temperature dependence of E_0 is described by

$$E_0(T, X) = K \frac{\hbar \omega_0}{\omega_0^2 M} \left(\frac{1}{2} \coth \frac{\hbar \omega_0}{2kT} + \frac{X}{2}\right) \qquad (3.106)$$

with $\hbar \omega_0 \simeq 34$ meV, i.e. of the order of typical phonon energies. This is, except for the constant, but sample-dependent term $X/2$, equal to the intrinsic temperature dependence of the optical gap, as expressed in (3.104). *Cody*

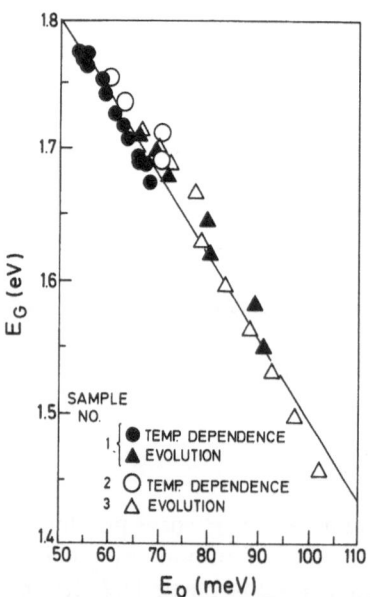

Fig. 3.57. Optical gap $E_G(T, X)$ as a function of $E_0(T, X)$, the width of the Urbach tail, for three samples of glow discharge a-SiH$_{0.13}$. The circles represent measurements at different temperatures and the triangles are for room temperature measurements on films which have been annealed between 425 and 624 °C (variable X) [3.280]

et al. indeed found that $E_0(T, X)$ and $E_G(T, X)$ are proportional to each other. The proportionality is not limited to the temperature dependence of E_0 and E_G, however, but it also includes samples annealed between 425 and 624 °C, as illustrated in Fig. 3.57. In this way a unique linear relationship between E_0 and E_G was established that covers almost the whole range of E_0 values observed so far.

Exponential absorption edges that obey both (3.105, 106) are called Urbach edges because they were first observed by *Urbach* in the absorption spectrum of AgBr [3.282]. In the meantime, Urbach edges have been identified in many crystalline semiconductors and insulators [3.283] as well as in a number of glasses [3.5, 284, 285] and amorphous selenium [3.286]. They are interpreted as absorption edges that are exponentially broadened by some mechanism involving phonons as suggested by the form of (3.106) [3.283]. The logarithmic slopes of all Urbach edges are very similar and cover the range from about 50 to 150 meV [3.283].

The temperature-independent term in (3.106), present only in amorphous materials, takes into account the contribution of disorder to the Urbach edge. *Tauc* [3.287] has suggested that this aspect of disorder could be described by "frozen-in" phonons so that (3.106) can be cast into a form analogous to (3.103):

$$E_0(T, X) = K(\langle u^2 \rangle_T + \langle u^2 \rangle_x) , \tag{3.107}$$

where $\langle u^2 \rangle_x$ is the average displacement of the atoms due to the frozen-in phonons. *Cody* et al. [3.280] extended this idea of equivalence of structural ($\langle u^2 \rangle_x$) and thermal ($\langle u^2 \rangle_T$) disorder to the band gap E_G and modified (3.103) accordingly:

$$E_G(T, X) = E_G(0, 0) - D(\langle u^2 \rangle_T + \langle u^2 \rangle_x - \langle u^2 \rangle_0) . \tag{3.108}$$

Combining (3.107) with (3.108) we obtain the observed linear relationship between E_G and E_0:

$$E_G(T, X) = E_G(0, 0) - D \langle u^2 \rangle_0 - \frac{D}{K} E_0(T, X) . \tag{3.109}$$

A fit of the data in Fig. 3.57 to (3.109) yields 2.10 eV for $E_G(0, 0) - D\langle u^2 \rangle_0$ in good agreement with the focal energy of the Urbach edges (2.2 eV), and 6.1 for the ratio D/K.

In lieu of the appropriate values for silicon, we take D for crystalline germanium from the recent calculation of *Allen* and *Cardona* [3.278]. The deformation potential is somewhat temperature dependent. For 20 °C, $D_{\mathrm{germ}} \simeq 10$ eV Å^{-2} and therefore $K = 1.6$ eV Å^{-2}. Similarly, $\langle u^2 \rangle_0^{1/2} = 0.08$ Å yields for $E_G(0, 0)$ 2.16 eV. $E_G(0, 0)$ represents, according to the model of *Cody* et al. [3.280], the upper limit for the zero-degree optical band gap in the a-Si:H family of materials. It should be noted that the magnitude of

$E_G(0,0)$ is in remarkable, though possibly fortuitous agreement with the value of 2.1 ± 0.1 eV for the mobility gap of a-Si:H at 2K obtained recently by *Lang* et al. from DLTS measurements [3.154]. One consequence of the present model is in contradiction to the previously accepted view that the incorporation of hydrogen removes states from the top of the valence bands and thereby increases the optical gap. Now the optical band gap $E_G(T, X)$ is determined by the degree of disorder in the network as described by the parameter $\langle u^2 \rangle_x$. The hydrogen reduces $\langle u^2 \rangle_x$ and thus affects E_G only indirectly. It remains to be seen how this view can be reconciled with the photoemission results and the E_G versus c_H relationship discussed earlier, and also with the electronic structure calculations which yield an increase in E_G with increasing hydrogen concentration, even with no change in disorder [3.288].

Just which aspect of disorder (defects, strain, internal fields, etc.) is operational in the model of *Cody* et al. in determining E_G and E_0 remains unspecified, however, as does the role of hydrogen to reduce the disorder. It might, therefore, be useful to examine a few of the many theories that have been developed to explain the Urbach tails in crystalline insulators and semiconductors in light of the situation at hand [3.289]. Our choice is limited to theories which rely on the electron-phonon interaction to explain the Urbach tail and we expect a value for the coupling constant K in reasonable agreement with experiment. These requirements are met by a model proposed by *Skettrup* to explain the Urbach edge in crystalline semiconductors [3.290]. The model yields directly the proportionality between E_0 and E_G as a function of T and therefore also implicitly as a function of disorder when the concept of frozen-in phonons is employed. The basic idea is to treat the optical absorption in the Born-Oppenheimer or adiabatic approximation. This implies that the transition between valence and conduction bands takes place in a time much shorter than the atomic motion due to the thermal phonons. As a consequence, the optical gap of relevance is not the time and spatial *average* $E_G(T)$ related to the average phonon occupation number $\langle n \rangle$ according to (3.103, 104) but rather a local and instantaneous gap $E_{G,n}$ which is obtained when $\langle n \rangle$ in (3.103, 104) is replaced by a particular phonon occupation number n:

$$E_n = E_G(0) - A\,(n_x + n_y + n_z)\ , \tag{3.110}$$

where $A = (D/3)[\hbar\omega_0/\omega_0^2 M]$ in the Einstein approximation to the phonon spectrum.

Bose-Einstein fields exhibit large fluctuations. At a temperature T the standard deviation of the occupation number n is given by

$$\sigma_n = (\langle n^2 \rangle - \langle n \rangle^2)^{1/2} = \langle n \rangle \left(1 + \frac{1}{\langle n \rangle}\right) . \tag{3.111}$$

Hence, σ_n is never less than the average occupation number $\langle n \rangle$ and at low temperatures σ_n will even be considerably larger than $\langle n \rangle$. As a result, the

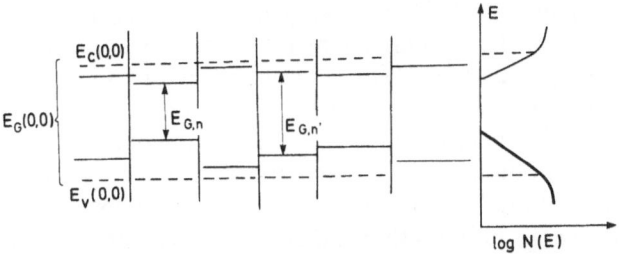

Fig. 3.58. Schematic representation of the positions of the band edges E_v and E_c in a disordered solid according to the model of *Skettrup* [3.290]. $E_G(0,0)$ is the band gap in the absence of disorder and at $T = 0$ and $E_{G,n}$ are the local and instantaneous gaps due to the excitation of n phonons (thermal or frozen-in). Also indicated is the shape of the densities of states $N(E)$

observed absorption spectrum is the superposition of optical edges $E_{G,n}$, each edge weighted with the probability $P_n(T)$ of finding an occupancy n:

$$\langle a(h\nu)\rangle_T = \sum_n P_n(T)\, a(h\nu - E_{G,n}) \ . \tag{3.112}$$

The situation is schematically sketched in Fig. 3.58 which is meant to represent the spatial distribution of valence and conduction band edges at a particular moment. Following *Skettrup* [3.290], the solid is divided into cells (one-dimensional in Fig. 3.58), each cell having a local gap $E_{G,n}$. The size of the cells is equal to the coherence length of the phonon modes, i.e., 10–100 times their wavelength. This is sufficiently large to define local band edges. The probability $P_n(T)$ that a particular cell contains n_x, n_y, n_z phonons is given by the distribution function for bosons [3.291]:

$$P_n(T) = \exp[-(n_x + n_y + n_z)\hbar\omega_0/kT][1 - \exp(-\hbar\omega_0/kT)]^3 \ . \tag{3.113}$$

Inserting (3.113) into (3.112) and making the additional assumption that $a(h\nu - E_{G,n})$ is a simple step function, one obtains for the spatial ($\equiv$ time) average of a at a temperature T:

$$\langle a(h\nu)\rangle_T = a_0 \exp\left\{ -[E_G(0) - h\nu]\frac{\hbar\omega_0}{AKT}\right\}, \tag{3.114}$$

where a_0 is a weakly temperature-dependent pre-factor. This is the correct exponential energy dependence. The temperature dependent logarithmic slope

$$E_0(T) = \frac{AKT}{\hbar\omega_0} = \frac{D}{3\omega_0^2 M}\, kT \tag{3.115}$$

agrees with the experimental results of *Cody* et al. [3.280] in the high temperature limit (3.106) and a comparison of factors in (3.106, 115) yields $D = 3K$

compared to the experimental result $D = 6.1K$. Whether this discrepancy is significant or just a result of the simplifications made, for instance, in the phonon spectrum remains to be seen. *Skettrup* has extended the simple picture given here to a particular case and also obtains the correct low temperature behaviour of E_0 [3.290]. An implication of this derivation of the Urbach edge is that the transitions in the exponential absorption region are between band states and should therefore have transition matrix elements comparable to those above E_G. *Dunstan* [3.292] has shown that this is indeed the case for a wide class of amorphous semiconductors lending further support to Skettrup's model.

We close this section with a few remarks about the relationship between the Urbach edge and the valence and conduction band tails in light of the model just discussed.

The variation of E_G with temperature and disorder as expressed in (3.103, 108) is the result of changes in the energy of valence and conduction states due to the electron-phonon interaction. In general, the states at the valence band edge are pushed up and those of the conduction band edge are pushed down, decreasing E_G. The temperature and disorder induced shifts in the band edges are related to the phonon occupation numbers in a way completely analogous to $E_G(T, X)$, provided the deformation potential D is replaced by the potentials D_v and D_c appropriate for valence and conduction band edges, respectively. Applying Skettrup's arguments to the band edges, exponential tails are obtained with slopes which are determined by D_v and D_c.

Allen and *Cardona* have calculated D_v and D_c in c-Ge [3.278]. Their results indicate that the observed temperature coefficient of the band gap is well reproduced by modifying the pseudopotential band structure in a manner analogous to the way in which Debye-Waller factors are introduced in the theory of temperature-dependent x-ray scattering. Under these conditions band energies can be referred to the temperature-independent average pseudopotential V_0, and it turns out that the valence band edge shifts 50–100 times more than the conduction band edge. The Urbach edge in a-Si:H thus essentially represents the valence band tail and the conduction band tail is much steeper (a result in agreement with the generally held view [3.293]), provided the deformation potentials in Si are not too different from those in Ge.

The direct connection between the Urbach edge and band tails is a specific property of Skettrup's model. Another class of models ascribes the Urbach edge to an electric field induced broadening of the band edge or exciton states via the Franz-Keldysh effect [3.287, 294–298]. Static electric fields are set up by charged defects or by fluctuating charges on the atoms, as we have shown for a-Si in connection with the core level broadening in Sect. 3.3.14. Temperature-dependent fields arise through the action of LO phonons on the charged atoms. Connected with the fields are, of course, potential fluctuations ΔV which modulate the position of the band edges so that

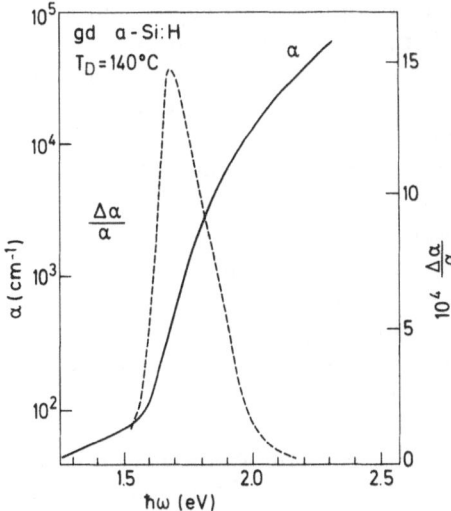

Fig. 3.59. Electroabsorption spectrum obtained on glow discharge a-Si : H prepared at 140 °C. Plotted is the absorption spectrum in the absence of an electric field and the relative change $\Delta\alpha/\alpha$ measured with an applied field of order of 10^4 Vcm^{-1} [3.281]

they move in parallel. Thus, we have the situation where the band edges are broadened in proportion to ΔV^2, whereas the broadening of the absorption edge is determined by fluctuations in the *gradient* of V. Band-tail slopes and Urbach slopes are decoupled and for long wavelength fluctuations, i.e., ΔV finite and $\nabla V \rightarrow 0$, we would have band-edge tails and a sharp absorption edge.

Al Jalali and *Weiser* investigated the influence of an *external* electric field on the absorption spectrum of glow discharge a-Si : H [3.281]. Their results, shown in Fig. 3.59, demonstrate that external electric fields affect the absorption spectrum only in the region of the Urbach edge; the relative change in α, $\Delta\alpha/\alpha$ drops off as $\hbar\omega$ approaches transitions above E_G and also for transitions from defect states below 1.4 eV. A theory of electroabsorption in the presence of random internal fields [3.299] fails to explain these results, and *Al Jalali* and *Weiser* conclude that random fields are not responsible for the Urbach edge. One should stress, however, that an exponential distribution of local energy gaps yields a mere shift in the absorption edge ($\Delta\alpha/\alpha$ = const) upon application of a field [3.294] – a result that is also not observed.

It thus appears that further measurements are necessary. In particular, a combination of photoemission studies of the valence band edges similar to those of Fig. 3.17 with optical measurements show some promise of shedding new light on the origin of the Urbach edge.

3.4.5 Defect Induced Transitions

At energies below ~ 1.4 eV the absorption coefficient $\alpha(h\nu)$ begins to deviate from its exponential shape and a shoulder develops, as indicated in region C of Fig. 3.49. The shoulder is a common feature in photoconductivity spectra [3.228, 229 a, 266, 271, 300, 301]. Its presence has been confirmed by photoacoustic [3.222, 302] and photothermal deflection spectroscopy [3.223, 303, 304], both techniques which measure the absorption constant $\alpha(h\nu)$ reliably down to small absorption coefficients as explained earlier (Sect. 3.4.1 a). The simultaneous measurements of photoconductivity and direct optical absorption by *Al Jalali* and *Weiser* [3.281] on very thick films ($d = 100$ μm) down to 0.7 eV suggest that α derived from photoconductivity may overestimate the defect induced absorption below ~ 1.4 eV, however.

The magnitude of the shoulder depends on the preparation conditions. The shoulder increases with increasing rf power for glow-discharge a-Si : H [3.303, 304] and with decreasing deposition temperature [3.222, 303]. For films deposited at 100 °C, annealing leads at first to a decrease in the absorption coefficient around 1.2 eV until $\alpha(1.2)$ increases again above 250–300 °C as hydrogen evolves [3.222, 303] (Fig. 3.53).

These results suggest that the shoulder is an extrinsic property of the a-Si specimens related to defects. For the best films the absorption edges are indeed simple exponentials down to $\alpha \simeq 1$ cm^{-1} [3.222, 304]. The strength $\Delta\alpha$ of the defect induced absorption band may therefore be obtained by subtracting the Urbach tail from the measured absorption spectrum:

$$\Delta\alpha(h\nu) = \alpha(h\nu)_{\text{meas}} - \alpha_0 \exp(h\nu/E_0) , \tag{3.116}$$

where α_0 and E_0 are adjusted to fit the exponential part of the spectrum well above the shoulder. It turns out that the integrated absorption strength $\int \Delta\alpha(\nu)d\nu$ is proportional to the spin density N_s(ESR) determined from ESR measurements [3.222, 304]:

$$N_s = A \int \Delta\alpha(\nu)d\nu .$$

Yamasaki et al. [3.222] obtained $A \simeq 1.2 \times 10^{16}$ cm^{-2} eV from an annealing series. *Jackson* and *Amer* [3.304], on the other hand, calculated A using the same expression that was employed to calculate the oscillator strength of the Si–H infrared modes including the local field corrections and the "fudge factor" 1/2 [3.305]. The result, $A = 7.9 \times 10^{15}$ cm^{-2} eV, yields values for the defect density N_s (absorption) which are in perfect agreement with N_s (ESR) over three orders of magnitude in N_s as shown in Fig. 3.60. This leaves little doubt that the extra absorption involves directly dangling bonds and not just defects that occur under the same preparation conditions as the dangling bonds. The evidence that the spin carrying defects are indeed dangling bonds is that the ESR signal used to calculate N_s(ESR) has $g = 2.0055$ [3.306, 307]. A positive correlation between the ESR signal and the 1.3 eV shoulder in the photoconductivity spectrum has first been observed by *Stuke* [3.308] in a-Si : H. In unhydrogenated evaporated a-Si, a contribution in the sub-band-gap absorption proportional to N_s^2(ESR) has been reported [3.309].

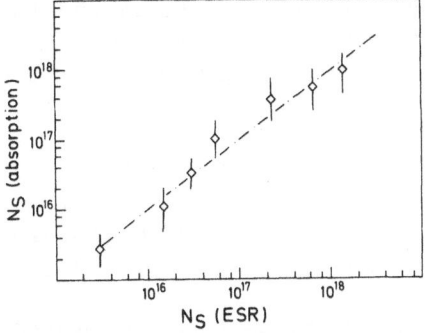

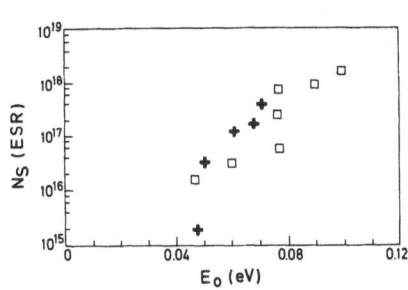

Fig. 3.60. Number of defects deduced from absorption spectrum versus number of spins measured by ESR [3.304]

Fig. 3.61. Logarithmic slope of the Urbach tail E_0 versus equilibrium spin density for a number of a-Si:H samples ($\square$) data from [3.303]; (+) data from [3.315]

Phosphorus and boron doping also introduce a strong extrinsic absorption band at virtually the same energy [3.240, 300–302, 304]. There is no $g = 2.0055$ spin signal for doped a-Si:H because the shift in E_F fills the defect states with two electrons (P-doping) or empties them completely (B-doping) [3.306]. The density of defects may, nevertheless, be determined from light induced ESR (LESR) or from the quenching of luminescence [3.310]. The result is again in excellent agreement with estimates based on the integrated strength of $\Delta\alpha$ yielding defect densities between 10^{16} and 10^{18} cm^{-3}. For a PH$_3$ concentration of 3×10^{-4}, *Jackson* and *Amer* [3.304] obtained $N_s = 3.7 \times 10^{17}$ cm^{-3}.

The g-factor of 2.0055 in LESR and the similar energy of the absorption band strongly suggest that the defect states introduced through P and B are identical to the dangling-bond states in undoped a-Si:H. The photoexcitation cross section of these states is 1.2×10^{-16} cm^2 [3.304].

There is general agreement that the transitions in the shoulder C take place from the defect states to empty states above the conduction band edge E_c [3.229 a]. Depending on the analysis of the absorption spectrum, the defect states are therefore situated 1.3 eV [3.304] or between 0.8 and 1.1 eV [3.229 a, 311] below E_c. *Collins* et al. [3.311] estimate a width (FWHM) of ~0.3 eV for the defect band. These estimates place the defect states between 0.3 and ~0.5 eV above E_v ($E_G \simeq 1.6$ eV), a position which agrees with the photoemission results discussed earlier for P-doped a-Si:H samples (Fig. 3.19) and with the results of DLTS measurements which also yield defect densities in good agreement with the absorption data [3.312]. The same assignment for B-doped samples poses problems because the absence of equilibrium spins means that the defect states are empty and cannot contribute to transitions into the conduction band.

The position of dangling-bond defects in the lower half of the gap is in agreement with the arguments summarized by *Voget-Grote* et al. [3.313]. The same defect quenches the band-to-band luminescence at 1.2 eV and it is thought to give rise to the 0.9 eV luminescence which increases with increasing ESR signal. *Street,* however, places the corresponding defect in the upper half of the gap [3.314].

Spin-carrying defects also affect the slope of the Urbach tail as shown in Fig. 3.61. The figure summarizes data from *Jackson* and *Amer* [3.303] and *Abeles* [3.315]. E_0 is seen to increase with the defect density. Unfortunately, this result sheds no new light on the question what causes the Urbach tails in a-Si : H. It would do so if the defect states were charged and could therefore cause Urbach tails by the field broadening mechanism discussed in the previous section. It is, however, expected that only spinless defects are charged. We have to accept, consequently, a mechanism where spins and disorder are created under the same deposition conditions without further specification.

3.5 Concluding Remarks

It was the aim of this chapter to present photoemission and optical data so as to provide, as far as possible, a reasonably consistent picture of the experimental situation. There is a wealth of optical data on amorphous silicon now available and firm outlines of some of the basic properties such as the position and shape of the absorption edge begin to appear despite the enormous variability in the preparation conditions. The situation is somewhat less favourable in photoemission where the data are largely limited to one source. What appears to be most promising for the future is a combination of both techniques to elucidate such unsolved questions as the origin of the Urbach tail in a-Si : H and the relationship between hydrogen concentration and band gap. Photoemission measurements will certainly have an impact on the investigation of surface and interface properties of a-Si : H in the near future.

One aspect of optical spectroscopy that has been omitted from this review is the field of photoinduced changes in the absorption coefficient. Employing picosecond techniques, the thermalization of hot carriers has been followed in this way and new insight into the energy dissipation mechanism gained. This area has been comprehensively reviewed at the time of writing by *Tauc,* the initiator of the technique applied to a-Si : H [3.316].

Acknowledgement. I am indebted to my colleagues at the Max-Planck-Institut who contributed considerably to the material presented here: K. J. Gruntz, R. Johnson, R. Kärcher, H. Richter, and B. von Roedern. I would also like to thank J. Tauc, M. Cardona, and P. Allen for a number of illuminating discussions and M. Cardona for a critical reading of the manuscript.

Note Added in Proof

Recently, *Griep* and *Ley* [3.317] succeeded in measuring the distribution of valence tail states and occupied gap states in a-Si : H employing the technique of photoelectric yield spectroscopy. They observed an exponential valence band tail over two to three orders of magnitude in $N_V(E)$ and a band of defect states located 0.3 to 0.5 eV above the valence band maximum. The logarithmic slope of the valence band tail varies between 75 and 50 meV depending on deposition temperature in quantitative agreement with the observed slopes of the Urbach tails. These measurements are thus in perfect agreement with the ideas about the origin of the Urbach tail developed on pages 156 and 157.

References

3.1   D. L. Greenaway, G. Harbeke: *Optical Properties of Semiconductors* (Pergamon, Oxford 1968)

3.2   M. Cardona: "Optical Properties and Band Structure of Germanium and Zincblende Type Semiconductors", in *Atomic Structure and Properties of Solids*, ed. by E. Burstein (Academic, New York 1972)

3.3   J. C. Phillips: *Solid State Physics*, Vol. 18 (Academic, New York 1966)

3.4   J. Tauc: In *Optical Properties of Solids*, ed. by F. Abeles (North-Holland, Amsterdam 1972) p. 277

3.5   J. Tauc: In *Amorphous and Liquid Semiconductors*, ed. by J. Tauc (Plenum, New York 1974) p. 159

3.6   M. Závetová, B. Velický: In *Optical Properties of Solids, New Developments*, ed. by B. O. Seraphin (North-Holland, Amsterdam 1976) p. 379

3.7   G. A. N. Connell: In *Amorphous Semiconductors*, ed. by M. H. Brodsky, Topics Appl. Phys., Vol. 36 (Springer, Berlin, Heidelberg, New York 1979) p. 73

3.8   N. F. Mott, E. A. Davis: *Electronic Processes in Non-Crystalline Materials* (Clarendon, Oxford 1979)

3.9   F. C. Brown: In *Solid State Physics*, Vol. 29 (Academic, New York 1974) p. 1

3.10 C. Kunz: In [Ref. 3.6, p. 473]

3.11 M. Cardona, L. Ley (eds.): *Photoemission in Solids I*, Topics Appl. Phys., Vol. 26 (Springer, Berlin, Heidelberg, New York 1978)

3.12 L. Ley, M. Cardona (eds.): *Photoemission in Solids II*, Topics Appl. Phys., Vol. 27 (Springer, Berlin, Heidelberg, New York 1979)

3.13 L. Ley, M. Cardona, R. A. Pollak: In [Ref. 3.12, p. 11]

3.14 M. Altarelli, D. L. Dexter, H. M. Nussenzweig, D. Y. Smith: Phys. Rev. **B 6**, 4502 (1972)

3.15 D. T. Pierce, W. E. Spicer: Phys. Rev. **B 5**, 3017 (1972)

3.16 J. Tauc, R. Grigorovici, A. Vancu: Phys. Stat. Sol. **15**, 627 (1966)

3.17 N. K. Hindley: J. Non-Cryst. Solids **5**, 17 (1970)

3.18 E. A. Davis, N. F. Mott: Phil. Mag. **22**, 903 (1970)

3.19 J. Tauc: Mater. Res. Bull. **5**, 721 (1970)

3.20 F. Stern: Phys. Rev. **B 3**, 2636 (1971)

3.21 H. Fritzsche: J. Non-Cryst. Solids **6**, 49 (1970)

3.22 J. D. Dow, J. J. Hopfield: J. Non-Cryst. Solids **8–10,** 664 (1972)
3.23 H. R. Philipp, H. Ehrenreich: Phys. Rev. **129,** 1550 (1963)
3.24 M. H. Brodsky, A. Lurio: Phys. Rev. **B 9,** 1646 (1974)
3.25 H. Raether: *Excitations of Plasmons and Interband Transitions by Electrons,* Springer Tracts Mod. Phys. **88** (1980)
3.26 A. H. Clark: Phys. Rev. **154,** 750 (1970)
3.27 T. M. Donovan, W. E. Spicer, J. M. Bennet, E. J. Ashley: Phys. Rev. **B 2,** 397 (1970)
3.28 J. C. Knights, G. Lucovsky: CRC Critical Rev. **9,** 211 (1980)
3.29 F. W. Lyttle: In *Physics of Non-Crystalline Solids,* ed. by J. A. Prins (North-Holland, Amsterdam 1965) p. 12
3.30 H. Richter, A. Rukwied: Z. Physik **160,** 473 (1960)
3.31 S. C. Moss, J. F. Graczyk: Phys. Rev. Lett. **23,** 1167 (1969)
3.32 T. M. Donovan, K. Heinemann: Phys. Rev. Lett. **27,** 1794 (1971)
3.33 I. Ohdomari, M. Ikeda, H. Yoshimoto: Phys. Lett. A **64,** 253 (1977)
3.34 J. C. Knights: J. Non-Cryst. Solids **35/36,** 159 (1980)
3.35 Y. Katayama, T. Shimada, K. Usami: Phys. Rev. Lett. **46,** 1146 (1981)
3.36 J. C. Phillips: Rev. Mod. Phys. **42,** 290 (1970)
3.37 J. A. van Vechten: Phys. Rev. **182,** 891 (1969)
3.38 D. Penn: Phys. Rev. **128,** 2093 (1962)
3.39 M. Cardona, F. H. Pollak: In *Physics of Opto-Electronic Materials,* ed. by W. A. Alberts (Plenum, New York 1971) p. 81
3.40 S. H. Wemple, M. DiDominico: Phys. Rev. **B 3,** 1338 (1971)
3.41 S. H. Wemple: Phys. Rev. **B 7,** 3767 (1973)
3.42 A. G. Revesz, S. H. Wemple: (private communication)
3.43 W. E. Spicer: Phys. Rev. **112,** 114 (1958)
3.44 W. L. Schaich: In [Ref. 3.11, p. 105]
3.45 G. D. Mahan: Phys. Rev. **B 2,** 4334 (1970)
3.46 C. Caroli, D. Lederer-Rozenblatt, B. Roulet Saint-James: Phys. Rev. **B 8,** 4552 (1973)
3.47 I. Lindau, W. E. Spicer: J. Electron Spectrosc. **3,** 409 (1974)
3.48 C. M. Garner, I. Lindau, C. Y. Su, P. Pianetta, W. E. Spicer: Phys. Rev. **B 19,** 3944 (1979)
3.49 S. Brennan, J. Stöhr, R. Jaeger, J. E. Rowe: Phys. Rev. Lett. **45,** 1414 (1980)
3.50 J. Reichardt, L. Ley, R. L. Johnson: to be published
3.51 E. O. Kane: Phys. Rev. **159,** 624 (1967)
3.52 H. Ibach: In *Electron Spectroscopy for Surface Science,* ed. by H. Ibach, Topics Current Phys., Vol. 4 (Springer, Berlin, Heidelberg, New York 1977) p. 1
3.53 F. Forstmann: In *Photoemission from Surfaces,* ed. by B. Feuerbacher, B. Fitton, R. F. Willis (Wiley, New York 1977) Chap. 8
3.54 J. A. Appelbaum, D. R. Hamann: Rev. Mod. Phys. **48,** 479 (1976); Crit. Rev. Sol. State Sci. **6,** 357 (1978)
3.55 P. J. Feibelman, D. E. Eastman: Phys. Rev. **B 10,** 4932 (1974)
3.56 T. Grandke, L. Ley, M. Cardona: Phys. Rev. **B 18,** 3847 (1978)
3.57 W. D. Grobman, D. E. Eastman, J. L. Freeouf: Phys. Rev. **B 12,** 4405 (1975)
3.58 M. Cardona, L. Ley: In [Ref. 3.11, p. 1] and references therein
3.59 W. Bambynek, B. Crasemann, R. W. Fink, H.-U. Freund, H. Mark, C. D. Swift, R. E. Price, P. Venugopala Rao: Rev. Mod. Phys. **44,** 716 (1972)
3.60 S. T. Manson, J. W. Cooper: Phys. Rev. **165,** 126 (1968)
3.61 U. Gelius: In *Electron Spectroscopy,* ed. by D. A. Shirley (North-Holland, Amsterdam 1972) p. 311
3.62 K. J. Gruntz, L. Ley, M. Cardona, R. Johnson, G. Harbeke, B. von Roedern: J. Non-Cryst. Solids **35/36,** 453 (1980)
3.63 V. I. Nefedov, N. P. Sargushin, Y. V. Salyn, I. M. Band, M. B. Trzhaskovskaya: J. Electr. Spectrosc. **2,** 383 (1973); **7,** 175 (1975)
3.64 J. H. Scofield: J. Electron. Spectrosc. **8,** 129 (1976)
3.65 S. M. Goldberg, C. S. Fadley, S. Kono: J. Electron. Spectrosc. **21,** 285 (1981)

3.66 R. C. G. Leckey: Phys. Rev. A 13, 1043 (1976)
3.67 R. G. Cavell, S. P. Kowalczyk, L. Ley, R. A. Pollak, B. Mills, D. A. Shirley, W. Perry: Phys. Rev. B 7, 5313 (1973)
3.68 H. R. Philipp, H. Ehrenreich: Phys. Rev. 129, 1550 (1963)
3.69 C. Gähwiller, F. C. Brown: Phys. Rev. B 2, 1918 (1970)
3.70 K. J. Gruntz: Dissertation, University of Stuttgart (1981) unpublished
3.71 Y. Katayama, K. Usami, T. Shimada: Phil. Mag. B 43, 283 (1981)
3.72 K. Siegbahn et al.: ESCA Applied to Free Molecules (North-Holland, Amsterdam 1969)
3.73 U. Gelius: Physica Scripta 9, 133 (1974)
3.74 F. J. Grunthaner, P. J. Grunthaner, R. P. Vasquez, B. F. Lewis, J. Maserjian, A. Madhukar: Phys. Rev. Lett. 43, 1683 (1979)
3.75 L. Pauling: The Nature of Chemical Bonds (Cornell Univ. Press, New York 1967)
3.76 J. C. Phillips: Bonds and Bands in Semiconductors (Academic, New York 1973)
3.77 C. S. Fadley, S. B. M. Hagström, M. P. Klein, D. A. Shirley: J. Chem. Phys. 48, 3779 (1968)
3.78 A. D. Baker, C. R. Brundle (eds.): Electron Spectroscopy, Vols. 1, 2 (Academic, New York 1977, 1978)
3.79 C. Kunz (ed.): Synchrotron Radiation, Topics Current Phys., Vol. 10 (Springer, Berlin, Heidelberg, New York 1979)
3.80 J. Reichardt, L. Ley, R. L. Johnson: In Proc. Xth Intl. Conf. Amorphous and Liquid Semiconductors, Tokyo (1983)
3.80 a M. Tanielian, H. Fritzsche, C. C. Tsai, E. Symbalisty: Appl. Phys. Lett. 33, 353 (1978)
3.81 See, e.g., E. Kay: Advan. Electron. Electron Phys. 17, 3 (1962)
3.82 R. H. Williams, R. R. Varma, W. E. Spear, P. G. LeComber: J. Phys. C 12, L 209 (1979)
3.83 B. von Roedern, L. Ley, M. Cardona, F. W. Smith: Phil. Mag. B 40, 433 (1979)
3.83 a S. R. Herd, P. Chaudhuri, M. H. Brodsky: J. Non-Cryst. Solids 7, 309 (1972)
3.84 T. M. Donovan, W. E. Spicer, J. Bennett: Phys. Rev. Lett. 22, 1058 (1969)
3.85 T. M. Donovan, W. E. Spicer: Phys. Rev. Lett. 21, 1571 (1968)
3.86 D. T. Pierce, W. E. Spicer: Phys. Rev. B 5, 3017 (1972)
3.87 D. E. Eastman, W. D. Grobman: In Proceed. Intern. Conf. on Phys. Semic. (PWN-Polish Scientific Publishers, Warsaw 1972) p. 889
3.88 L. Ley, S. P. Kowalczyk, R. Pollak, D. A. Shirley: Phys. Rev. Lett. 29, 1088 (1972)
3.89 G. Wiech, E. Zöpf: In Band Structure Spectroscopy of Metals and Alloys, ed. by D. J. Fabian, L. M. Watson (Academic, New York 1973) p. 637
3.90 J. D. Joannopoulos, M. L. Cohen: Phys. Rev. B 7, 2644 (1973)
3.91 D. Weaire, M. F. Thorpe: Phys. Rev. Lett. 27, 1581 (1971)
3.92 F. Herman, J. P. van Dyke: Phys. Rev. Lett. 21, 1575 (1968)
3.93 B. Kramer: Phys. Stat. Sol. (b) 47, 501 (1971)
3.94 D. Brust: Phys. Rev. 186, 768 (1969)
3.95 D. Weaire, M. F. Thorpe: Phys. Rev. B 4, 2548, 3517 (1971)
3.96 J. D. Joannopoulos, M. L. Cohen: In Solid State Physics, Vol. 31 (Academic, New York 1976) p. 71
3.97 R. Alben, S. Goldstein, M. F. Thorpe, D. Weaire: Phys. Stat. Sol. (b) 53, 545 (1972)
3.98 I. B. Ortenburger, W. E. Rudge, F. Herman: J. Non-Cryst. Solids 8–10, 653 (1972)
3.99 J. D. Joannopoulos, M. L. Cohen: Phys. Rev. B 8, 2733 (1973) and references therein
3.100 W. Y. Ching, C. C. Ling, D. L. Huber: Phys. Rev. B 14, 620 (1976)
3.101 W. Y. Ching, C. C. Ling, L. Guttmann: Phys. Rev. B 16, 5488 (1977)
3.102 M. J. Kelly, D. W. Bullett: J. Non-Cryst. Solids 21, 155 (1976)
3.103 P. E. Meek: J. Phys. C 10, L 59 (1977)
3.104 P. Steinhardt, R. Alben, M. G. Duffy, D. E. Polk: Phys. Rev. B 8, 6021 (1973)
3.105 D. Henderson: J. Non-Cryst. Solids 16, 317 (1974)
3.106 H. Richter, G. Breitling: Z. Naturforsch. 13 a, 988 (1958)
3.107 M. V. Coleman, D. J. D. Thomas: Phys. Stat. Sol. 22, 593 (1967)
3.108 R. Grigorovici, R. Manaila: Thin Solid Films 1, 343 (1968)
3.109 S. Mader: J. Vac. Sci. Technol. 8, 247 (1971)

3.110 F. Yonezawa, M. H. Cohen: In *Fundamental Physics of Amorphous Semiconductors,* ed. by F. Yonezawa, Springer Ser. Solid-State Sci., Vol. 25 (Springer, Berlin, Heidelberg, New York 1981) p. 11
3.110a J. Singh: Phys. Rev. B **23,** 4156 (1981)
3.111 B. von Roedern, L. Ley, M. Cardona: Phys. Rev. Lett. **39,** 1576 (1977)
3.112 K. C. Pandey, T. Sakurai, H. D. Hagstrum: Phys. Rev. Lett. **35,** 1728 (1975)
3.113 T. Sakurai, H. D. Hagstrum: Phys. Rev. B **14,** 1593 (1976)
3.114 K. Fujiwara: Phys. Rev. B **24,** 2240 (1981)
3.115 K. C. Pandey: Phys. Rev. B **14,** 1593 (1976)
3.116 K. M. Ho, M. L. Cohen, M. Schlüter: Phys. Rev. B **15,** 3888 (1977)
3.117 W. Y. Ching, D. J. Lam, C. C. Lin: private communication
3.118 W. Y. Ching, D. J. Lam, C. C. Lin: Phys. Rev. Lett. **42,** 805 (1979)
3.119 W. Y. Ching, D. J. Lam, C. C. Lin: Phys. Rev. B **21,** 2378 (1980)
3.120 E. N. Economou, D. A. Papaconstantopoulos: Phys. Rev. B **23,** 2042 (1981)
3.121 W. E. Picket: Phys. Rev. B **23,** 6603 (1981)
3.122 P. Lemaire, J. P. Gaspard: J. Phys. (Paris) C 4 **42,** 765 (1981)
3.123 D. C. Allan, J. D. Joannopoulos, W. B. Pollard: Phys. Rev. B **25,** 1065 (1982)
3.124 J. C. Phillips: Phys. Rev. Lett. **42,** 1151 (1979)
3.125a D. P. DiVincenzo, J. Bernholc, M. H. Brodsky, N. O. Lipari, S. T. Pantelides: AIP Conf. Proc. **73,** 156 (1981)
3.125b D. P. DiVincenzo, J. Bernholc, M. H. Brodsky: J. Phys. (Paris) C 4 **42,** 137 (1981)
3.126 H. R. Shanks, C. J. Fang, L. Ley, M. Cardona, F. J. Demond, S. Kalbitzer: Phys. Stat. Sol. (b) **100,** 43 (1980)
3.127 J. A. Reimer, R. W. Vaughan, J. C. Knights: Solid State Commun. **37,** 161 (1981)
3.128 B. J. Waclawski, J. W. Gadzuk, J. F. Herbst: Phys. Rev. Lett. **41,** 583 (1978)
3.129 K. J. Gruntz: Thesis (Stuttgart 1981), unpublished
3.130 J. A. Appelbaum, G. A. Baraff, D. R. Hamann, H. D. Hagstrum, T. Sakurai: Surf. Sci. **70,** 654 (1978)
3.131 G. Allie, C. Lauroz, A. Chenevas-Paule: J. Non-Cryst. Solids **35/36,** 267 (1980)
3.132 F. C. Brown, O. P. Rustgi: Phys. Rev. Lett. **28,** 497 (1972)
3.133 W. Gudat, C. Kunz: Phys. Rev. Lett. **29,** 169 (1972)
3.134 E. O. Kane: Phys. Rev. **146,** 558 (1966)
3.135 M. Altarelli, D. L. Dexter: Phys. Rev. Lett. **29,** 1100 (1972)
3.136 R. S. Bauer, R. Z. Bachrach, J. C. McMenamin: Nuovo Cimento **39,** 409 (1977)
3.137 W. Eberhardt, G. Kalkoffen, C. Kunz, D. Aspnes, M. Cardona: Phys. Stat. Sol. (b) **88,** 135 (1978)
3.138 M. Altarelli: J. Phys. (Paris) C 4 **39,** 95 (1978)
3.139 A similar DOS was suggested by Spicer and Donovan for a-Ge based on photoemission spectra taken at various photon energies; W. E. Spicer, T. M. Donovan: J. Non-Cryst. Solids **2,** 66 (1970)
3.140 B. von Roedern, L. Ley, F. W. Smith: *The Physics of Semiconductors,* ed. by L. H. Wilson (Institute of Physics, London 1978) p. 701
3.141 J. A. Reimer, R. W. Vaughan, J. C. Knights: Phys. Rev. Lett. **44,** 1936 (1980)
3.142 J. C. Knights, R. A. Lujan: Appl. Phys. Lett. **35,** 244 (1979)
3.143 D. C. Allan, J. D. Joannopoulos: Phys. Rev. Lett. **44,** 43 (1980)
3.144 A. Madan, P. G. LeComber, W. E. Spear: J. Non-Cryst. Solids **20,** 239 (1976)
3.145 P. G. LeComber, A. Madan, W. E. Spear: J. Non-Cryst. Solids **11,** 219 (1972)
3.146 A. R. Moore: Appl. Phys. Lett. **31,** 766 (1977)
3.147 M. H. Brodsky: Solid State Commun. **36,** 55 (1980)
3.148 W. Beyer, J. Stuke, H. Wagner: Phys. Stat. Sol. (a) **30,** 231 (1975)
3.149 J. E. Rowe, H. Ibach, H. Froitzheim: Surf. Sci. **48,** 44 (1975)
3.150 I. Solomon, T. Dietl, D. Kaplan: J. Phys. (Paris) **39,** 1241 (1978)
3.151 N. B. Goodman: Phil. Mag. B **45,** 407 (1982)
3.152 M. Grünewald, K. Weber, W. Fuhs, P. Thomas: J. Phys. (Paris) C 4, **42,** 523 (1981)
3.153 M. Hirose, T. Suzuki, G. H. Döhler: Appl. Phys. Lett. **34,** 234 (1979)

3.154a D. V. Lang, J. D. Cohen, J. P. Harbison: Phys. Rev. Lett. **31**, 292 (1977)

3.154b D. V. Lang: In *Thermally Stimulated Relaxation in Solids,* ed. by P. Bräunlich, Topics Appl. Phys., Vol. 37 (Springer, Berlin, Heidelberg, New York 1979)

3.155 R. F. Willis, L. D. Laude, B. Fitton: Surf. Sci. **37**, 395 (1973)

3.156 L. D. Laude, R. F. Willis, B. Fitton: Solid State Commun. **12**, 1007 (1973)

3.157 C. W. Peterson, J. H. Dinan, T. E. Fischer: Phys. Rev. Lett. **25**, 861 (1970)

3.158 D. T. Pierce, W. E. Spicer: Phys. Rev. Lett. **27**, 1217 (1971)

3.159 T. E. Fischer, M. Erbudak: Phys. Rev. Lett. **18**, 1220 (1971)

3.160 M. Erbudak, T. E. Fischer: J. Non-Cryst. Solids **8–10**, 965 (1972)

3.161 L. D. Laude, R. F. Willis, B. Fitton: Phys. Rev. Lett. **29**, 472 (1972)

3.162 B. von Roedern, G. Moddel: Solid State Commun. **35**, 467 (1980)

3.163 B. von Roedern, L. Ley, M. Cardona: Solid State Commun. **29**, 415 (1979)

3.164 P. D. Persans: J. Non-Cryst. Solids **35/36**, 369 (1980)

3.165 G. H. Döhler: J. Non-Cryst. Solids **35/36**, 363 (1980)

3.166 H. Overhof, W. Beyer: J. Non-Cryst. Solids **35/36**, 375 (1980)

3.167 W. E. Spear, P. G. LeComber: Phil. Mag. **33**, 935 (1976)

3.168 D. G. Ast, M. H. Brodsky: J. Non-Cryst. Solids **35/36**, 611 (1980)

3.169 J. H. Thomas: J. Vac. Sci. Technol. **17**, 1306 (1980)

3.170 D. I. Jones, P. G. LeComber, W. E. Spear: Phil. Mag. **36**, 54 (1977)

3.171 L. Guttman, W. Y. Ching, J. Rath: Phys. Rev. Lett. **44**, 1513 (1980)

3.172 D. Adler: Phys. Rev. Lett. **41**, 1755 (1978)

3.173 D. Haneman: In *Surface Physics of Phosphors and Semiconductors,* ed. by C. G. Scott, C. E. Reed (Academic, London 1975) p. 2

3.174 M. H. Brodsky, D. Kaplan: J. Non-Cryst. Solids **32**, 431 (1979)

3.175 S. Brennan, J. Stöhr, R. Jaeger, J. E. Rowe: Phys. Rev. Lett **45**, 1414 (1980)

3.176 F. J. Himpsel, P. Heimann, T. C. Chiang, D. E. Eastman: Phys. Rev. Lett. **45**, 1112 (1980)

3.177 L. Ley, J. Reichardt, R. L. Johnson: Phys. Rev. Lett. **49**, 1664 (1982)

3.178 K. Usami, T. Shimada, Y. Katayama: Jpn. J. Appl. Phys. **19**, L 389 (1980)

3.179 B. Kramer, H. King, A. Mackinnon: In *Proc. XVIth Intern. Conf. on Physics Semiconductors,* ed. by M. Averous (North-Holland, Amsterdam 1983) p. 944

3.180 S. C. Shen, C. J. Fang, M. Cardona, L. Genzel: Phys. Rev. B **22**, 2913 (1980)

3.181 Y. Katayama, K. Usami, T. Shimada: Phil. Mag. B **43**, 283 (1981)

3.182 R. Kärcher, L. Ley: In *Proc. Xth Intl. Conf. Amorphous and Liquid Semicond.,* Tokyo (1983)

3.183 A. Madan, S. R. Ovshinsky, E. Benn: Phil. Mag. B **40**, 259 (1979)

3.184 A. Matsuda, S. Yamasaki, K. Nakagawa, H. Okushi, K. Tamaka, S. Itzima, M. Matsumura, H. Yamamoto: Jpn. J. Appl. Phys. **19**, L 305 (1980)

3.185 K. J. Gruntz, L. Ley, R. L. Johnson: Phys. Rev. B **24**, 2069 (1981)

3.186 L. Ley, K. J. Gruntz, R. L. Johnson: In *Tetrahedrally Bonded Amorphous Semiconductors,* ed. by R. A. Street, D. K. Biegelsen, J. C. Knights (American Inst. Phys., New York 1981) p. 161

3.187 C. J. Fang, L. Ley, H. R. Shanks, K. J. Gruntz, M. Cardona: Phys. Rev. B **22**, 6140 (1980)

3.188 A. E. Jonas, G. K. Schweitzer, F. A. Grimm, T. A. Carlson: J. Electron Spectrosc. Relat. Phenom. **1**, 29 (1972/73)

3.189 W. Y. Ching: J. Non-Cryst. Solids **35/36**, 61 (1980)

3.190 W. Y. Ching: private communication

3.191 L. Ley, H. Richter, R. Kärcher, R. L. Johnson, J. Reichardt: J. Phys. (Paris) C 4 **42**, 753 (1981)

3.192 R. Kärcher, L. Ley: unpublished

3.193 G. Lucovsky, R. J. Nemanich, J. C. Knights: Phys. Rev. B **19**, 2064 (1979)

3.194 G. Müller, F. Demond, S. Kalbitzer, H. Damjantschitsch, H. Mannsperger, W. E. Spear, P. G. LeComber, R. A. Gibson: Phil. Mag. B **41**, 571 (1980)

3.195 F. J. Kampas, R. W. Griffith: Sol. Cells **2**, 385 (1980)

3.196 F. J. Kampas, R. W. Griffith: In [Ref. 3.186, p. 1]
3.197 B. A. Scott, M. H. Brodsky, D. C. Green, P. B. Kirby, R. M. Plecenik, E. E. Simonyi: Appl. Phys. Lett. **37**, 725 (1980); [Ref. 3.186, p. 6]
3.198 N. Kasupke, M. Henzler: Surf. Sci. **92**, 407 (1980)
3.199 J. N. Miller, I. Lindau, W. E. Spicer: Phil. Mag. B **43**, 273 (1981)
3.200 *The Physics of SiO₂ and its Interfaces*, ed. by S. Pantelides (Pergamon, New York 1978)
3.201 F. J. Grunthaner, P. J. Grunthaner, R. P. Vasquez, B. F. Lewis, J. Maserjian, A. Madhukar; J. Vac. Sci. Technol. **16**, 1443 (1979)
3.202 R. Kärcher, L. Ley: Solid State Commun. **43**, 415 (1982)
3.203 C. M. Garner, I. Lindau, J. N. Müller, P. Pianetta, W. E. Spicer: J. Vac. Sci. Technol. **14**, 372 (1977)
3.204 J. E. Rowe: Appl. Phys. Lett. **25**, 576 (1974)
3.205 T. H. DiStefano, D. E. Eastman: Phys. Rev. Lett. **27**, 1561 (1971)
3.206 D. E. Eastman, W. D. Grobman: Phys. Rev. Lett. **28**, 1378 (1972)
3.207 L. F. Wagner, W. E. Spicer: Phys. Rev. B **9**, 1512 (1974)
3.208 G. D. Cody, B. Abeles, C. R. Wronski, R. B. Stephens, B. Brooks: Sol. Cells **2**, 227 (1980)
3.209 H. Wolter: In *Handbuch der Physik*, Vol. XXIV, ed. by S. Flügge (Springer, Berlin, Göttingen, Heidelberg 1956) p. 461
3.210 O. S. Heavens: *Optical Properties of Thin Solid Films* (Dover, New York 1965)
3.211 E. Abeles, M. L. Theye: Surf. Sci. **5**, 325 (1966)
3.212 T. M. Donovan, W. E. Spicer, J. M. Bennett, E. J. Ashley: Phys. Rev. B **2**, 397 (1970)
3.213 R. M. A. Azzam, N. M. Bashara: *Ellipsometry and Polarized Light* (North-Holland, Amsterdam 1977)
3.214 D. E. Aspnes: In *Optical Properties of Solids, New Developments*, ed. by B. O. Seraphin (North-Holland, Amsterdam 1976) p. 799
3.215 B. Drevillon, J. Huc, A. Lloret, J. Perrin, G. de Rosny, J. P. M. Schmitt: In [Ref. 3.186, p. 31]
3.216 D. E. Aspnes, B. G. Bagley, A. A. Studna, A. C. Adams, F. B. Alexander Jr.: In [Ref. 3.186, p. 307]
3.217 G. A. N. Connell, A. Lewis: Phys. Stat. Sol. (b) **60**, 291 (1973)
3.218 M. Kerker: *The Scattering of Light* (Academic, New York 1969)
3.219 D. L. Wood, J. Tauc: Phys. Rev. B **5**, 3144 (1972)
3.220 A. Bubenzer, S. Hunklinger, K. Dransfeld: J. Non-Cryst. Solids **40**, 605 (1980)
3.221 Shi-fu Zhao, S. Hunklinger: To be published
3.222 S. Yamasaki, N. Hata, T. Yoshida, H. Oheda, A. Matsuda, H. Okushi, K. Tanaka: J. Phys. (Paris) C 4 **42**, 297 (1981)
3.223 W. B. Jackson, N. H. Amer: In [Ref. 3.186, p. 263]
3.224 W. B. Jackson, N. M. Amer, A. C. Boccara, D. Fournier: J. Appl. Opt. **20**, 1333 (1981)
3.225 R. C. Chittick, J. H. Alexander, H. F. Sterling: J. Electrochem. Soc. **116**, 77 (1969)
3.226 R. C. Chittick: J. Non-Cryst. Solids **3**, 255 (1970)
3.227 A. Rose: In *Concepts in Photoconductivity and Allied Problems* (Wiley, New York 1963)
3.228 R. J. Loveland, W. E. Spear, A. Al-Sharbaty: J. Non-Cryst. Solids **13**, 55 (1973/74)
3.229 B. Abeles, C. R. Wronski, T. Tiedje, G. D. Cody: Solid State Commun. **36**, 537 (1980)
3.229 a G. Moddel, D. A. Anderson, W. Paul: Phys. Rev. B **22**, 1918 (1980)
3.229 b J. Kočka, M. Vaneček, J. Stuchlik, O. Štika, E. Sipek, H. T. Ha, A. Triska: In *Proc. IVth EC Photovoltaic Solar Energy Conf.* Stresa, Italy (1982)
3.230 W. E. Spear, R. J. Loveland, A. Al-Sharbaty: J. Non-Cryst. Solids **15**, 410 (1974)
3.230 a R. S. Crandall: Sol. Cells **2**, 319 (1980)
3.230 b H. M. Welsch, W. Fuhs, K. H. Greeb, H. Mell: J. Phys. (Paris) C 4 **42**, 567 (1981)
3.231 D. Beaglehole, M. Zavetova: J. Non-Cryst. Solids **4**, 272 (1970)
3.232 K. Maschke, P. Thomas: Phys. Stat. Sol. **41**, 743 (1970)
3.233 M. H. Brodsky, R. S. Title, K. Weiser, G. D. Petit: Phys. Rev. B **1**, 2632 (1970)
3.234 See, e.g., B. Kramer: In *Advances in Solid State Physics*, ed. by O. Madelung (Vieweg, Braunschweig 1972) p. 133 ff.

3.235 G. Weiser, D. Ewald, M. Milleville: J. Non-Cryst. Solids **35/36**, 447 (1980)
3.236 D. Ewald, M. Milleville, G. Weiser: Phil. Mag. B **40**, 291 (1979)
3.237 R. S. Bauer: In *Proc. Vth Intern. Conf. Amorphous and Liquid Semiconductors,* ed. by J. Stuke, W. Brenig (Taylor and Francis, London 1974) p. 595
3.238 A. Donnadieu, J. P. Ferraton, J. M. Berger, A. Divrechy, C. Raisin, J. Robin, D. Booth: Sol. Energy Mat. **2**, 201 (1979/80)
3.239 A. Donnadieu, G. Weiser, J. Beichler: Sol. Energy Mat. **4**, 455 (1981)
3.240 E. C. Freeman, W. Paul: Phys. Rev. B **20**, 716 (1979)
3.241 C. C. Tsai, H. Fritzsche, M. H. Tanielian, P. J. Gaczi, P. D. Persans, M. A. Vesaghi: In *Proc. 7th Intern. Conf. Amorphous and Liquid Semiconductors, Edinburgh 1977,* ed. by W. E. Spear (University of Edinburgh 1977) p. 339
3.242 P. J. Zanzucchi, C. R. Wronski, D. E. Carlson: J. Appl. Phys. **48**, 5227 (1977)
3.243 J. C. Knights: Japan. J. Appl. Phys. Suppl. **18-1**, 101 (1979)
3.244 C. C. Tsai, H. Fritzsche: Sol. Energy Mat. **1**, 11 (1979)
3.245 M. H. Brodsky, R. S. Title, K. Weiser, G. D. Pettit: Phys. Rev. B **1**, 2632 (1970)
3.246 G. K. M. Thutupalli, S. G. Tombin: J. Phys. C **10**, 467 (1977)
3.247 R. Grigorovici, A. Vancu: Thin Solid Films **2**, 105 (1968)
3.248 J. E. Fischer, T. M. Donovan: J. Non-Cryst. Solids **8**, 202 (1972)
3.249 S. K. Bahl, S. M. Bhagat: J. Non-Cryst. Solids **17**, 409 (1975)
3.250 M. Hirose, M. Taniguchi, Y. Osaka: In [Ref. 3.241, p. 352]
3.251 G. A. N. Connell, J. R. Pawlik: Phys. Rev. B **13**, 787 (1976)
3.252 R. H. Klazes, M. H. L. M. van den Broek, J. Bezemer, S. Radelaar: Phil. Mag. B **25**, 377 (1982)
3.253 R. S. Crandall: J. Non-Cryst. Solids **35/36**, 381 (1980)
3.254 R. S. Crandall: Phys. Rev. Lett. **44**, 749 (1980)
3.255 R. Hulthén: Phys. Scr. **12**, 342 (1975)
3.256 J. Tauc, R. Grigorovici, A. Vancu: Phys. Stat. Sol. **15**, 627 (1966)
3.257 A. Matsuda, M. Matsumura, K. Nakagawa, T. Imura, H. Yamamoto, S. Yamasaki, H. Okushi, S. Iizima, K. Tanaka: In [Ref. 3.186, p. 192]
3.258 F. R. Jeffrey, H. R. Shanks, G. C. Danielson: J. Non-Cryst. Solids **35/36**, 261 (1980)
3.259 M. H. Brodsky, R. S. Title: In *Structure and Excitations of Amorphous Solids* (American Inst. Phys., New York 1976) p. 97
3.260 H. Fritzsche, M. Tanielian, C. C. Tsai, P. J. Gaczi: J. Appl. Phys. **50**, 3366 (1978)
3.261 K. Tanaka, S. Yamasaki, K. Nakagawa, A. Matsuda, H. Okushi, M. Matsumura, S. Iizima: J. Non-Cryst. Solids **35/36**, 475 (1980)
3.262 A. Deneuville, A. Mini, J. C. Bruyère: J. Phys. C **14**, 4531 (1981)
3.263 D. K. Biegelsen, R. A. Street, C. C. Tsai, J. C. Knights: J. Non-Cryst. Solids **35/36**, 285 (1980)
3.264 N. Aspley, E. A. Davis, A. P. Troup, A. D. Yoffe: In [Ref. 3.241, p. 447]
3.265 J. Chevallier, H. Wieder, A. Onton, C. R. Guarnieri: Solid State Commun. **24**, 867 (1977)
3.266 G. Moddel, J. Blake, R. W. Collins, P. Viktorovitch, D. K. Paul, B. von Roedern, W. Paul: In [Ref. 3.186, p. 25]
3.267 P. G. LeComber, R. J. Loveland, W. E. Spear, R. A. Vaughn: In [Ref. 3.237, p. 245]
3.268 M. Janai, D. D. Aldred, D. C. Booth, B. O. Seraphin: Sol. Energy Mat. **1**, 11 (1979)
3.268a I. Solomon, J. Perrin, B. Bourdon: Inst. Phys. Conf. Ser. **43**, 689 (1979)
3.269 C. Tsang, R. A. Street: Phys. Rev. B **19**, 3027 (1979)
3.270 J. Chmel, H. Fritsche: unpublished
3.271 Landoldt-Börnstein: New Series, Vol. 17, *Semiconductors,* ed. by O. Madelung (Springer, Berlin, Heidelberg, New York 1982) p. 43
3.272 G. A. N. Connell, W. Paul: J. Non-Cryst. Solids **8–10**, 215 (1973)
3.273 B. Welber, M. H. Brodsky: Phys. Rev. B **16**, 3660 (1977)
3.274 S. Minomura, K. Tsuji: J. Non-Cryst. Solids **35/36**, 513 (1980)
3.275 S. Minomura: J. Phys. (Paris) C 4 **42**, 181 (1981)
3.276 B. A. Weinstein: unpublished

3.277 D. L. Camphausen, G. A. N. Connell, W. Paul: Phys. Rev. Lett. **26**, 184 (1971)
3.277a H. Y. Fan: Phys. Rev. **82**, 900 (1951)
3.278 P. B. Allen, M. Cardona: Phys. Rev. B**23**, 1495 (1981); B**24**, 7479 (E) (1981)
3.279 M. Olivier, P. Bouchut: J. Phys. (Paris) C4 **42**, 305 (1981)
3.280 G. D. Cody, T. Tiedje, B. Abeles, B. Brooks, Y. Goldstein: Phys. Rev. Lett. **47**, 1480 (1981)
3.281 S. Al Jalali, G. Weiser: J. Non-Cryst. Solids **41**, 1 (1980)
3.282 F. Urbach: Phys. Rev. **92**, 1324 (1953)
3.283 M. V. Kurik: Phys. Stat. Sol.(a) **8**, 9 (1971)
3.284 K. L. Chopra, S. K. Bahl: Thin Solid Films **11**, 377 (1972)
3.285 R. A. Street, T. M. Searle, I. G. Austin, R. S. Sussmann: J. Phys. C**7**, 1582 (1974)
3.286 K. J. Siemsen, E. W. Fenton: Phys. Rev. **161**, 632
3.287 J. Tauc: Mat. Res. Bull. **5**, 721 (1970)
3.288 D. A. Papaconstantopoulos, E. N. Economou: Phys. Rev. B**24**, 7233 (1981)
3.289 For reviews on the theories of the Urbach edges see, e.g., [3.5–8, 283]
3.290 T. Skettrup: Phys. Rev. B**18**, 2622 (1978)
3.291 R. C. Tolman: *The Principles of Statistical Mechanics* (Oxford, London 1938)
3.292 D. J. Dunstan: J. Phys. C**30**, L419 (1982)
3.293 See, e.g., T. Tiedje, J. M. Cebulka, D. L. Morel, B. Abeles: Phys. Rev. Lett. **46**, 1425 (1981)
3.294 W. Franz: Z. Naturforsch. **13a**, 484 (1958)
3.295 D. Redfield: Phys. Rev. **130**, 916 (1963)
3.296 D. L. Dexter: Phys. Rev. Lett. **19**, 1383 (1967)
3.297 J. D. Dow, D. Redfield: Phys. Rev. B**5**, 594 (1972)
3.298 V. L. Bronch-Bruevich: Phys. Stat. Sol. **42**, 35 (1970)
3.299 B. Esser: Phys. Stat. Sol.(b) **51**, 735 (1972)
3.300 D. A. Anderson, G. Moddel, W. Paul: J. Non-Cryst. Solids **35/36**, 345 (1980)
3.301 T. D. Moustakas: Solid State Commun. **35**, 745 (1980)
3.302 S. Yamasaki, K. Nakagawa, H. Yamamoto: AIP Conf. Proc. **73**, 258 (1981)
3.303 W. B. Jackson, N. M. Amer: J. Phys. (Paris) C4 **42**, 293 (1981)
3.304 W. B. Jackson, N. M. Amer: Phys. Rev. B**25**, 5559 (1982)
3.305 M. H. Brodsky, M. Cardona, J. J. Cuomo: Phys. Rev. B**16**, 3556 (1977)
3.306 H. Dersch, J. Stuke, J. Beichler: Phys. Stat. Sol.(b) **105**, 265 (1981)
3.307 R. A. Street, D. K. Biegelson: Solid State Commun. **33**, 1159 (1980)
3.308 J. Stuke: In *Proc. 6th Intern. Conf. Amorph. Liq. Semiconductors,* ed. by B. T. Kolomiets (Leningrad 1976) p. 193
3.309 M. H. Brodsky, D. M. Kaplan, J. F. Ziegler: In *Proc. 11th Intern. Conf. Physics of Semiconductors* (Warsaw 1972) p. 529
3.310 R. A. Street: Phys. Rev. B**24**, 969 (1981)
3.311 R. W. Collins, M. A. Paesler, G. M. Moddel, W. Paul: J. Non-Cryst. Solids **35/36**, 681 (1980)
3.312 J. D. Cohen, D. V. Lang, J. P. Harbison: In (Ref. 3.186, p. 217]
3.313 U. Voget-Grote, W. Kümmerle, R. Fischer, J. Stuke: Phil. Mag. B**41**, 127 (1980)
3.314 R. A. Street: J. Phys. (Paris) C4 **42**, 283 (1981)
3.315 B. Abeles: private communication
3.316 J. Tauc: In *Festkörperprobleme: Advances in Physics,* Vol. XXII ed. by P. Grosse (Vieweg, Braunschweig 1982) p. 85
3.317 S. Griep, L. Ley: In *Proc. Xth Intl. Conf. on Amorphous and Liquid Semiconductors,* Tokyo (1983)

4. Conductivity, Localization, and the Mobility Edge

Sir Nevill Mott

With 8 Figures

The many forms of amorphous silicon, hydrogenated to a greater or lesser extent, appear to have widely differing structures and electrical properties. A major aim of theory in the field of amorphous materials must be to relate electrical, magnetic and optical properties to structure. The purpose of this chapter is to describe the extent to which this can be done. We shall ask, whether the concept of a mobility edge is theoretically justified and experimentally observed, what are the roles of short-range variations of potential, such as might exist in a continuous random network, relative to those of long-range fluctuations, caused either by charged defects or fluctuations in composition. We describe the meaning of a "defect" in an amorphous material and give possible explanations of the double sign anomaly in the Hall effect, found in all amorphous silicons investigated up till now, and discuss the applications of the concept of a polaron to this material.

4.1 Background

Much of our understanding of the conduction bands of amorphous materials comes from the classical paper of *Anderson* [4.1] on "The Absence of Diffusion in Certain Random Lattices". In this and subsequent sections we shall see what can be deduced about conductivity from the model of Anderson's paper, and later in this chapter apply these results to amorphous silicon. In Anderson's paper a crystalline array of potential wells is considered, as shown in Fig. 4.1, with depths spread in a random way over a range of energies V_0. If B is the tight-binding bandwidth in the absence of disorder, then Anderson showed that, if V_0/B exceeds a certain critical value $(V_0/B)_{\text{crit}}$, all states in the band are localized; that is to say, following later analysis [4.2], their wave functions may be written

$$\Psi = [\Sigma \, C_n \, e^{i\phi_n} \, \psi_n] \, e^{-\alpha r} \, , \tag{4.1}$$

where ψ_n is an atomic wave function on the well n, ϕ_n a random phase and c_n a coefficient which, in its turn, will vary in some random way from site to site. Each eigenstate Ψ is localized at some point in space, falling off exponentially with distance from it. The quantity varies with energy in the band and

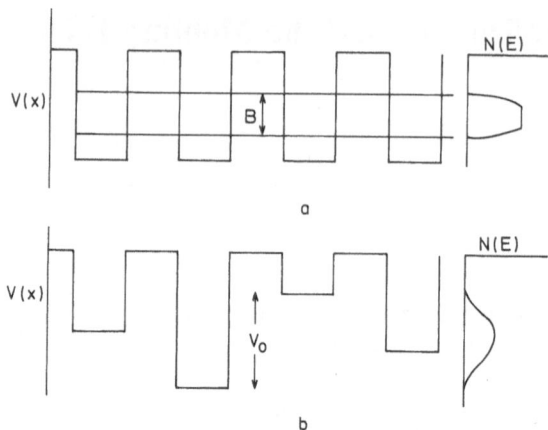

Fig. 4.1. Potential energy $V(x)$ of an electron in a lattice with diagonal disorder. (**a**) With $V_0 = 0$; (**b**) With a finite value of V_0. The density of states $N(E)$ is also shown. B is the bandwidth without disorder

tends to zero as V_0/B tends to the critical value. This depends on the coordination number z; for $z = 6$ it is probably [4.3] about 2, differing considerably from that in Anderson's original paper.

If states are localized, then an electron can move from one state to another only through thermal activation, for instance, by phonons. The present author [4.4] first pointed out that if V_0/B were less than the critical value, a tail to the band would none the less exist and states in the band tail would still be localized, and for a rigid lattice, localized states would be separated from nonlocalized ones by a sharp[1] energy E_c; electrons with energies below E_c can move only through thermal activation, while those with energies above will have an unactivated mobility. For this reason, the energy E_c has been named the "mobility edge" [4.5].

A great deal of our understanding of the electrical properties of amorphous and liquid semiconductors has been obtained through the application, to the conduction bands of these materials, of results obtained from the Anderson model of Fig. 4.1, although of course, the potential in real materials is very different; the experimental situation nearest to the Anderson model is that of an impurity band in a doped and fairly heavily compensated ($\sim 50\%$) semiconductor, though here there is the additional complication that the wells of Fig. 4.1 (the donors) are random in space. We shall then first outline the predictions of the Anderson model, with references when appropriate to impurity conduction.

The position of the mobility edge in this model has been calculated by *Abou-Chacra* and *Thouless* [4.6]. If V_0/B lies some way below the critical value, we expect a tail of localized states, as shown in Fig. 4.2, which illustrates only the lower part of the band. The tail extends over a range of energies V_0, and these authors show that if $V_0/B \ll 1$, practically the whole tail is localized. This is because no fluctuation of depth V_0 will itself produce a

---

1 If interaction with phonons is taken into account, a lifetime τ will lead to a broadening h/τ.

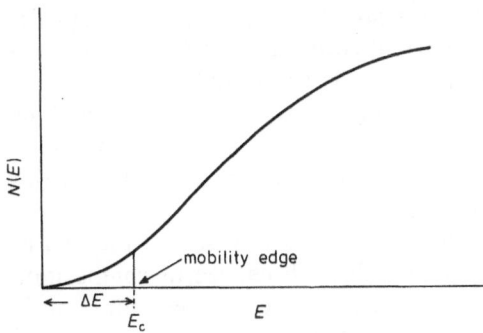

Fig. 4.2. Density of states in the conduction band of a noncrystalline material showing the mobility edge (E_c)

trap, but only groups of deep wells near together. Such groupings are rare and so extended states cannot be formed from them. Only when $V_0/B \sim 1$ does E_c move down into the tail.

As regards the conductivity, we may consider two cases:

a) The Fermi energy lies in the band of localized and extended states; this case is appropriate to impurity bands. Then if E_c lies below E_F, the conductivity will depend little on phonon scattering and be roughly constant except at high temperatures. On the other hand, if E_F lies below E_c, we expect two forms of conduction. These are: at high temperatures (by excitation to a mobility edge) when the conductivity will be of the form

$$\sigma = \sigma_{\min} \exp[- (E_c - E_F)/kT] , \tag{4.2}$$

(the significance of $\sigma_{\min}$ being discussed below), and at low T (variable-range hopping by electrons with energies near the Fermi energy) where an *approximate* formula for the conductivity is

$$\sigma = A \, e^{-B/T^{1/4}} . \tag{4.3}$$

This is discussed in Sects. 4.5, 12.

b) The Fermi energy lies below the band. Here again, two forms of conduction are expected, one given by (4.2) and the other, at low tempera-

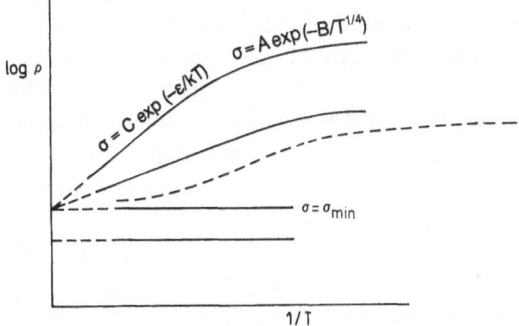

Fig. 4.3. Typical behaviour of the resistivity ϱ of a degenerate electron gas showing a metal-insulator transition of the Anderson type. ε denotes $E_c - E_F$. The dotted line shows the behaviour if there is no minimum metallic conductivity

tures, resulting from hopping between band-edge localized states. This is a form of variable-range hopping, treated theoretically by *Grant* and *Davis* [4.7], but does *not* lead to a conductivity of the form (4.3).

It may be noted that in case b, any lifetime broadening of the mobility edge will be a consequence of interaction with phonons, while in case a, the stronger Auger interaction with the degenerate electron gas in states below E_F will be important, except when E_F lies at E_c.

Figure 4.3 shows the resistivity to be expected in case a if the relative positions of E_F and E_c are moved, either through change of composition, magnetic field or in other ways. For comparison with experiment for impurity bands and inversion layers, see [4.2, 8–10]. A metal-insulator transition is observed.

4.2 The Minimum Metallic Conductivity

This quantity was introduced by the present author [4.11] and estimated to be

$$\sigma_{min} \sim C e^2/\hbar a_E . \tag{4.4}$$

Here C is a numerical constant, which depends both on the coordination number and on the square of V_0/B for the Anderson localization criterion, and is probably in the range of 0.025–0.05. If the mobility edge lies near mid-gap, $a_E \simeq a$, where a is the distance between wells; otherwise [4.12],

$$(a/a_E)^3 = \int_{-\infty}^{E} N(E)dE / \int_{-\infty}^{\infty} N(E)dE . \tag{4.5}$$

The quantity σ_{min} may be used in two ways. a) As the pre-exponential factor in (4.2). b) A smallest value of the unactivated conductivity.

Early considerations by *Cohen* and *Jortner* [4.13] based on classical percolation theory, and a more recent approach due to *Götze* [4.14] suggest that there is no discontinuous change in the mobility at E_c; the present author believes these to be incorrect in the limit of high T, and that (4.2) is then valid. At the time of writing, however, there is considerable doubt [4.14–17] about whether σ_{min} exists in the limit of low T. In other words, do values of $(E_F - E_c)$ exist for which ϱ behaves like the dotted line in Fig. 4.3, tending as $T \rightarrow 0$ to a finite value much above $1/\sigma_{min}$, and to infinity when $E_F = E_c$? The present author believes that the answer to this question, though of great theoretical interest and likely to be relevant to impurity bands at low temperatures, will not affect the properties of amorphous silicon in any practical case. In [4.17] he related it to the properties of the coefficients c_n in (4.1); σ will tend to zero as $E \rightarrow E_c$ only if these show long-range fluctuations with changing position (i.e., with n), with a correlation length tending to infinity as $E \rightarrow E_c$. This, in its turn, depends on the way $\alpha \rightarrow 0$ in (4.1) as $E \rightarrow E_c$; if this is as $(|E - E_c|)^s$, these fluctuations will occur only if $s < 2/3$. Theoretical estimates of s are 0.6 or 2/3, so the question remains open at the time of writing.

The conclusion that (4.2) is valid for conduction at a mobility edge is queried by some authors (e.g., *Götze* [4.14], and *Belize* and *Götze* [4.18] pointed out how difficult it is to use experimental data to test this. According to their analysis, $\sigma \sim \sqrt{(E - E_c)}$ near a mobility edge, and they show this to be consistent with some of the earlier experimental work on the temperature-dependence of the conductivity.

In what follows we shall, however, assume (4.1) to be valid for charge transport at a mobility edge.

4.3 Hall Mobility and Thermopower for Charge Transport at a Mobility Edge

The Hall mobility μ_H has been calculated for electrons or holes with energies at a mobility edge by *Friedman* [4.19], using a method derived from the theory of polarons (Sect. 4.4). According to this analysis,

$$\mu_H \simeq 0.1 \, ea_E^2/\hbar \tag{4.6}$$

and the sign should be negative (n-type), whether for electrons or holes. This is smaller than the "conductivity mobility" μ_c; if we write

$$\sigma = e\mu_c N(E_c)kT \, e^{-W/kT} \tag{4.7}$$

and equate the pre-exponential term to $\sigma_{\min}$ ($\simeq 0.05 \, e^2/\hbar a_E$), we see that

$$\mu_c \sim 0.05 \, ea_E^2 \, \Delta E/kT \, , \tag{4.8}$$

or if $N(E_c) \sim 1/\Delta E \, a_E^3$,

$$\mu_c \sim 0.05 \, ea_E^2 \, \Delta E/kT \, . \tag{4.9}$$

Here ΔE is the range of localized states; if $\Delta E/kT \gg 1$, μ_c should be the larger.

The thermopower when charge transport is by electrons at a mobility edge is given by [4.2, 20]

$$S = \frac{k}{e} \left(\frac{E_c - E_F}{kT} + 1 \right) . \tag{4.10}$$

If the Fermi energy shifts linearly with temperature, we may write

$$E_c - E_F = \varepsilon - \gamma T . \qquad (4.11)$$

Then (4.2) becomes

$$\sigma = \sigma_0 e^{-\varepsilon/kT} , \qquad \sigma_0 = \sigma_{min} e^{\gamma/k} . \qquad (4.12)$$

There has been some dispute in the literature as to whether the same substitution can be made in (4.12), but this appears to be the case [4.2]. The problem of the shift in the Fermi energy is discussed further in Sect. 4.10.

4.4 Polarons

The extent to which polarons play a role in charge transport in noncrystalline semiconductors has been a matter of controversy, and a few lines on their nature are added here [4.2, 21]. In crystalline materials, they are of two kinds.

a) Dielectric polarons, limited to ionic crystals. An electron or hole will always polarize its surroundings outside a sphere having a "polaron radius" r_p, creating a potential well for itself. In principle, this is found by minimizing the sum of the potential energy $(- e^2/r_p)(1/\kappa_\infty - 1/\kappa)$ and the kinetic energy $h^2/2m_e r_p^2$. Here κ_∞ and κ are the high-frequency and static dielectric constants. If r_p is much greater than the lattice parameter a, a "large polaron" is formed which can increase the effective mass but is of little importance here. For large m_{eff} (e.g., in d-bands), r_p can be comparable with a, and the polaron is then called "small".

b) Acoustic or molecular (small) polarons are formed when the carrier allows a bond to form between two adjacent atoms, or in molecular crystals when the presence of the carrier deforms a molecule. As shown in different ways by *Toyozawa* [4.22], *Emin* [4.23], and *Mott* and *Stoneham* [4.24], no polaron of this type is formed at all unless the coupling between the carrier and the localized phonon is strong enough, and the effective mass large enough.[2]

Small polarons of both types move at high temperatures ($T > \Theta_D/2$) by "hopping" with mobility of the form

$$\mu = \mu_0 \exp(- W_H/kT) . \qquad (4.13)$$

---

2 *Towozawa* [4.22], and also *Mott* and *Stoneham* [4.24], have shown that there must be a delay in the formation of this kind of polaron involving an activation energy, which has been estimated numerically by *Mott* [4.25]. *Mott* and *Stoneham* [4.24] described evidence for such a delay from the migration of excitons (an electron in the field of a hole which can be self-trapped). The first observation for holes was due to *Laredo* et al. [4.26] in AgCl.

At low temperatures, band motion is possible, with effective mass greatly enhanced, of order

$$m_p \sim 5\, m \exp\left[W_H \bigg/ \left(\frac{1}{2}\, \hbar\omega \right) \right]^{-1}. \tag{4.14}$$

It is doubtful if this has been observed; such high values of the effective mass would lead to Anderson localization [4.27] if there were a very small random field, caused, for instance, by charged impurities. What has been observed, however, is a drop in W_H to very low values (for instance, in vanadium phosphate glasses [4.2, 25]) where disorder does give some localization but the main activation energy for conduction is of polaron type.

Some peculiarities of the acoustic polaron are worth mentioning. In one-dimensional systems a polaron will *always* form (because in one dimension any well, however shallow, will trap an electron). In three dimensions one expects an equilibrium between free and trapped carriers so that as the temperature rises, charge transport would be due to free carriers [4.29].

For polarons in the range of temperatures for hopping, the Hall effect is explained by a mechanism quite different from that for free carriers. *Friedman* and *Holstein* [4.30] were the first to show, using a three-site model and assuming s-like wave functions, that

a) the Hall effect would be n-type both for electrons and holes, and

b) the activation energy in μ_H is $W_H/3$.

We see therefore that, in contract to the behaviour of free carriers in a conduction band, there are different activation energies for conduction, thermopower and Hall mobility. The only material in which to our knowledge all three have been observed is slightly reduced $LiNbO_3$ ([4.31] and [Ref. 4.2, p. 84]).

A feature of a-Si-H is the double sign reversal of the Hall effect, electrons showing p-type behaviour and holes of n-type; this occurs also in amorphous III–V compounds (references are in Sect. 4.8). With a view to understanding this, *Emin* [4.32] considered polarons formed on bonds (rather than atoms) and showed that the anomaly could be explained by supposing that hopping was predominantly round the rings and that odd-membered rings predominated. Some criticisms and refinements of the theory were given by *Grünewald* et al. [4.33]. Whether these ideas can be related to amorphous silicon will be discussed in Sect. 4.8.

The application of the polaron concept to noncrystalline materials in general is as follows. For band-edge localized states, as also for any bound states such as donors in crystals, there must be some deformation of the network or lattice by the carrier. If, however, the criterion for (acoustic) polaron formation in the corresponding crystal is not met by some margin, it seems to us unlikely that there will be polaron effects for carriers above the mobility edge. If, however, polarons are formed in the crystal (or would be if some softening of the phonons due to amorphicity were taken into account), it will be appropriate to take the polaron in the noncrystalline material as a heavy quasiparticle and consider its behaviour in whatever random field exists in the material. If W_H is large enough ($> \sim 0.1$ eV) for hopping motion to be observed when $T > \Theta_D/2$, the effective mass (4.9) for low temperatures will probably be large enough to ensure Anderson localization throughout the band, with no mobility edge; the mobility will behave like

$$\exp\left[-\left(W_H + \frac{1}{2} w\right)/kT\right] \qquad T > \frac{1}{2}\Theta_D$$

$$\exp(-w/kT), \qquad T \ll \frac{1}{2}\Theta_D$$

where w is a small hopping energy due to disorder. However, systems probably exist where polaron formation gives some mass enhancement, but there is still a mobility edge; according to the present author, $La_{1-x}Sr_xVO_3$ is one [Ref. 4.2, p. 144].

4.5 Variable-Range Hopping

Two forms must be distinguished.

a) Hopping by electrons excited into localized states at a band edge; here an analysis was given by *Grant* and *Davis* [4.7].

b) Hopping by electrons with energies near the Fermi level; this is a phenomenon observed in impurity bands as well as in some amorphous semiconductors. For this phenomenon, a theory which neglects electron-electron interaction can be shown by various methods [4.34–36] (see also [Ref. 4.2, p. 33]) to yield the law

$$\sigma = A \exp[-(T_0/T)^{1/4}] . \qquad (4.15)$$

However, if electron-electron interaction is taken into account, major changes in this behaviour are predicted, especially at low temperatures, which have not with certainty been observed and for which several theories exist [4.37, 38].

The thermopower S has been discussed by several authors [Ref. 4.2, p. 55]. It behaves like

$$S = (k/e)(W^2/kT)(d \ln N/dE)_{E = E_F} . \qquad (4.16)$$

Here W is the hopping energy given by $W^2/kT = k(T_0T)^{1/2}$ for variable-range hopping, so that S is proportional to $T^{1/2}$. According to *Whall* [4.39], (4.13) is also valid for nearest-neighbour hopping, so S should *decrease* at high T.

The contribution of the spins to the thermopower is believed [4.10] to add a term

$$(k/e)\ln 2$$

if these are random.

The Hall effect is expected to be small; recent theoretical approaches are not in complete agreement [4.41–43], but give the same order of magnitude.

Since variable-range hopping is between localized states, there must always be some distortion of the surroundings when the occupation of a site changes, so effects of polaron type are present. According to *Mott* and *Davis* [Ref. 4.2, p. 87], in the limit of low T this will simply decrease the pre-exponential factor by $\exp(-4W_H/\hbar\omega)$ but will not greatly affect the term within the exponential. As the temperature is raised, however, an activation energy of the type of W_H will gradually be introduced.

In view of all there complications, it is hardly to be expected that a $T^{1/4}$ law will always be observed, and it is perhaps surprising how accurately it represents experimental data in some cases [4.2] and the corresponding $T^{1/3}$ law for two dimensions [4.8].

4.6 Application of Theory to Amorphous Silicon

In seeking to apply theoretical models to amorphous silicon, the following are some of the relevant considerations:

a) There is a wide variety in the properties of silicon films prepared by various methods, both as regards hydrogen content, concentration of defects and probably structure, and some at least cannot be regarded as homogeneous.

b) All films show the double-sign anomaly in the Hall effect; that is, when they are *n*-type (according to thermopower measurements), the Hall effect is positive, and when *p*-type (for instance, through doping), the Hall effect is *n*-type. For Si, results are due to *Le Comber* et al. [4.44], *Beyer* et al. [4.45] and *Dresner* [4.46]; for Ge, *Seager* et al. [4.47]; and *Beyer* and *Mell* [4.48] for some amorphous III–V compounds.

c) There is a wide variety in the pre-exponential factors σ_0 of (4.8) in the conductivity; σ_0, in general, increases with activation energy.

d) In general, the activation energies for conduction (E_σ) and thermopower (E_S) are not equal, but $E_\sigma > E_S$, though they approach each other for "good" specimens. Whether this is so or not is conveniently tested by a plot of $\ln \sigma + (e/k)S$ versus $1/T$. The temperature dependence of the activation energy, namely $-\beta T$, cancels out in this expression, as was first pointed out by *Beyer* et al. [4.49]. Figure 4.4, reproducing results from *Jones* et al. [4.50] and later work by Jones shows that specimens can be obtained in

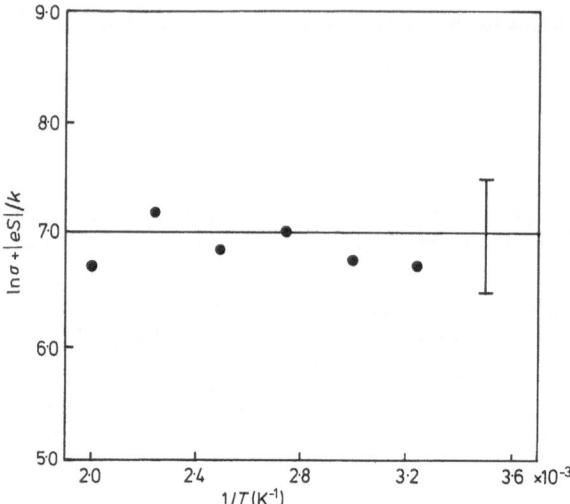

Fig. 4.4. Plot of ln σ + eS/k versus 1/T for a specimen of a-Si [4.59]. No difference between E_σ and E_S is apparent, but if the high temperature point were in error, a difference of up to 0.1 eV would be consistent with the data

which, according to these workers, $E_s \simeq E_\sigma$. On the other hand, if, as Beyer maintains, the high temperature results are not reliable, even the flattest curves may have a slope of $(E_\sigma - E_S)$ in the range from 0.05 to 0.1 eV. In any case, a larger slope is observed when hydrogen is driven off, and indeed is seen frequently in other work. Thus, *Beyer* et al. [4.49] investigated glow-discharge-deposited silicon doped with lithium and found $E_\sigma - E_S$ in the range 0.1 to 0.2 eV, with no evidence for two-channel conduction above 200 K.

This last fact could, in principle, be explained in two ways. One is to assume that the carriers are small polarons, and this has been proposed. However, if this were so it is difficult to see why $(E_\sigma - E_S)$ should vary from specimen to specimen and increase, for instance, on bombardment. The alternative explanation, developed particularly by *Overhof* and *Beyer* [4.51, 52], is that silicon films normally contain *long-range* fluctuations of potential, caused either by charged defects, lack of homogeneity or fluctuations in the concentration of deep states, and that these should be treated as opaque to tunnelling, so that electrons travel along potential valleys and E_σ is determined by the height of the passes separating them, while E_S is determined by the height of the bottoms of the valleys above E_F. We know of no other model which will explain the facts and shall adopt it in this chapter.

If this is so, it might be tempting to abandon altogether the concept of a mobility edge treated as a consequence of Anderson localization; long-range fluctuations will, of course, lead to an activated mobility and as shown, for instance, by *Pistolet* [4.53] and by *Dusseau* [4.54], most of the electrical properties can be accounted for on this model; earlier models of this kind are by *Fritzsche* [4.55] and *Shklovskii* and *Efros* [4.56]. In our view, the strongest argument for not doing this is the double sign anomaly in the Hall effect. As

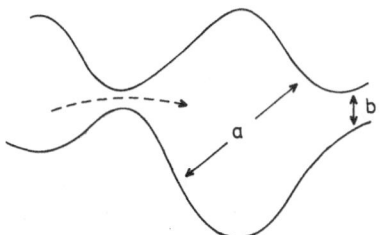

Fig. 4.5. Showing schematically a percolation path (---) in a material with long-range fluctuations of potential

shown below, although in terms of charge transport at a mobility edge we have no fully accepted the explanation of this phenomenon, it is predicted for polarons and is perhaps not unexpected; certainly classical behaviour in long-range fluctuations will not yield it. We shall, therefore, in this chapter, try to combine the idea of a mobility edge with that of long-range fluctuations in potential.

We suppose also that the position of the Fermi energy E_F is determined by a low density of gap states and is highly T-dependent, thus accounting for the abnormal values[3] of σ_0. These and other features of the gap states will be discussed in Sect. 4.10.

4.7 Long-Range Fluctuations of Potential

If a material contains long-range fluctuations of potential *and no other scattering mechanism*, then in our view the concept of a minimum metallic conductivity should be applicable with σ_{min} given by (4.4) and with a equal to the linear dimensions of the fluctuations. There is evidence that this is so from the experiments of *Abeles* et al. [4.51] on the conductivity of metallic particles of Ni, Pt and Au of size less than 100 Å prepared by co-sputtering with SiO_2 or Al_2O_3 [Ref. 4.2, p. 157].

If, on the other hand, there is, in addition, a scattering mechanism giving a mean-free path small compared with a, or polaron hopping, then classical percolation theory should apply; in the former case, the mobility should increase as $(E-E_c)^{1.6}$ as the energy rises above the height of the passes. If this scattering leads to a mobility edge at an energy ΔE above E_A, then E_σ should be the sum of ΔE and $(E-E_F)$ where E lies at the top of the passes. Although the model should lead to some decrease in the pre-exponential factor below σ_{min}, we do not think this should be important. If, in Fig. 4.5, a is the radius of the long-range fluctuations and b the width of the passes, then if ϱ is the resistivity in the neighbourhood of the pass, the resistance of each pass is ϱb and the bulk resistivity is $\varrho b/a$. We might suppose that b is given by

$$\Delta W(b/a)^2 = \Delta E , \qquad (4.17)$$

---

3 See, however, Sect. 4.10 for an alternative explanation due to *Spear*.

where ΔW is the magnitude of the long-range fluctuations, so b/a is unlikely to be small.

We suggest then that the activation energy appearing in the drift mobility and conductivity is the sum of two terms: ΔE, as illustrated in Fig. 4.2; and ΔW, the difference in energies between valleys and passes.

ΔW may have its origin in charged-point defects and fluctuations in defect density near E_F (Sect. 4.9). It should be highly sensitive to the method of preparation. On the other hand, we believe that ΔE is characteristic of the disorder in a continuous random network, that is the result of random dihedral angles and perhaps stretched bonds. Obviously it will depend on composition, but as we shall see below, it is rather insensitive to moderate hydrogen content.

We next ask, what evidence is there for a mobility edge in other noncrystalline materials. In vitreous silica the mobility of electrons has been measured and increases with temperature ([4.58]; cf. [Ref. 4.2, p. 572]), so if a mobility edge exists, ΔE is small compared with kT down to liquid air temperatures. The same is true of the liquid rare gases [4.59]. The present author believes that this is to be expected if the wave function is s-like at the bottom of the band [4.2]. For holes in chalcogenides, the thermopower and Hall effect have been analysed first by *Nagels* et al. [4.60] and in detail by *Mytilineou* and *Davis* [4.61] in terms of a mobility edge with $\Delta E \sim 0.1$ eV. An equally satisfactory analysis was given by *Emin* and coworkers [4.62] in terms of the hypothesis that holes form small polarons. For the present author's arguments against this hypothesis, see [Ref. 4.2, Chap. 10].

In silicon various calculations exist; we first describe one by *Davies* [4.63]. The essence of this method is to calculate the scattering, and hence the mean free path in midband, to be expected from a random orientation of the silicon tetrahedra; that is to say, random values of the dihedral angle. It is then assumed that the relationship between the mean free path L and ΔE is the same as in the Anderson model, with [4.2]

$$L \sim 4\pi b(B/V)^2 , \qquad (4.18)$$

and ΔE, according to *Abou-Chacra* and *Thouless* [4.6], given by

$$\Delta E/V = (12\,b/L)^{1/2} + 4\pi b/3\,L . \qquad (4.19)$$

The first term is the important one, giving the displacement of the band edge due to disorder. In calculating L, a value of the effective mass m_{eff} in midband must be assumed; reasons are given for taking $m_e = m_{eff}$. The calculated value of ΔE is 0.28 eV.

The observed effect of hydrogen is to widen the gap from 1.8 to 2.2 eV for 15% hydrogen [4.64] but this, as shown in [4.71], is due to a shift of the valence band, the conduction band being little affected. By considering the maximum change in the level of the conduction band due to hydrogen and estimating the scattering by a potential hill (or well) that would produce this displacement, Davies found that a 10% concentration of hydrogen would increase ΔE by 0.02 eV or less. It is known that the hydrogen is not uniformly distributed. We believe, however, that our conclusion (that the effect on the conduction band is small) is correct.

The treatment given here neglects any effect on the mobility edge due to stretched bonds; the localized states are at regions in the CRN where variations in the dihedral angle are abnormally small. The assumption that stretched bonds do not play a major role in determining the mobility edge is not proved [4.65]. What is clear, as we shall see, is that states (E_Y in Fig. 4.7) exist above the valence band with no states of corresponding density below the conduction band, so that one assumption could be that there are about 10^{19} cm^{-3} very stretched bonds which trap holes, but not electrons. Other models are possible, which on the whole we prefer (cf. [4.71] and Sect. 4.9).

As regards other treatments of the band edge, *Yonezawa* and *Cohen* [4.66] using a tight-binding Hamiltonian found 0.1 eV for the tail; thus, if we accept from the work of Abou-Chacra and Thouless that the range of localized states fills most of the tail, the results agree as regards the order of magnitude with those of Davies. *Bonch-Bruevich* [4.67] found, however, a much smaller value.

There is much evidence that in silicon the mobility is activated; Spear's work estimates the range of tail states to be ~ 0.2 eV [4.00]. In future work, an attempt to separate ΔE from ΔW, for instance, by determining $(E_\sigma - E_S)$, would be valuable, as would also work on CVD material. *Tiedje* et al. [4.68] in recent work found that the mobility in hydrogenated silicon ranges between 0.05 and 0.8 cm^2/V s, that its temperature-dependence shows only minor variations, while the dispersion parameter ranged from 0.1 to 0.7 with increasing hydrogen.

By no means are all interpretations of the mobility based on the concept of a mobility edge. An exponential tail of localized states is assumed, for instance, by *Döhler* [4.60] and by *Tiedje* et al. [4.70].

4.8 The Hall Effect in Amorphous Silicon

We turn next to the Hall effect and the double-sign anomaly described in Sect. 4.6. We have first of all the polaron hypothesis of *Emin* [4.32], namely, that the effect can be explained if small polarons of acoustic type are formed on Si–Si bonds and odd membered rings predominate. Against this it can be argued [4.71]:

a) the phenomenon is also observed in III–V compounds [4.45] for which odd membered rings should be few;

b) no argument is given to show why polarons should form in the amorphous material and not in crystals;

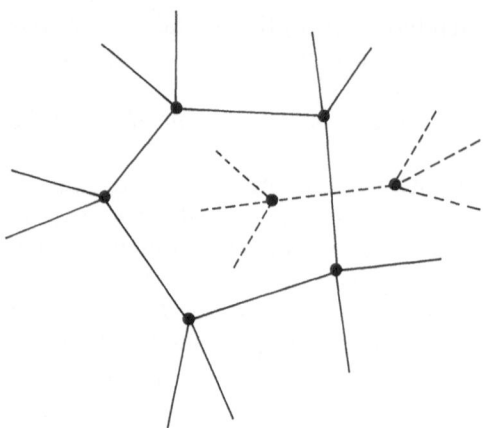

Fig. 4.6. Suggested form of bonds giving rise to the Hall effect in a-Si. The dotted line represents another bond in a ring above the plane of the paper

c) the activation energy for hopping can be only a small fraction of the whole, since this is structure-dependent. If so, the arguments of *Sumi* [4.26] show that at high temperatures, transport should be by free carriers which would give a normal sign to the Hall constant. This has not been observed.

Another possibility is to assume a random change of phase from bond to bond, as assumed by *Friedman* [4.72], which would allow the application of Emin's argument. However, this in our view would imply a zero bandwidth and would also necessitate the assumption of odd-membered rings. The present author [4.73] has suggested that Emin's argument could be combined with Friedman's discussion of conduction at a mobility edge by linking three sites at a distance a_E with wave functions $f(x,y,z)\exp(-\alpha r)$. The Hall anomaly would follow if f had odd parity and it was argued that for the conduction band this should be so. This suggestion is, however, highly speculative and we attempt another one here.

In the sense of the tight-binding approximation, we may suppose that the phase ϕ_n of the wave function changes slowly in moving from one bond to another along the rings, but that the Hall effect arises from three-site coincidence where there is sufficient overlap between antibonding orbitals on *different* rings. For the latter, we may well assume a random phase. We suppose, then, that such configurations are as illustrated in Fig. 4.6. We think that, in the spirit of Friedman's work of 1971 and that of Emin, the anomaly in the Hall constant might thus be explained[4].

If long-range fluctuations exist, then according to *Fritzsche* [4.74], the Hall coefficient should be determined by electrons in the valleys rather than at the passes. Thus, the activation energy in the Hall mobility should be the same as $(E_\sigma - E_S)$. *Le Comber* et al. [4.75] found that, for lightly doped n-type samples, μ_H seems virtually independent of T, as in earlier work in chal-

---

4 For an alternative discussion, see [4.78]

cogenides [4.76]; in silicon on doping an activation energy appeared. Further investigation to see if any relation exists between the $(E_\sigma - E_S)$ and the Hall mobility would be useful.

That the sign anomaly depends on short-range disorder seems to be substantiated by the work of *Spear* et al. [4.77] who have investigated the Hall effect in microcrystalline silicon films as a function of crystallite size. It will be seen that the effect regains its normal sign at about 20 Å. Although we are far from having a satisfactory theory of this effect, we conjecture that it shows that $a_E \sim 20$ Å for amorphous silicon and that the sign anomaly depends on having disorder within a smaller distance [4.78].

4.9 Defects and States in the Gap

It has been argued in the literature that the concept of a point defect in an amorphous substance is not acceptable [4.79]. While we agree for a material such as a "metallic glass", for glasses and for deposited films for which the structure can be represented by a continuous random network, which we believe to be the case for amorphous silicon and oxide and chalcogenide glasses, defects seem to have a real existence. A true defect, in our view, will have *either* a spin (e.g., a dangling bond) *or* a charge (e.g., an unoccupied or doubly charged dangling bond). Examples are the singly (C_1^-) and triply coordinated (C_3^+) chalcogens in the chalcogenide glasses, the threefold coordinated silicons in SiO_2 which are positively charged or neutral (E^1 centres), and the nonbridging oxygens. In silicon, a continuous random network will contain as part of the disorder of the "perfect" network odd and even-numbered rings, a random distribution of dihedral angles and doubtless stretched bonds.

Defects analogous to vacancies and divacancies may exist, without spin in the neutral state. These are not to be *sharply* distinguished from fluctuations in density.

Whether the concept of valence alternation pairs, with positive and negative charges (negative Hubbard U), is applicable to a-Si is uncertain, though it has been suggested [4.80]. It should however, be noted that in crystalline silicon, *Watkins* and *Troxell* [4.81] have given evidence that the positively charged vacancy has this property; if it is denoted by V^+, the reaction

$$2V^+ \rightarrow V^{++} + V$$

is found to be exothermic. A similar possibility exists for the amorphous material.

The inhomogeneity of amorphous silicon films has been discussed by many authors [4.82] and will not be described in detail here. An important point made particularly by *Revesz* et al. [4.83] is that hydrogen renders the network more flexible with a reduction in number of the stretched bonds and voids. We believe, however, that a strained network, containing dangling bonds, is a useful model with which to start.

In amorphous silicon hydrogenated and otherwise, the esr signal shows a line with the g-value 2.0055. This is clearly to be identified with the dangling bond and has been observed by *Stuke* and coworkers [4.84] at the interface between crystalline silicon and SiO_2. Its strength is enhanced when hydrogen is driven off [4.85]. Other lines observed under illumination are assigned by *Street* and *Biegelson* [4.86] as follows:

$g = 2.0043$ localized electrons in a conduction band tail state

$g = 2.010–2.013$ hole in the valence band tail states.

Voget-Grote et al. [4.87] identified spin signals from positively and negatively charged vacancies in crystals with $g = 2.011$ and 2.0045, respectively.

No density-of-states curve which neglects correlation (the Hubbard U) can be meaningful in a-Si (or in doped crystalline silicon for that matter). The fact that centres are singly occupied and thus show an esr signal and Curie behaviour is essentially dependent on correlation, as emphasized by *Schweitzer* et al. [4.88]. Thus, in a curve such as that of Fig. 4.7, A is the band of levels for dangling bonds, broadened by disorder; B is the band of levels for doubly occupied bonds of which only a few in specially low states will normally be occupied. If the zero-temperature Fermi energy is determined by the overlapping of these bands, we see that if conductivity is observable, $\beta = (E_c - E_F)/kT$ must be such that $e^{-\beta}$ is not negligible, but that if new dangling bond states are introduced, for instance, by irradiation, E_F may change.

The E_Y state is though to be a hole trap associated with hydrogen [4.89].

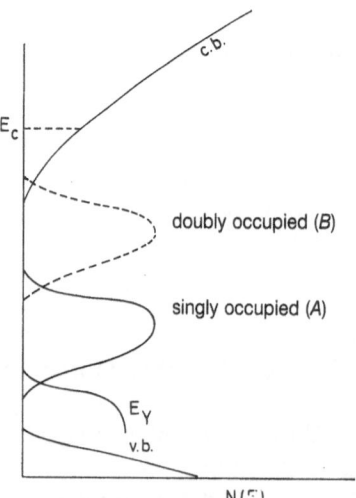

Fig. 4.7. Density of states in hydrogenated amorphous silicon (schematic)

It has already been emphasized that for any trap, particularly a deep trap, some Stokes shift is to be expected. For the E_Y state, *Tsang* and *Street* [4.90] gave some evidence for a Stokes shift in the photoluminescence line at 1.6 eV, ascribed to transitions between band edge localized states in the conduction band and holes at E_Y; the magnitude of the shift was found to be 0.2 eV.

Stokes shift, namely, a distortion of the surroundings, is, of course, relevant to the magnitude of the Hubbard U; for instance, that required to shift an electron from one occupied dangling bond to another, creating two charged states. This will always reduce U, and in chalcogenides it is normally negative. In crystalline silicon, as we have seen, *Watkins* and *Troxell* [4.81] show that a charged vacancy has this property though there is no firm evidence that this is so for any defect in amorphous silicon. However, if the Hubbard U is to be small enough for doubly charged dangling bond states to lie in the gap, then some deformation must be present to lower its value. *Chakraverty* [4.91] estimates this, without deformation, to be ~ 1.7 eV. This is too large to allow the doubly occupied dangling bond state to lie in the gap. *Dersch* et al. [4.84], however, concluded from the doping dependence of the spin density that the effective correlation energy of the dangling bond amounts to 0.4 to 0.5 eV. The model set out in great detail by *Street* [4.92] to describe luminescence also puts the doubly occupied dangling bond below the conduction band.

The model of dangling bond or other states with energies overlapping so that positive and negatively charged states exist, is reinforced by much evidence, such as that of *Anderson* and *Spear* [4.93] on recombination in doped material. Further evidence comes from the effects of removing hydrogen from a-Si:H annealing. *Fritzsche* [4.85] and *Tsai* et al. [4.94] showed that annealing increases the density of gap states and the paramagnetic susceptibility. The same process leads to a reduction in the photo-current [4.87] and in the luminescence [4.95]. It appears then that spin states act as recombination centres, producing states at the Fermi energy to which electrons can fall. *Hasegawa* and *Imai* [4.96] traced the relationship between photoconductivity and ESR linewidth.

We might add that pinning of the Fermi level is not necessarily due to overlap between the positively and negatively charged dangling bond states. Thus, *Voget-Grote* et al. [4.87] took into account both these and vacancy-like states. *Chakraverty* [4.91], as we have seen, considered that U is too large for the negative state to exist, so the pinning is due to a balance between neutral and positively charged dangling bond states and charged vacancy states. Such a model implies neutral and positively charged dangling bonds and neutral and negatively charged vacancies (V states). *Stuke* [4.84a], on the other hand, found evidence that overlap between neutral and negatively charged states is responsible for pinning, and that this fits his evidence on the Staebler-Wronski effect which he ascribes to an enhancement of the dangling bond density [4.11].

Street ([4.92] and private communication) considers that the best evidence for pinning by dangling bond states is that the ESR line at $g = 2.0025$ occurs in undoped material, but that in doped material in the dark it disappears but can be induced by illuminating at low temperatures. He also emphasizes the strong evidence that the line at $g = 2.0055$ is indeed from a dangling bond because the same g-value is observed at the interface between crystalline silicon and SiO_2.

4.10 The Pre-Exponential Factor in the Conductivity

We have seen that the pre-exponential factor for a material without long-range fluctuations in potential should be

$$\sigma_0 = \sigma_{min} e^{\beta/k} , \tag{4.20}$$

with $\sigma_{min} \sim 200\ \Omega^{-1}\ cm^{-1}$. In fact, very large ranges of σ_0 are observed, from 1 to $10^6\ \Omega^{-1}\ cm^{-1}$, and this will now be discussed.

Solomon et al. [4.97, 98] have claimed that this variation is the result of the production of an accumulation layer by surface changes, giving some band bending downwards. They found that when corrections are made for this, σ_0 is sensibly independent of preparation conditions and of order 10^4 $\Omega^{-1}\ cm^{-1}$. Other workers [4.99] claim to avoid these effects and still find a large variation of σ_0, and in certain cases that σ_0 obeys the Meyer-Nelden rule [4.100,102] in the form

$$\sigma_0 = const\ e^{E\sigma/kT} \tag{4.21}$$

when E_σ is varied by doping.

As regards the variation of $(E_c - E_F)$ with temperature and the term $exp(\beta/k)$, it has long been recognised [4.2] that the optical band gap changes with temperature, that this is probably due to a shift of the valence band and that if the Fermi energy is pinned to the valence band, a similar shift of $E_c - E_F$ should occur, linear above the Debye temperature. Such a shift could give at most $exp(\beta/k) \sim 100$, but would account nicely for a value of order 10^4 $\Omega^{-1}\ cm^{-1}$, since $100\ \Omega^{-1}\ cm^{-1}$ is a reasonable estimate for σ_{min} with a_E $\sim 10\text{--}20\ \text{Å}$. To account for larger variations, and particularly for large values of the order $10^6\ \Omega^{-1}\ cm^{-1}$, there are two possibilities.

a) *Spear* and co-workers [4.99, 103] have given evidence to show that the quantity $\Delta E = E_c - E_A$, that is, the height of the mobility edge above the bottom of the band, decreases with temperature and disappears (or nearly so) at ~ 400 K. This *only* occurs for undoped specimens for which ε_σ is in the range 0.65–0.8 eV and for which σ_0 has these high values ($10^5\text{--}10^6\ \Omega^{-1}\ cm^{-1}$). It does not occur for doped specimens for which the conduction channel

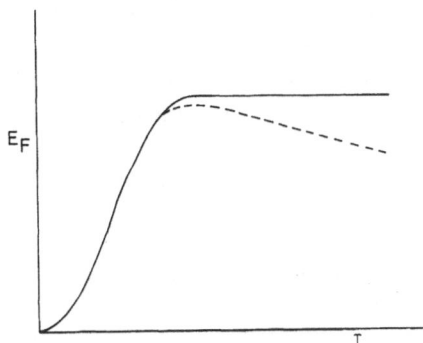

Fig. 4.8. Suggested temperature-dependence of the Fermi energy in amorphous silicon

seems to be in an impurity band. From the theoretical viewpoint, however, we have no satisfactory model to show why either ΔE or ΔW would be so sensitive to the temperature; if this behaviour is established, perhaps quite a new description of the mobility edge may be needed.

b) *Overhof* and *Beyer* [4.104], on the other hand, pointed out the possibility of a strong temperature-dependence of the Fermi energy under certain conditions. Perhaps this is most easily seen by considering a semiconductor in which the Fermi energy is "pinned" by deep donors (E_D below E_c) with low compensation. Then at zero temperature, $E_c - E_F = E_D$. As the temperature is raised, E_F will increase as $(kT)^2 N'(E_F)$. At higher temperatures, however, it will move towards $\frac{1}{2} E_D$, as in Fig. 4.8; it should then drop on account of thermal expansion as shown by the dotted line[5].

Determinations of the position of the Fermi energy by ultraviolet photoemission spectroscopy (UPS) which compare its energy with that of inner levels [4.105] will reveal shifts due to thermal expansion but *not* the temperature dependent effect shown here.

Such a model does reproduce one feature of the Meyer-Nelden rule, that σ_0 is large for a large value of the gap. However, the temperature at which a linear regime should set it must depend on the degree of compensation. We think this could depend on the homogeneity of the specimen. Neither the hydrogen content nor the concentration of dangling bonds are thought to be homogeneous, so the Fermi energy would vary from one region to another if each region remained neutral. Charge will be transferred from one region to another, setting up fields and giving another kind of long-range potential fluctuation in the conduction band. This could well increase any overlap between the dangling-bond band E_x and the doubly charged band of Fig. 4.8, thus raising the temperature at which the linear regime occurs.

In any case, then, we expect σ_0 to be very sensitive to homogeneity.

Possibly the Meyer-Nelden rule is normally to be accounted for by shifts in the Fermi energy of this type. There are, however, other models. Thus,

---

5 An analysis of this behaviour was given by *Roberts* [4.101]

Roberts [4.101] postulated a conduction band with an exponential tail but constant mobility and shows that a Meyer-Nelden form can be reproduced. We think this explanation is open to the criticism that in an exponential tail, Anderson localization should occur and the mobility should be a strong function of energy.

Deep level transient spectroscopy in the hands of *Cohen* et al. [4.106] and *Lang* et al. [4.107] has given evidence for a very low density of states (10^{14}–10^{15} eV^{-1} cm^{-3}) and a rapid variation with energy, of the kind which would produce a large change of Fermi energy with temperature. The results are not, however, confirmed by determination of $N(E_F)$ from the space-charge limited current ([4.108] and Chap. 3), found values some 3–10 times smaller than that deduced from the field effect and thus of order $\sim 10^{16}$ eV^{-1} cm^{-3} and little dependent on energy. If these are correct, the large values of σ_0 cannot be due to a shift in the Fermi energy and the only explanation available is a dependence of E_c on temperature [4.109].

4.11 The Staebler-Wronski Effect

A photoinduced ESR signal which is stable at low temperatures is found in many amorphous semiconductors; it was observed by *Bishop* et al. [4.110] in chalcogenides and interpreted in terms of the trapping of electrons in defects already existing, namely, the charged valence alternation pairs C_1^- and C_3^+. In principle, any electron or hole trap could give rise to photoinduced ESR.

Staebler and *Wronski* [4.111] report that when glow-discharge deposited a-Si is exposed at room temperature to broad band illumination (1.4 to 2.1 eV), a decrease in the conductivity by four orders of magnitude is observed which can be annealed out by heating to 150 °C with an activation energy of 1.52 eV. The metastable state contains about double the density of unpaired spins. The effect was observed only in lightly doped samples. Completely undoped films have been observed to show an inverse effect, while heavily doped films do not show it.

What has to be emphasized is that the trapped charges which produce the effect do not communicate with the conduction band, as do electrons at states near the Fermi energy at any temperature where the conduction is observed. It is also to be noted that the activation energy for annealing is greater than the band gap. We have to ask, then, whether electrons are trapped in a deep state stabilized by a large Stokes shift, or whether defects are formed.

One interpretation is that light induces two T_3^0 states (dangling bonds). This, as stated in Sect. 4.9, is the view of the Marburg group. A model of this type is that of *Elliott* [4.112]. These models were criticised by *Adler* [4.78]. The energy of the exciting light seems much less than that required to break two bonds and the lack of any exchange narrowing in the photoinduced ESR line indicates that any photoinduced T_3^0 centres must be more than 10 Å apart. It seems to us likely that spinless traps exist, for instance, vacancies which do give very stable trapping. Adler postulates states with negative U, but this does not seem essential though we note that in crystalline silicon, two V^+ (where V is a vacancy) should dissociate into V^{++} and V.

As regards the mechanism of the effect, Adler suggests that more point charges of one sign than the other will be produced so that to preserve charge neutrality, the Fermi energy will be pushed one way or the other. The effect will thus only be significant if $N(E_F)$ is small.

Hirabayashi et al. [4.113] and *Morigaki* et al. [4.114, 115] reported that prolonged irradiation produces $\sim 10^{17}$ cm^{-3} ESR centres; they claim that this is evidence for defects because the g-value indicates that these are dangling bonds. If illumination creates such bonds, e.g., by driving off hydrogen, it is not clear that they would produce the charge needed to give a considerable shift in the Fermi energy. They might, if $N(E_F)$ is very low, produce a shift for other reasons, namely, by changing the density of states. Strong evidence for this model was given by *Dersch* [4.84b, 116].

4.12 Variable-Range Hopping in Silicon

Our discussion in Sect. 4.5 has shown that a $T^{1/4}$ law is deduced on the assumption that $N(E)$ is constant in the neighbourhood of E_F and that the "Coulomb gap" caused by the interaction between carriers is neglected.

If, as is generally supposed, variable-range hopping, when observed in insufficiently hydrogenated material, is a phenomenon in which electrons jump from a singly occupied dangling-bond state to a charged one (empty or doubly occupied), then Fig. 4.8 shows

a) no relationship between $N(E_F)$ and the number of spins is to be expected, and

b) $N(E_F)$ may well be far from constant.

In the latter case, though calculations with particular models are missing, we think that a hopping formula of the form

$$\sigma = A \exp(- T_0/T)^S \tag{4.22}$$

with $\frac{1}{4} < S < \frac{1}{2}$ is to be expected; the lower limit $\frac{1}{2}$ comes from the calculations of *Hamilton* [4.117] for the form for the density of states

$$N(E) \propto (E - E_F)^2 . \tag{4.23}$$

Observations seem to give, particularly in a-Ge, a fair approximation to the $T^{1/4}$ law, but with a much higher pre-exponential factor than can be expected from interaction with phonons. It may well be that this is sometimes a consequence of forcing the observations to a $T^{1/4}$ law and that with a higher index, the prefactor would be reasonable.

Kalbitzer and co-workers [4.118, 119], have made detailed investigations of hopping transport in ion-bombarded amorphous silicon, determining the index in the expression $\exp[(T_0/T)^n]$ by a method due to *Hill* [4.120]. A value of $n \sim \frac{1}{2}$ is frequently found. A mechanism similar to that of *Abeles* et al. [4.57] for metallic particles dispersed in an insulator is proposed.

The thermopower for materials in which variable-range hopping is observed appears independent of T and can be explained *qualitatively* by (4.12), though the value is normally smaller. A rise in S at low temperatures observed by *Lewis* et al. [4.121] has been discussed by the present author [4.35] in terms of the Coulomb gap.

The Hall effect for hopping conduction has not been observed; for discussions of its magnitude see [4.41–43].

A relationship between the width of the ESR line with $g \sim 0.0055$ (due to dangling bonds) and hopping conduction has been found by the Marburg School, and *Movaghar* et al. [4.122] ascribe this to a locally fluctuating magnetic field due to the hopping, giving a relation of the form

$$\delta H(T) = C[\sigma(T)]^n ;$$ (4.24)

for further developments see *Dersch* et al. [4.84 b] and *Grünewald* et al. [4.123] and *Overhof* [4.124] who related the temperature-dependence of the ESR line with $g = 2.01$ to the density of states that he assumes in order to interpret the shift in the Fermi level.

Acknowledgements. The author is indebted to discussions with many colleagues, particularly E. A. Davis, H. Overhof, W. E. Spear, R. A. Street and J. Stuke.

References

4.1   P. W. Anderson: Phys. Rev. **109,** 1492 (1958)
4.2   N. F. Mott, E. A. Davis: *Electronic Processes in Non-Crystalline Materials,* 2nd ed. (University Press, Oxford 1979)
4.3   J. T. Edwards, D. J. Thouless: J. Phys. C. **5,** 807 (1972); see also [Ref. 4.2, p. 20]
4.4   N. F. Mott: Adv. Phys. **16,** 49 (1967)
4.5   M. H. Cohen, H. Fritzsche, S. R. Ovshinsky: Phys. Rev. Lett. **22,** 1069 (1969)
4.6   R. Abou-Chacra, D. J. Thouless: J. Phys. C **7,** 65 (1974)
4.7   A. G. Grant, E. A. Davis: Solid State Commun. **15,** 569 (1974)
4.8   N. F. Mott, M. Pepper, S. Pollitt, R. H. Wallis, C. J. Adkins: Proc. R. Soc. A. **345,** 169 (1975)
4.9   H. Fritzsche: In *The Metal-Insulator Transitions in Disordered Solids,* Proc. 19th Scottish Summer School in Physics, ed. by L. R. Friedman, D. P. Tunstall (1978) p. 193

4.10 Articles in Phil. Mag. Vol. **42**, No. 6, 723–1003 (1980), Proc. Würzburg Conference on Impurity Bands in Semiconductors
4.11 N. F. Mott: Phil. Mag. **26**, 1015 (1972)
4.12 N. F. Mott: Phil. Mag. B **43**, 941 (1981)
4.13 M. H. Cohen, J. Jortner: Phys. Rev. Lett. **30**, 699 (1973)
4.14 W. Götze: Solid State Commun. **27**, 1393 (1978); J. Phys. C. **12**, 1279 (1979); Phil. Mag. B **43**, 219 (1981)
4.15 E. Abrahams, P. W. Anderson, D. C. Licciardello, T. W. Ramakrishnan: Phys. Rev. Lett. **42**, 693 (1979)
4.16 T. F. Rosenbaum, K. Andres, G. A. Thomas, R. N. Bhatt: Phys. Rev. Lett. **43**, 1723 (1980)
4.17 N. F. Mott: Phil. Mag. B **44**, 265 (1981)
4.18 D. Belize, W. Götze: Phil. Mag. B **43**, 517 (1981)
4.19 L. Friedman: J. Non-Cryst. Solids **6**, 321 (1971)
4.20 M. Cutler, N. F. Mott: Phys. Rev. **181**, 1336 (1969)
4.21 I. G. Austin, N. F. Mott: Adv. Phys. **18**, 41 (1969)
4.22 Y. Toyozawa: Prog. Theor. Phys. **26**, 29 (1963)
4.23 D. Emin: Adv. Phys. **22**, 51 (1973)
4.24 N. F. Mott, A. M. Stoneham: J. Phys. C **10**, 3391 (1977)
4.25 N. F. Mott: Mat. Res. Bull. **13**, 1389 (1978)
4.26 A. P. Schmid: J. Appl. Phys. **39**, 3140 (1968)
4.27 Y. Nagaoka, H. Fukuyama (eds.): *Anderson Localization,* Springer Ser. Solid-State Sci., Vol. 39 (Springer, Berlin–Heidelberg, New York 1982)
4.28 E. Laredo, L. G.Rowan, L. Slifkin: Phys. Rev. Lett. **47**, 384 (1981)
4.29 A. Sumi: J. Phys. Soc. Jpn. **33**, 327 (1972); J. Chem. Phys. **70**, 3775 (1979); **71**, 3403 (1979)
4.30 L. Friedman, T. H. Holstein: Ann. Phys. N.Y. **24**, 494 (1963)
      T. H. Holstein, L. Friedman: Phys. Rev. **165**, 1019 (1968)
4.31 P. Nagels: *The Hall Effect and Its Applications,* ed. by C. L. Chien, C. R. Westgate (Plenum, New York 1980), p. 253
4.32 D. Emin: Phil. Mag. **35**, 1189 (1977)
4.33 M. Grünewald, R. Müller, P. Thomas, D. Würz: J. Phys. (Paris) C **4**, 99 (1981)
4.34 N. F. Mott: Phil. Mag. B **19**, 835 (1969)
4.35 B. I. Shklovskii, A. L. Efros: Sov. Phys. JETP **33**, 468 (1971)
4.36 V. Ambegaokar, S. Cochran, J. Kurkijarvi: Phys. Rev. B **8**, 3682 (1973)
4.37 M. Pollak: Phil. Mag. B **42**, 781 (1980)
4.38 A. L. Efros, B. I. Shklovskii: J. Phys. C **8**, L 49 (1975)
      B. I. Shklovskii, A. L. Efros: *Electronic Properties of Doped Semiconductors,* Springer Ser. Solid-State Sci., Vol. 45 (Springer, Berlin–Heidelberg, New York, Tokyo 1984)
      J. H. Davies, P. Lee, T. M. Rice: Phys. Rev. Lett. **49**, 1958 (1982)
4.39 T. E. Whall: J. Phys. C. **14**, L 887 (1981)
4.40 P. M. Chaikin, G. Beni: Phys. Rev. B **13**, 647 (1978)
4.41 L. Friedman, M. Pollak: Phil. Mag. B **44**, 487 (1981)
4.42 M. Pollak: J. Phys. (Paris) C **4**, 141 (1981)
4.43 P. Butcher: J. Phys. (Paris) C **4**, 91 (1981); P. Butcher, A. A. Kumar: Phil. Mag. B **42**, 201 (1980)
4.44 P. G. Le Comber, D. I. Jones, W. E. Spear: Phil. Mag. **35**, 1173 (1977)
4.45 W. Beyer, H. Mell, H. Overhof: Proc. 7th Conf. Amorphous and Liquid Semiconductors, ed. by W. E. Spear (CICL, University of Edinburgh 1977), p. 229
4.46 J. Dresner: Appl. Phys. Lett. **37**, 742 (1980)
4.47 C. H. Seager, M. L. Knotek, A. H. Clark: Proc. 5th Intern. Conf. on Amorphous and Liquid Semiconductors, ed. by J. Stuke, W. Brenig (Taylor and Francis, London 1974), p. 1173
4.48 W. Beyer, H. Mell: Solid State Commun. **39**, 375 (1981)
4.49 W. Beyer, R. Fischer, H. Overhof: Phil. Mag. B **39**, 205 (1979)

4.50 D. I. Jones, P. G. Le Comber, W. E. Spear: Phil. Mag. **36**, 541 (1977)
4.51 H. Overhof, W. Beyer: J. Non-Cryst. Solids **35/36**, 377 (1980)
4.52 H. Overhof, W. Beyer: Phil. Mag. B **43**, 433 (1981)
4.53 B. Pistolet, J. L. Robert, J. M. Dusseau, L. Ensuque: J. Non-Cryst. Solids **29**, 29 (1978)
4.54 J. M. Dusseau: Thesis Montpellier (1980)
4.55 H. Fritzsche: J. Non-Cryst. Solids **6**, 49 (1971)
4.56 B. Shklovskii, A. Efros: Sov. Phys. JETP **33**, 468 (1971)
4.57 B. Abeles, Ping Shen, M. D. Coutts, Y. Arie: Adv. Phys. **24**, 407 (1975); see also [Ref. 4.2, p. 157]
4.58 R. C. Hughes: Phys. Rev. Lett. **30**, 1333 (1973)
4.59 W. E. Spear: Adv. Phys. **26**, 811 (1977)
4.60 P. Nagels, R. Callaerts, M. Denayer: Proc. 5th Intern. Conf. on Amorphous and Liquid Semiconductors, ed. by J. Stuke, W. Brenig (Taylor and Francis, London 1974), p. 867
4.61 E. Mytilineou, E. Davis: Proc. 7th Conf. on Amorphous and Liquid Semiconductors, ed. by W. E. Spear (CICL University of Edinburgh 1977), p. 632
4.62 D. Emin, C. H. Seager, R. K. Quinn: Phys. Rev. Lett. **28**, 813 (1972)
4.63 J. H. Davies: Phil. Mag. B **41**, 373 (1980)
4.64 T. D. Moustakas: J. Elect. Mat. **8**, 391 (1979)
4.65 A. Deneuville, J. C. Bruyere, A. Mini, M. Kahil, R. Danielou, E. Ligeon: J. Non-Cryst. Solids **35/36**, 469 (1980)
4.66 F. Yonezawa, M. H. Cohen: In *Fundamental Physics of Amorphous Semiconductors,* ed. by F. Yonezawa, Springer Ser. Solid-State Sci., Vol. 25 (Springer-Verlag, Berlin, Heidelberg, New York 1981) p. 119
4.67 V. L. Bonch-Bruevich: J. Non-Cryst. Solids, **35/36**, 95 (1980)
4.68 T. Tiedje, T. D. Moustakas, D. L. Morel, J. M. Cebulka, B. Abeles: J. Phys. (Paris) **4**, 155 (1981)
4.69 G. H. Döhler: J. Non-Cryst. Solids **35/36**, 363 (1980)
4.70 T. Tiedje, A. Rose, J. M. Cebulka: AIP Conf. Proc. **73**, 197 (1981)
4.71 N. F. Mott: J. Phys. C **13**, 5433 (1980)
4.72 L. Friedman: Phil. Mag. B **41**, 347 (1980)
4.73 N. F. Mott: Phil. Mag. B **37**, 594 (1978)
4.74 H. Fritzsche: In *Fundamental Physics of Amorphous Semiconductors,* ed. by F. Yonezawa, Springer Ser. Solid-State Sci., Vol. 25 (Springer-Verlag, Berlin, Heidelberg, New York 1980) p. 1
4.75 P. G. Le Comber, D. I. Jones, W. E. Spear: Phil. Mag. **35**, 1173 (1977)
4.76 J. C. Male: Brit. J. Appl. Phys. **18**, 1543 (1967)
4.77 W. E. Spear, G. Willeke, P. G. Le Comber, A. G. Fitzgerald: J. Phys. (Paris) **4**, 257 (1981)
4.78 W. E. Spear: Proc. Int. Conf. on Amorphous and Liquid Semiconductors, Tokyo (1983, in press)
4.79 P. W. Anderson: J. Phys. (Paris) **4**, 339 (1976)
4.80 D. Adler: J. Phys. (Paris) **4**, 3 (1981)
4.81 G. D. Watkins, J. R. Troxell: Phys. Rev. Lett. **44**, 593 (1980)
4.82 H. Fritzsche: Solar Energy Mat. **3**, 447 (1980)
4.83 A. G. Revesz, S. H. Wemple, G. V. Gibbs: J. Phys. (Paris) **4**, 217 (1981); A. G. Revesz: Thin Solid Films **52**, L 29 (1978)
4.84 P. J. Caplan, E. H. Poindexter, B. E. Deal, R. R. Razonk: J. Appl. Phys. **50**, 5847 (1979);
      H. Dersch, J. Stuke, J. Beichler: Appl. Phys. Lett., **38**, 456 (1981), and Phys. Stat. Sol. (b) **105**, 265 (1981); **107**, 307 (1981)
4.85 H. Fritzsche: Proc. 7th Int. Conf. on Amorphous and Liquid Semiconductors, ed. by W. E. Spear (CICL, University of Edinburgh 1977), p. 3
4.86 R. A. Street, D. K. Biegelsen: Solid State Commun. **33**, 1159 (1980)
4.87 U. Voget-Grote, W. Kummerle, R. Fischer, J. Stuke: Phil. Mag. B **41**, 127
4.88 L. Schweitzer, M. Grünewald, H. Dersch: J. Phys. (Paris) **4**, 827 (1981)

4.89 J. D. Joannopoulos: J. Non-Cryst. Solids **35/36,** 781 (1980)
4.90 C. Tsang, R. A. Street: Phys. Rev. B **19,** 3027 (1979)
4.91 B. Chakraverty: J. Phys. (Paris) C **4,** 741 (1981)
4.92 R. A. Street: Adv. Phys.: **30,** 593 (1981); J. Phys. (Paris) C **4,** 283 (1981); see also Chap. 5 of this volume.
4.93 D. A. Anderson, W. Spear: Phil. Mag. B **36,** 695 (1977)
4.94 C. C. Tsai, H. Fritzsche, M. H. Tanielian, P. J. Gaczi, P. D. Persans, M. A. Vesaghe: Proc. 7th Intern. Conf. on Amorphous and Liquid Semiconductors, ed. by W. E. Spear (CICL, (University of Edinburgh 1977), p. 339
4.95 D. K. Biegelsen, R. A. Street, C. C. Tsai, J. C. Knights: Phys. Rev. B **20,** 4839 (1979)
4.96 S. Hasegawa, Y. Imai: Phil. Mag. (in press)
4.97 I. Solomon, T. Dietle, D. Kaplan: J. Phys. (Paris) **39,** 124 (1978)
4.98 I. Solomon, J. Perrin, B. Bonden: Intern. Conference on Physics of Semiconductors (1978), p. 689
4.99 W. E. Spear, D. Allen, P. G. Le Comber, A. Gaith: Phil. Mag. B **41,** 419 (1980)
4.100 W. Meyer, H. Nelden: Z. tech. Phys. **18,** 588 (1937)
4.101 G. G. Roberts: J. Phys. C **4,** 3167 (1971)
4.102 G. G. Roberts: Electronic and Structural Properties of Amorphous Semiconductors, Scottish University Summer School, ed. by P. G. Le Comber, J. Mort (Academic, London 1973) p. 409
4.103 W. E. Spear, Haifa Al-Ani, P. G. Le Comber: Phil. Mag. B **43,** 781 (1981)
4.104 H. Overhof, W. Beyer: J. Non-Cryst. Solids **35/36,** 663 (1980); Phys. Stat. Sol. (b) **107,** 207 (1981)
4.105 B. Von Roedern, L. Ley, M. Cardona, F. W. Smith: Phil. Mag. B **40,** 433 (1979)
4.106 J. D. Cohen, D. V. Lang, J. B. Harbisson: Phys. Rev. Lett. **45,** 197 (1980)
4.107 D. V. Lang, J. D. Cohen, J. B. Harbisson: Phys. Rev. (1981)
4.108 J. Den Boer: J. Phys. (Paris) C **4,** 451 (1981)
4.109 H. Overhof, W. Beyer: Phil. Mag. B **47,** 377 (1983)
4.110 A. R. Bishop, U. Strom, P. C. Taylor: Phys. Rev. Lett. **34,** 1346 (1975)
4.111 D. L. Staebler, C. R. Wronski: Appl. Phys. Lett. **31,** 292 (1977)
4.112 S. R. Elliott: Phil. Mag. **36,** 1291 (1977)
4.113 I. Hirabayashi, K. Morigaki, S. Nitta: Tech. Rpt. of Inst. of Solid State Phys. Tokyo, Ser. A, No 1053 (1980)
4.114 K. Morigaki, I. Hirabayashi, S. Nitta: J. Phys. (Paris) C **4,** 335 (1981)
4.115 K. Morigaki, Y. Sano, I. Hirabayashi: Solid State Commun. **39,** 947 (1981)
4.116 H. Dersch: Dissertation, University of Marburg, FR Germany (1983); H. Dersch, J. Stuke, J. Beichler: Phys. Stat. Sol. (b) **105,** 265 (1981)
4.117 E. M. Hamilton: Phil. Mag. **26,** 1043 (1972)
4.118 R. Pfeilsticker, S. Kalbitzer, G. Müller: Nucl. Instr. Meth. **182/183,** 603 (1981)
4.119 G. Müller, S. Kalbitzer, R. Pfeilsticker: Z. Physik. B **39,** 21 (1980)
4.120 R. M. Hill: Proc. 7th Intern. Conf. on Amorphous and Liquid Semiconductors, ed. by W. E. Spear (CICL, University of Edinburgh 1977), p. 229
4.121 A. J. Lewis, G. A. Connell, W. Paul, J. R. Pawlik, R. J. Temkin: AIP Conf. Proc. **20,** 27 (1974)
4.122 B. Movaghar, L. Schweitzer, H. Overhof: Phil. Mag. B **37,** 638 (1978)
4.123 M. Grünewald, P. Thomas, T. D. Würz: J. Phys. C **14,** 4083 (1981)
4.124 H. Overhof: Phil. Mag. B **47,** 377 (1983)

5. The Spectroscopy of Localized States

R. A. Street and D. K. Biegelsen

With 50 Figures

In preceding chapters the fundamental theoretical concepts related to amorphous silicon (a-Si) have been developed. In this chapter we focus on the experimental findings concerning the nature of localized electronic states in the gap. The characteristics and distribution (both in energy and space) of the states dominate the electrical and optical properties vis-à-vis their role as trapping and recombination centers. In order to understand and tailor the electronic properties of real materials, an accurate assessment of the states in the gap must be achieved. We shall show here that much has been learned, but many fundamental aspects remain both qualitatively as well as quantitatively uncertain. In the introductory section we briefly restate the essential theoretical notions. In Sect. 5.2 we describe the results of various spectroscopies which have been used to ascertain the characteristics of the states. In Sect. 5.3 we cover material variations and extrinsic material properties insofar as they relate to gap states.

5.1 Introduction

In the idealized picture of a-Si as a continuous random network, the local structure is much like that of crystalline silicon. All atoms are covalently bonded to a nearly regular tetrahedral array of nearest neighbors. Therefore, chemically, the two states share many properties. In the amorphous state the spatial disorder of atoms gives rise to a distribution of fluctuations about the average in the potentials acting on electrons. Thus, electronic bands, associated with bonding and antibonding states in the crystal, are broadened by tails corresponding to the greatest perturbations (i.e., particularly strong or weak bonding and antibonding states). The potential fluctuations manifest themselves within the bands by reducing the coherence length of the phase of the electronic wave functions to distances of the order of the interatomic spacing. Strong electronic scattering and low microscopic mobilities result.

As first discussed by *Anderson* [5.1] and *Mott* [5.2] strong quantitative disorder can lead to localization of the electronic carriers (Chap. 4). The nature of the transition region between extended and localized states (the mobility edge) is a controversial subject which has not been satisfactorily resolved in a-Si. However, the results of most of the spectroscopic measurements described here do not depend on the sharpness of the mobility edge. The mobility edges delineate qualitatively different states. Between them (within the mobility gap) lie localized band-tail states and other localized deep states arising from structural defects and impurities. The investigation of these states is the subject of this chapter.

The primary characteristics sought are the energy distribution of band-tail, defect and impurity-related states, the spatial features (e.g., degree of electronic localization and symmetry) and the electron-phonon coupling and electron-electron correlation strengths. We also wish to determine the specific roles of these states, e.g., as traps and/or recombination centers. And finally we would like to develop the etiology of these states, e.g., doping, native defects, doping and impurity-related defects, deposition-related variations, inhomogeneities and surface states.

A priori, one could expect insurmountable complexities arising from the presence of a dense distribution of localized tail states and a large variety of defects. Fig. 5.1 shows a general density-of-states diagram for such a system. The solid curve represents the states of the continuous random network tailing between the valence and conduction bands. The dashed line in Fig. 5.1 indicates a continuous distribution of possible defect or impurity levels coexisting with the network states.

The nature of the localized states can also be viewed in the context of some other general expectations. A particuar spatial fluctuation can give rise to different energy shifts in the valence and conduction band tails because of differences in electronic symmetry at the two band edges. For example, if the

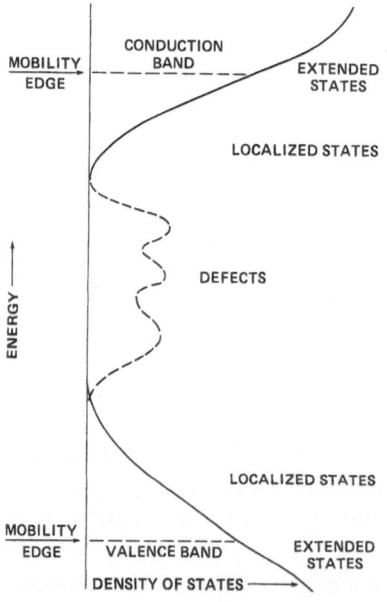

Fig. 5.1. Schematic density-of-states diagram of an amorphous semiconductor showing the extended states and localized band tails of the ideal network and defect states in the gap

wave functions of the conduction band edge electrons are predominantly
s-like (spherically-symmetric), the electronic energies would be expected to
be fairly insensitive to differences in angular orientation of nearest neighbors. Radial distribution functions derived from x-ray scattering experiments
show that the distribution of nearest-neighbor spacings is very small, whereas
bond angle variations are considerable. The fluctuations in electronic potentials might thus be expected to be small for the conduction band-tail states. If
the valence band edge states, on the other hand, are more nearly p-like [5.3],
the increased sensitivity to bond angle variations would lead to a wider
distribution in potentials (a wider band tail) and therefore stronger localization. Stronger localization in turn may lead to stronger electron-phonon
coupling (lattice relaxation) and increased electronic correlation energy.

A characteristic feature of amorphous materials is a relaxation of structural constraints relative to the crystalline state. The distribution in atomic
configurations and bond strengths enables bond reconstruction and/or defect
formation, network termination (e.g., Si–H bonding) and creation of internal
surfaces (microstructure). Of particular interest are the resulting strong electron-phonon coupling and presence of isolated dangling bonds (threefold
coordinated Si atoms), neither of which are allowed in the crystalline state.

The picture which is evolving for amorphous silicon is in fact surprisingly
more simple than that of Fig. 5.1. When a-Si was first prepared by sputtering
high hopping conductivities, pinning of the Fermi level and high electron spin
densities seemed to indicate a picture similar to that described above. However, when hydrogen was added to a-Si during deposition, an alloy, a-Si : H,
was created in which the density of gap states was dramatically reduced
[5.4 a]. In terms of the short-range chemical picture, strong Si–H bonds
replace weak Si–Si bonds and terminate dangling bonds (defects in the continuous random network) and thus remove states from the gap. It is believed
that the hydrogen adds no states in the gap and acts only to remove states.
The effectiveness of the hydrogenation process depends on the deposition
conditions and method (e.g., reactive sputtering, glow discharge, evaporation, etc.). Suffice it to say though that for most methods there exists a
reasonably large and controllable region of the parameter space for which the
a-Si : H has a density of states low enough to make many spectroscopic
probes viable. (A further consequence is that the Fermi level can also be
moved by dopant incorporation [5.4 b]).

We will demonstrate that two great simplifications in the gap state picture
occur for a-Si : H. The first is the relatively narrow extent of the band tails
(i.e., the density of tail states at midgap is negligible). The second arises from
the nature of electronically active defects. Essentially the only defect is the
dangling Si bond. The absence of other electronic defect states in the gap in
spite of the known presence of microstructural defects ([5.5] and [Ref. 5.6,
Chap. 2]) comes about from the ability of the a-Si structure to relax.
Extended defects are largely reconstructed (paired) and/or hydrogenated giving rise to tail states (e.g., weak or strong bonds) or states outside the gap.

Only the unhydrogenated, isolated dangling bond is left to act as a deep trap and nonradiative recombination center.

The ideal electronic spectroscopy would provide information about the number of states in given energy intervals in the gap. It would indicate the local anisotropy and spatial extent of the electronic states. Furthermore, it could be used to study the dynamics of charge carriers in nonequilibrium situations. From all these data, models of the microscopic origins of the states in the gap could be inferred. Such spectroscopic measures exist for gap states (impurities and defects) in crystalline silicon. However, the disorder inherent in a-Si makes most probes much less informative than the same measurements in crystalline silicon. Optical absorption and emission bands become very broad and nearly featureless. For electron-spin-resonance lines, the site-specific information about local anisotropy and environmental variations is averaged out. Broad bands of gap states (instead of discrete levels) make interpretations of electrical data ambiguous.

Nevertheless, by performing coordinated experiments, a consistent picture, as sketched above, does emerge. In this chapter then we will review both the experimental tools and the various results which have been useful in deducing the nature of the states in the gap of a-Si. Only those aspects of the spectroscopic measurements will be developed which help reveal the nature of the localized states. Furthermore, particular attention will be paid to a-Si:H, i.e., material with relatively low densities of states.

5.2 Spectroscopic Techniques

5.2.1 Electron Spin Resonance (ESR)

ESR is a very sensitive and selective probe of microscopic paramagnetic entities. A simple homogeneous resonance line is characterized by a Lorentzian shape with an intensity determined by the spin density, a position (usually resonant magnetic field for fixed microwave frequency) determined by the spin Hamiltonian couplings, and a width determined by the dominant lifetime mechanism of the excited spins. An inhomogeneous line consists of a distribution of related homogeneous lines which differ slightly throughout the sample. Examples of fluctuations which modify the spin energies and therefore resonance positions are locally varying stresses or a random distribution of distances to dissimilar spins (electronic or nuclear).

ESR has been a particularly powerful tool in the study of the structural identity of localized states in crystalline materials. For example, from measurements of the strength and symmetry of hyperfine shifts in the resonance due to interactions with ^{29}Si nuclei, the wave function of the unpaired, paramagnetic electron in the charged divacancy of crystalline silicon has been mapped at more than 100 neighboring lattice sites. The symmetry and local chemical configuration of many dopants, impurities and point defects have been unambiguously inferred from ESR measurements [5.7]. By varying E_F through doping or by excitation with subband gap optical illumination, the discrete energy levels of the charge states of such defects have been determined.

Historically, ESR was one of the earliest probes applied to the study of a-Si [5.9]. It vividly demonstrated the presence of a high density of paramagnetic states in undoped material. In sputtered samples [5.8, 9] a single asymmetric absorption was found at a g-value of 2.0055 with spin densities in the range of 10^{19}–10^{20} cm^{-3}. (The g tensor is the dimensionless linear coupling between external magnetic fields and local magnetic moments. In an amorphous material, the random occurrence of site directions leads to an observed isotropic line. A scalar g-value is assigned to the peak of the resonant absorption.) This signature has also been found in a-Si of many other origins; e.g., evaporated and glow discharge deposited samples or crystalline material driven to amorphicity by particle irradiation. The center is therefore intrinsic, i.e., not due to an impurity. An example of the $g = 2.0055$ resonance is shown in Fig. 5.2, along with the other characteristic ESR signals observed in a-Si : H.

In fact, in undoped a-Si, independent of preparation conditions, the *only* ESR line found in equilibrium is the $g = 2.0055$ spectrum. It is thus particularly important to understand the origin of this line as fully as possible. We will show that there is a great deal of evidence from spin resonance that the center is a silicon dangling bond.

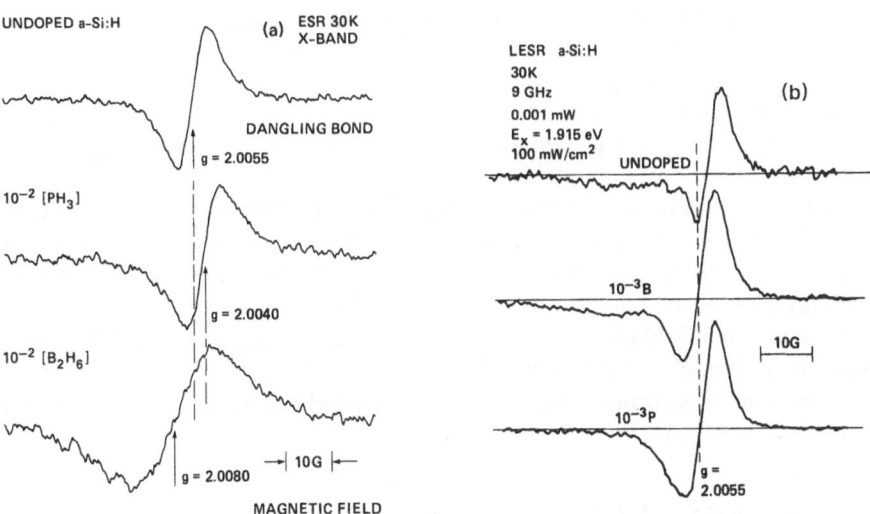

Fig. 5.2. a, b. Characteristic ESR line shapes in a-Si : H. (a) Equilibrium ESR in undoped, *n*-type and *p*-type material; (b) light-induced ESR in undoped and doped material [5.35]. All measurements are taken at 30 K

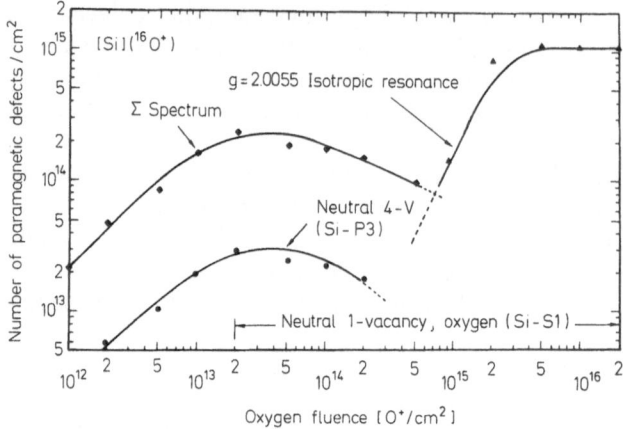

Fig. 5.3. Spin density versus oxygen fluence of bombarded crystalline Si. At low fluence the ESR is characteristic of crystalline Si defects. At high fluence these are replaced by the amorphous Si defect resonance [5.10]

The creation of amorphous silicon by increasingly damaging crystalline silicon is particularly revealing because the ESR studies illustrate that a-Si is qualitatively different from damaged crystalline silicon. When undoped crystalline silicon is bombarded with energetic ions there is a characteristic progression in damage regimes. Fig. 5.3, from a study by *Brower* and *Beezhold* [5.10], is typical. At low implant doses the number density of nonoverlapping small clusters of noninteracting point defects is observed to grow linearly with fluence. At higher doses the spin density saturates as ESR signals of more complex centers are detected. Above a cumulative spin density of $\sim 10^{18}$ cm^{-3}, the amorphous silicon ESR signature begins to dominate and all other lines disappear. The implication is that when a sufficient density of broken bonds exists, the remaining network can collapse to a lower energy state, the gross features of which are independent of history. A further point of interest learned from ESR measurements is the common property of most linear defects induced in crystalline silicon by radiation [5.7]. Jahn-Teller distortions lead to bond reconstruction along the length of the defect with unpaired spins remaining only at the ends. Thus, reconstruction into weak bonds and a relatively small number of dangling-bond-like paramagnetic centers is characteristic of relaxed, disordered silicon.

The spin Hamiltonian for a single unpaired electron spin associated with an anisotropic defect is given by [5.11]

$$H = \beta H_0 g S + \sum_i I_i A_i S . \tag{5.1}$$

Here the first term is the Zeeman energy of an $S = \frac{1}{2}$ spin in an external magnetic field H_0. The second term is the hyperfine interaction between the electron spin and the nuclear spins, ^{29}Si (4.7% natural abundance) and ^{1}H, both with nuclear spin $I = \frac{1}{2}$. The hyperfine interaction causes pairs of satellites, corresponding to $I = \pm \frac{1}{2}$ shifted from the Zeeman energy by amounts depending on the wave-function density at the nuclei. In a-Si hyperfine structure has not yet been identified. Presumably the hyperfine split-off signals are too weak and broad to be observable, most likely because of the low concentration of nuclear spins and because of the great sensitivity of the hyperfine interaction to the fluctuations in local environments.

The g tensor of the unpaired electron is shifted from the free electron g-value of 2.0023 due to a mixing of the electron spin with orbital angular momentum through the spin-orbit coupling [5.11]. To first order in the spin-orbit interaction V_{so},

$$\Delta g_{ij} = -2\sum_n \frac{\langle d|(V_{so})_i|n\rangle\langle n|L_i|d\rangle}{E_n - E_d} , \tag{5.2}$$

where the index d stands for the defect ground state and n varies over all excited states. Because the defect is a deep level and is strongly localized, molecular orbital (LCAO) calculations have been applied successfully to defects in crystalline silicon [5.7]. The same notions can reasonably be expected to explain the spectrum of dangling bonds (and paramagnetic tail states) in a-Si.

Assuming for simplicity that the defect *is* a dangling bond, LCAO calculations provide several pertinent insights. First, because of the axial symmetry of a dangling bond, $m_{II} = 0$ along the bond axis and matrix elements for angular momentum parallel to the bond axis are zero. Thus, $\Delta g_{II} \sim 0$. Furthermore, $\Delta g_\perp$ is approximately axially isotropic. The sign of $\Delta g_\perp$ is calculated and measured to be positive for many dangling-bond-like defects [i.e., defects for which the paramagnetic electron resides predominantly in a (111)-directed orbital localized on a single atom], independent of their position in the band gap. The phosphorus-vacancy complex, boron-vacancy complex and negative divacancy are examples [5.12]. The origin of the sign of $\Delta g_\perp$ is a much stronger spin-orbit matrix element between the unpaired electron wave function and the bonding states with the nearest neighbors sp^3-like (valence band) than with the antibonding states [5.13] (sp-like conduction band states). From (5.2), because $(E_n - E_d)$ is negative for the bonding excited states (and because the sum of products of matrix elements is positive-definite along the principal axes), Δg is positive. In principle then, states lying closest to the valence band should have the greatest positive contribution to the shift from the energy denominator. As can be seen in Fig. 5.2a and as will be discussed below, the localized paramagnetic states in a-Si : H qualitatively agree with this spin-orbit shift.

For an ensemble of axially symmetric sites (all with identical g-tensors) randomly oriented in an amorphous material, all g-values will be observed from g_{II} to $g_\perp$ (a "powder pattern"). Therefore, centers with the greatest $\Delta g_\perp$ will also have the greatest widths. Similar calculated results and experimental findings of positive g-shifts occur for the paramagnetic charge states of weak bonds, e.g., the oxygen-vacancy center in crystalline silicon. Again this might be expected to be applicable to the band-tail states in amorphous silicon.

Turning again to the experimental results for a-Si, it has been found that the spin density N_s associated with the $g = 2.0055$ spectrum varies from $\sim 10^{20}$ to $< 10^{15}$ cm^{-3} depending on deposition conditions [5.14, 15] or post-deposition treatments such as annealing [5.15–18] or particle bombardment [5.19, 20]. For $N_s \leq 10^{18}$ cm^{-3}, the line shape is independent of N_s and temperature [5.19]. From measurements at different magnetic fields H_0, *Voget-Grote* et al. [5.21] have found that the line width is nearly proportional to H_0. This is indicative of a line inhomogeneously broadened by both the random orientation of sites having the same anisotropic g-tensor and random environmental variations in the components of g-tensors. From the first (dominant) term in (5.1) it can be seen that a given distribution in g-values will lead to a field-dependent scaling of the resonance width. Actual values of the intrinsic, homogeneous linewidth of individual sites can be estimated from saturation studies as a function of modulation rate [5.22]. Such measurements indicate a homogeneous width $\sim 1/100$ of the inhomogeneous width. The spin-spin relaxation time T_2 for localized spins in a-Si:H is expected to be very long. From the Anderson formula [5.23] for the dipolar broadening due to a random spatial distribution of identical spins, $T_2^{-1} \sim 1.6 \times 10^{-12} N_s$, which is independent of temperature. Thus, $T_2 \sim 1$ m s for $N_s \leq 10^{15}$ cm^{-3}. However, the coupling between spins is greatly decreased in amorphous solids because the random environmental variations shift the resonant frequencies of neighboring spins. This leads to values of T_2 greater than the spin-lattice relaxation time T_1. The homogeneous width is thus determined by T_1 (except for very high spin densities or possibly at very low temperatures where T_1 becomes long). The spin-lattice relaxation in silicon is mediated by the spin-orbit interaction. Because this is particularly weak, T_1's for defects in silicon are generally quite long [5.24]. The magnitude and temperature dependence of T_1 for the $g = 2.0055$ line in a-Si:H have been found to be very similar to the dangling bond centers in crystalline silicon [5.22, 25]. The $g = 2.0055$ line thus arises from an ensemble of isolated spins which vary from site to site in orientation and environmental details.

Because the dangling bond is a characteristic property of a-Si and plays such an important role (as will be shown below) as the dominant recombination center, a digression will be made here to discuss a more precisely characterizable defect in a related disordered silicon system. *Caplan* et al. [5.26] have used ESR to study clean interfaces between crystalline silicon and amorphous SiO$_2$. They have found that a single intrinsic interfacial defect exists. The beauty of this system is that the crystal substrate causes all defects to be oriented and thus the anisotropic properties of the center can be measured. The results show conclusively that the defects are dangling silicon

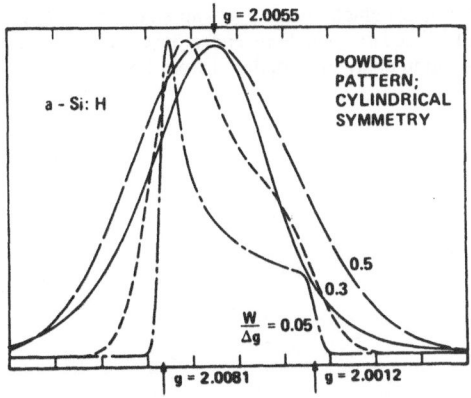

Fig. 5.4. Calculated powder pattern for the dangling bond using g values obtained from the Si/SiO$_2$ interface resonance and including various Gaussian broadenings W. The solid line is the observed shape in a-Si:H

bonds oriented along (111) axes. The g tensor is axially symmetric with $g_{\text{II}} = 2.0012$ and $g_{\perp} = 2.0081$. This system is one of the few in which dangling bonds are topologically allowed in the crystalline material. To compare the dangling bond resonance with the defect resonance in a-Si, we must first average the g tensor over all spatial orientations. The resultant spectrum, a powder pattern, is shown in Fig. 5.4. As discussed above, mechanisms other than g-value anisotropy exist in an amorphous material for inhomogeneously broadening the resonance response. In a-Si two other important contributions are (a) the random variations in g-tensor components arising from random structural variations from site to site and (b) unresolved hyperfine shifts due to nuclear moments on neighboring sites. The dashed curves in Fig. 5.4 give the powder pattern convolved with a Gaussian of differing widths. A more accurate treatment would include the notion that in a-Si, $g_{\perp}$ is expected to be broadened by lattice fluctuations much more than g_{II}. Thus, the line should be less steep at low fields and steeper at high fields. The solid curve in Fig. 5.4 is the defect signature in a-Si. The similarity of the two lines seems to us to be the most convincing piece of evidence that the defect associated with the $g = 2.0055$ line is a dangling bond. Further evidence is given shortly.

It is interesting to note that there are, in fact, many similarities between the disordered Si/SiO$_2$ interface and a-Si:H. Among these are the density and nature of the dominant defect, the existence of band tailing [5.27, 28] and the specific, salutory role of hydrogen passivation [5.28].

The last concept to be discussed before considering doped a-Si is the impact of the effective correlation energies of a group of states on the temperature dependence of the spin density. At $T = 0$ for states with $U_{\text{eff}} = 0$, the sites are doubly occupied and spin-paired below E_F and unoccupied above. At finite temperatures a range of states $\sim kT$ wide are singly occupied and paramagnetic. A temperature-dependent spin density $N_s \sim N(E_F)kT$ is then expected. If U_{eff}, on the other hand, is large ($\gg kT$), then states within U_{eff} below E_F will be paramagnetic independent of temperature. The dangling

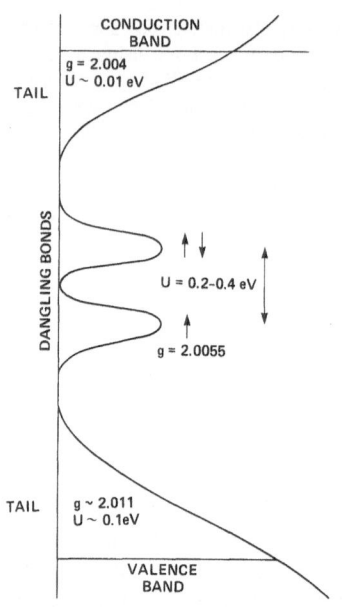

Fig. 5.5. Schematic density of gap states showing the band tail and dangling bond energy levels, g values and correlation energies as deduced from ESR data

bond resonance (for $N_s \le 10^{18}$ cm^{-3}) has been found to be independent of temperature below ~ 400 K [5.29]. We can thus infer that $U_{\text{eff}} > 0.1$ eV. ESR measurements in doped samples, for which E_F can be shifted, are discussed below. These show that several charge states of the dangling bond lie within the gap. Estimated values of U_{eff} of 0.2–0.4 eV are found.

When a-Si:H is doped n or p-type to a level comparable with N_s in undoped material prepared under similar conditions, the defect spin density becomes undetectable and the Fermi energy moves towards the band edges. This behavior is indicative of a density of neutral, spin = ½ defects lying just below the center of the gap. Positive and negative charge states (diamagnetic) must lie within the gap, as indicated in Fig. 5.5 and demonstrated further below. The simplest defect which has spin ½ when neutral is the dangling bond. For dangling-bond-like defects in crystalline silicon, 60–70% of the unpaired electron is localized on one atom [5.7]. The insensitivity of the a-Si defect signature to deposition conditions is further indication of the strong localization of the dangling-bond electron wave function. Until hyperfine measurements are made on ^{29}Si enriched material, we can conclude that the circumstantial evidence is strong that the dominant defect in a-Si is the simple dangling bond. In Sect. 5.2.2 we will review experiments which show that the dangling bond is also the dominant nonradiative recombination center in a-Si. The electrically active defects in a-Si are thus spin-marked and so, fortuitously, electron-spin-resonance spectroscopies are especially useful.

Many groups have observed the charge state transition of the dangling bond in various forms of a-Si:H. It is manifested directly in the reduction of

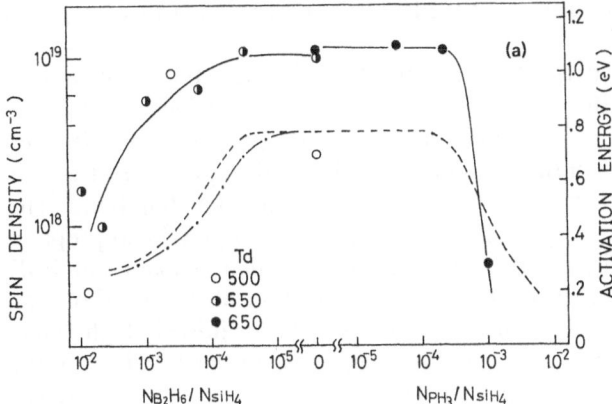

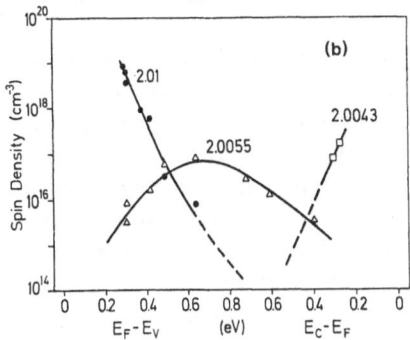

Fig. 5.6. (a) Dangling bond spin density (*solid line*) and conductivity activation energy (*dashed line*) for doped CVD a-Si:H [5.30]; (b) spin density of unpaired dangling bond and band-tail states as a function of conductivity activation energy in glow discharge a-Si:H [5.21]

the equilibrium dangling bond resonance with increased doping as shown in Figs. 5.6 a, b [5.21, 30]. In both cases the spin density in the undoped material is very high. However, when a dopant level approximately equal to the defect density is incorporated into the films, the spin density rapidly drops off and the Fermi level moves towards the band edge. The spin densities drop to half at ~ 0.2 eV above and 0.4 eV below midgap for n and p-type doping, respectively. This implies either (a) a chemical compensation effect during deposition (i.e., removal of dangling bonds by impurity pairing) followed by efficient electrical doping, or (b) initially efficient electrical doping with energy levels for negatively and positively charged (diamagnetic) dangling bond states at approximately $E_c - 0.6$ eV and $E_c - 1.2$ eV, respectively. We present LESR evidence below which strongly implies the latter mechanism. These results indicate that the dangling bond correlation energy is approximately 0.4–0.6 eV. However, the effects of doping are more complex than would appear from Fig. 5.6. As will be discussed in Sect. 5.3.2 b, defects are created concomitantly with doping and therefore, interpretation of the Fermi level variations within a fixed band model overestimates the true correlation energy.

When the doping level is sufficiently high to pull E_F into the conduction or valence band tails, different ESR lines are observed which increase in strength with dopant incorporation. The conduction (valence) band-tail line is narrower (broader) than the dangling bond line and its peak is shifted less (more) from the free electron g-value than that of the dangling bond. The linewidths and positions vary with dopant concentration [5.21] and preparation conditions. Fig. 5.2a shows the dangling bond and band tail ESR spectra. Fig. 5.6b, taken from the work of *Dersch* et al. [5.21], shows an example of the doping dependence of all three lines for initially high defect density material. Other workers have also seen these lines when E_F is in the band tails in glow discharge deposited [5.31, 32] and chemical vapor deposited material [5.30, 33]. That the lines are most probably intrinsic and not specific to the dopant is deduced from the results of light-induced ESR (LESR) discussed below.

Several further aspects of the two, band tail-related ESR lines are indicative of the relative degree of localization of the unpaired electrons giving rise to the resonances. First, *Dersch* et al. [5.21] have measured the variations in linewidth with doping and magnetic field and ascribe the variations to hyperfine interactions with the dopant nuclei. Although at the same doping concentration the hyperfine interaction should be smaller for phosphorus than for boron, they find the opposite to be true. They conclude that the wave function of the valence band-tail states are necessarily much more localized than the conduction band-tail states. Two other experimental features are interpreted to follow from a much weaker effective correlation energy for the conduction band-tail states. They are the much smaller ESR signal and the much larger (positive) temperature derivative of ESR spin density for conduction band-tail states than valence band-tail states at the same dopant incorporation level. Although these could possibly be explained by a higher valence band-tail density of states, the notion of a smaller U_{eff} for conduction band-tail states is consistent with the weaker localization. From the temperature dependence of equilibrium ESR strengths, the correlation energies for the conduction band tail, valence band tail and dangling bond states are estimated to be 0.01, 0.2 and 0.4 eV, respectively. *Hasegawa* et al. [5.30] have observed a similar increase in spin density with temperature in phosphorus-doped CVD a-Si: H. They argue that the increase in singly-occupied tail states arises from a statistical shift in E_F through a peak in the distribution. Because there is no evidence that the character of band-tail states changes qualitatively with energy and because the tails increase monotonically towards the band edges, such an explanation seems unlikely.

The microscopic nature of the tail states has not been established yet. Whether they can be thought of simply as the bonding and antibonding states of weak bond configurations, or whether the nature of the disorder is more complex, has not been resolved. Rough LCAO g-value calculations based on the picture of tail states as weak bonds are consistent with the ESR spectra in a-Si : H [5.34]. The experimentally-determined short-spin lattice relaxation times at temperatures above 100 K are also consistent with states more strongly coupled to mobile carriers than the deeper dangling bond states [5.21].

The discussion above has dealt only with ESR of spins in equilibrium. A very useful extension is the study of nonequilibrum states, either the steady-state or transient response. Illuminating thin a-Si films with band gap radiation excites electrons and holes which are paramagnetic until recombination. In doped material, for which the dangling bonds are charged (diamagnetic), a nonequilibrium response at $g = 2.0055$ can also be expected if recombination occurs via a Shockley-Read-like mechanism, i.e., by initial trapping of a minority carrier at the charged defect followed by the recombination event. Fig. 5.2 b shows the LESR spectrum in undoped a-Si : H with unmeasurably small equilibrium ESR response [5.35]. The signal can be shown to be the superposition of two lines by their different microwave saturation behaviors. The two lines in undoped material are very nearly the same as the equilibrium spectra found in heavily doped material. The lines are thus seen to be intrinsic to disordered a-Si : H and not directly related to impurity states. Fig. 5.2 b also contains LESR spectra for p-type and n-type material. The p-type sample can be deconvolved into the band-tail hole line and the dangling bond line. The n-type sample is most likely the band-tail electron and dangling bond lines. However, the proximity of spectral peaks ($\sim 2G$) relative to linewidths ($\sim 7G$) makes positive identification uncertain. It should be noted that the g-value of the LESR excited hole line in heavily boron doped samples lies at higher g-values than the equilibrium signal. More work is needed to understand this shift. It will be shown below that the correlated behavior of LESR and luminescence in undoped and doped samples substantiate the assignments of the ESR signatures.

At low temperatures ($T < 50$ K) where carrier diffusion is negligible, recombination of the light-induced states could occur via a direct, bimolecu-

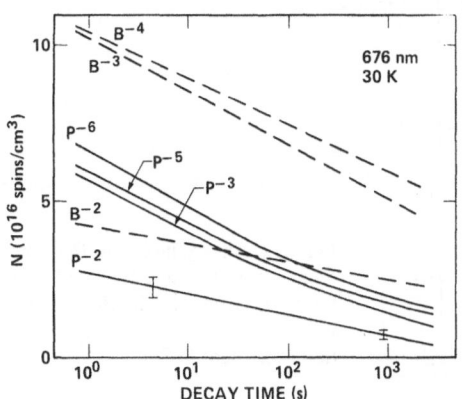

Fig. 5.7. Decay of dangling bond LESR in various doped a-Si : H samples at 30 K [5.36]

lar tunneling process or via monomolecular, multiple tunneling transitions in a relatively large density of states at or below the quasi Fermi levels. Fig. 5.7 shows recent results for n-and p-type samples at 30 K [5.36]. The very long lifetimes are in agreement with the direct tunneling picture and indicate densities of defect and tail states near the equilibrium values of $E_F \leq 10^{17}$ cm^{-3} eV^{-1}. Further discussion of the LESR recombination channels will be presented in the next section.

5.2.2 Optical Spectroscopy

a) Luminescence

Luminescence is one of the most widely used techniques for studying localized states in semiconductors. Generally one can obtain information about the energy of the states from the luminescence spectrum and about the recombination mechanisms from studies of the emission kinetics. As we shall see, in a-Si : H the second aspect has proved to be more informative than the first.

At low temperatures a-Si : H usually has a single broad featureless luminescence peak. The peak energy is reported in the range 1.25–1.45 eV depending on the deposition conditions and the band gap of the material [5.37]. The width of the peak is typically 0.3 eV, and an example of the spectrum is shown in Fig. 5.8. Since in luminescence the energy is usually emitted as a combination of phonons and photons, the relation between the peak energy and the density of states is not immediately obvious. Distinct

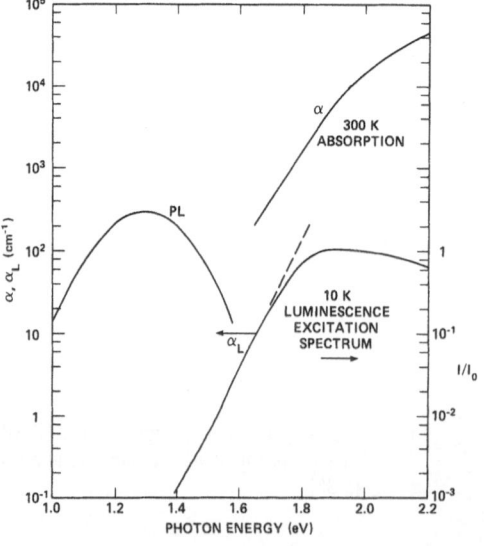

Fig. 5.8. An example of the low-temperature luminescence and excitation spectrum of a-Si : H compared with the room temperature absorption coefficient [5.38]

phonon sidebands have not been observed in a-Si : H and so the estimate of the phonon contribution has relied on a comparison of the luminescence and excitation spectrum [5.38], an example of which is also shown in Fig. 5.8. Detailed balance arguments have concluded that the absorption strength is too small by several orders of magnitude at the luminescence energy to interpret the luminescence in terms of zero-phonon transitions. Instead, a Stokes shift of 0.4–0.5 eV is deduced so that the luminescence position and line width is dominated by the phonon sidebands. The absence of observable structure in the spectrum makes it impossible to accurately deconvolve the electronic and phonon energies. However, the zero-phonon energies are estimated to be in a band near 1.6–1.8 eV.

The 1.4 eV luminescence has been almost universally attributed to a transition between band-tail states, for various reasons. Firstly, the zero-phonon energies are in the range expected for the band-tail states. Secondly, it is observed that the luminescence intensity is a maximum when the defect density is lowest [5.39], indicating that the recombination is in some way "intrinsic". Thirdly, the luminescence has been directly associated with the band-tail states observed in LESR [5.40]. This evidence is given in Fig. 5.9 which shows the temperature dependence of the LESR spin density for light totally absorbed in the sample. Also plotted is the temperature dependence of the lifetime of the 1.4 eV band edge luminescence measured under the same conditions. The LESR spin density in steady state is given by

$$N_s = G\tau , \tag{5.3}$$

where G is the optical generation rate and τ is an average lifetime. Thus, if the luminescence and LESR states are identical, then the temperature dependence of the LESR N_s should be the same as that of the luminescence lifetime. The experimental agreement in Fig. 5.9 then supports the assignment of the origin of the 1.4 eV luminescence.

Time-resolved luminescence decay measurements have been even more useful than cw measurements in mapping the distribution and character of the tail states. In high efficiency samples at low temperatures the radiative recombination rate, as shown in Fig. 5.10, varies over about six orders of magnitude [5.41]. This is the signature of a radiative tunneling process between two spatially separated localized states. The time-resolved shift of the luminescence band down in energy by ~ 0.15 eV is argued to be a direct manifestation of the width of the conduction band tail.

Figure 5.11 is a schematic representation of the luminescence process. Absorption of band gap radiation creates electron-hole pairs which thermalize in the order of 10^{-12} s to the band tails. At low temperatures, diffusion is negligible after thermalization and so the transitions from the band tails occur by tunneling. From the evidence for a Stokes shift and the much smaller drift mobility of holes than electrons, it is surmised that band-tail holes become self-trapped with a binding energy ~ 0.3–0.4 eV [5.41]. This

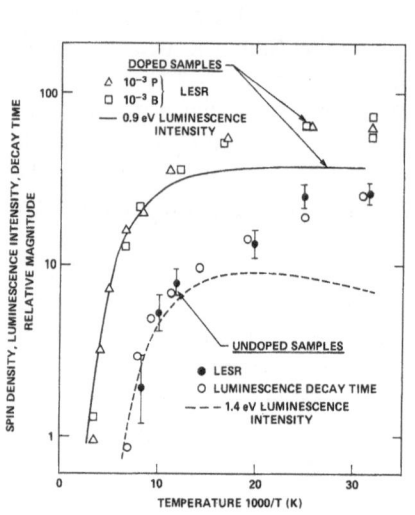

Fig. 5.9. Temperature dependence of luminescence intensity, decay times and LESR spin densities in doped and undoped a-Si:H [5.35]

Fig. 5.10. Examples of the luminescence decay after a short excitation pulse at various temperatures. The lower curves show the distribution of decay times corresponding to the data shown [5.42]

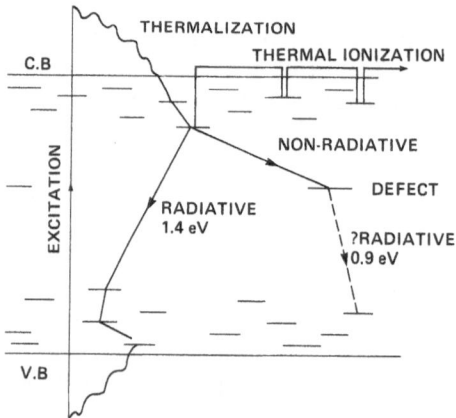

Fig. 5.11. Schematic diagram of the luminescence processes shows the radiative and nonradiative processes, as described in the text

occurs quickly, presumably at distorted sites in the silicon network corresponding to particularly large disorder potential fluctuations. Band-tail electrons, on the other hand, apparently have a much smaller electron-phonon coupling and greater spatial extent, in agreement with the ESR and mobility data. Radiative recombination then occurs by tunneling of band-tail electrons and holes. The lifetime of pairs with separation R is [5.42]

$$\tau(R) = \tau_0 e^{2R/R_0} \tag{5.4}$$

with $\tau_0 \sim 10^{-8}$ s. An exponentially broad distribution in lifetimes thus occurs with a median value found to be 10^{-3} s. As the excitation intensity is varied, the recombination is observed to change from a geminate to a nongeminate process (5.41). From the excited carrier density at which this occurs, one can obtain the distance scale of the recombination. In this way it is estimated that the effective Bohr radius R_0 is ~ 10 Å and so the separation of an electron-hole pair with a recombination lifetime of 10^{-3} s is about 50 Å.

The presence of dangling-bond defects has a major influence on the luminescence. Figure 5.12 shows that when the ESR spin density N_s exceeds about 10^{17} cm^{-3}, the luminescence is rapidly quenched [5.31]. The data imply that dangling bonds act as nonradiative centers and again the behavior is interpreted as electron tunneling, in this case nonradiatively to the nearest defect. The rate $\nu(R_D)$ to tunnel to a defect at distance R_D is

$$\nu(R_D) = \nu_0 e^{-2R_D/R_0} , \tag{5.5}$$

where the prefactor ν_0 is now a phonon frequency $\sim 10^{12}$ s^{-1}. Clearly then if $\nu(R_D) \tau(R) > 1$, the band-edge radiative recombination will be quenched. This occurs when R_D is less than R_c where

$$R_c = \tfrac{1}{2} R_0 \ln(\nu_0 \tau) . \tag{5.6}$$

Assuming a random distribution of nonradiative centers with density N_s, and an average value of the radiative lifetime τ, the luminescence efficiency y_L is given by [5.43]

$$y_L = \exp(-4\pi R_c^3 N_s/3) . \tag{5.7}$$

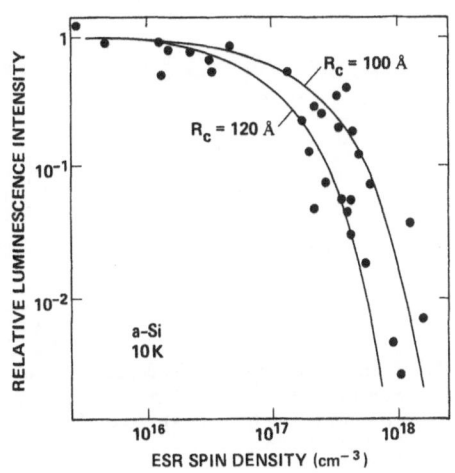

Fig. 5.12. Plot of luminescence intensity versus spin density for a variety of samples prepared under different deposition conditions. The solid line is the fit to the nonradiative tunneling model [5.39]

The solid curves in Fig. 5.12 are plots of (5.7) for R_c = 100 and 120 Å. Assuming a mean value of $\tau(R) = 10^{-3}$ s results in a value of R_0 of 10–12 Å which is in good agreement with the estimate made above. Thus, a simple random distribution of isolated dangling bonds is sufficient to explain the nonradiative recombination channel competing with the band edge luminescence.

At temperatures above about 50 K, the luminescence intensity decreases, being reduced by a factor $\sim 10^4$ at room temperature [5.44]. The thermal quenching is explained by the increased mobility of carriers, which can diffuse through the material to recombination centers. There is evidence that the dominant recombination centers are again dangling bonds [5.45]. However, the thermally activated diffusion process primarily yields information about the band-tail states. As shown in Fig. 5.13, the temperature dependence obeys a relation first observed in chalcogenide glasses [5.46]:

$$y_0/y(T) - 1 = e^{T/T_0} . \tag{5.8}$$

This relation can be understood in terms of the thermal release of carriers from an exponential distribution of band-tail states. As the temperature is raised, carriers can be excited from deeper states and a calculation of the fraction remaining which recombine radiatively leads directly to (5.8). As we shall see in Sect. 5.3.1, this model is similar to that developed to account for the dispersive drift mobility, and an almost identical distribution is deduced

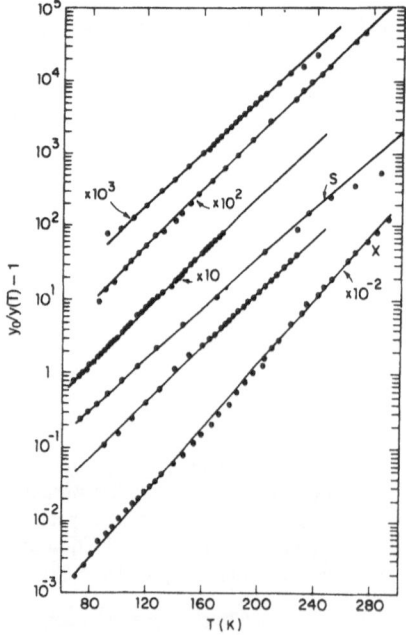

Fig. 5.13. Temperature dependence of luminescence of samples from different laboratories plotted to show the fit to (5.8) [5.44]

in both cases [5.47]. This model also explains the large shift of the luminescence peak with temperature shown in Fig. 5.14. The explanation is again
that the weakly-bound carriers are thermally excited and recombine nonradiatively so that the luminescence is increasingly dominated by deep states.
As before, the magnitude of the shift is consistent with the same exponential
tail. Thus, the luminescence and transport results can be understood both
qualitatively and quantitatively within a common band-tail model.

In many a-Si:H samples, a lower energy luminescence transition at 0.9
eV is found [5.48]. Again, this is a broad featureless peak with a width of
~0.35 eV. The 0.9 eV band has been observed in defective undoped material
[5.48], radiation damaged material [5.49] and material doped n- or p-type
when the dangling bond density is ~10^{18} cm^{-3} [5.50]. Because of its association with material containing a large dangling-bond density, it was suggested
that the recombination involved this defect. Further evidence for this
hypothesis is shown in Fig. 5.9 in which the LESR dangling bond spin density
and the luminescence intensity are plotted versus temperature. The 0.9 eV
luminescence is too weak to obtain good lifetime measurements. However,
because the luminescence intensity is given by

$$y_L = P_r/(P_r + P_{nr}) = P_r \tau , \qquad (5.9)$$

where P_r and P_{nr} are the radiative and nonradiative recombination rates, and
because usually P_r is not strongly temperature dependent, the temperature
dependence of y_L should approximate that of τ. The good agreement
between the LESR and luminescence indicates that the 0.9 eV luminescence
and the LESR in doped samples involve the same excited states. As can be
seen in Figs. 5.2b, the LESR spectra in doped a-Si:H consist of the band-tail
hole line and/or the dangling-bond line. The suggestion is, therefore, that the
origin of 0.9 eV luminescence is the tunneling of a conduction band-tail

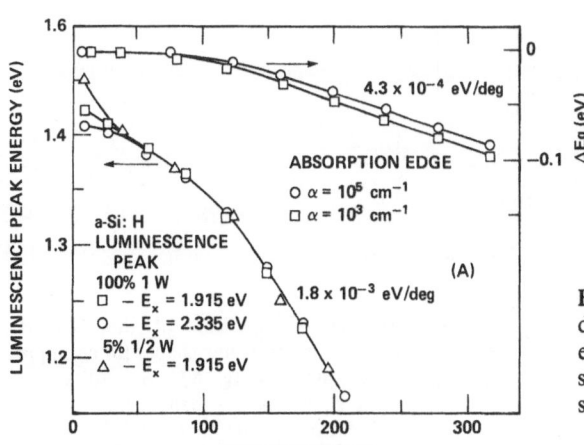

Fig. 5.14. Temperature dependence of the luminescence peak
energy compared to the much
smaller shift of the optical absorption edge [5.42]

electron into a neutral (paramagnetic) dangling bond followed by radiative recombination with a self-trapped band-tail hole. Consideration of the estimated Stokes shift for this transition then leads to the estimate that the dangling bond level is about 0.5 eV below the conduction band. However, further data are needed to confirm this result.

As we have seen, the luminescence and ESR data are closely related, because both radiative and nonradiative recombination involve paramagnetic states. For the same reason, the recombination is found to be dependent on the spin orientation of the centers. This effect is most easily observed by the technique of optically detected magnetic resonance (ODMR) in which the change of luminescence intensity is observed as the paramagnetic states are brought into microwave resonance [5.25, 51, 52]. These experiments can be a very precise determination of the recombination mechanisms, although the results in a-Si : H remain somewhat controversial.

Figure 5.15 shows ODMR results in samples of different dangling bond density [5.25]. The ODMR spectrum is essentially that of the dangling bond ESR, and the magnitude of the spin dependence increases with N_s. In addition, time-resolved measurements using microwave pulses demonstrate that the spin dependence is from a nonradiative process. These results therefore reaffirm that the dangling bonds are indeed nonradiative centers. The fact

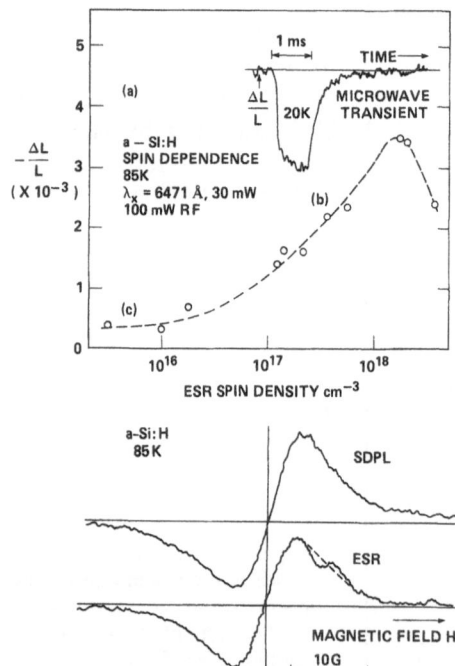

Fig. 5.15. Comparison of the ODMR (SDPL) and ESR line shapes for a sample with spin density $N_s = 3 \times 10^{17}$ cm^{-3}. Also shown is the dependence of the ODMR magnitude on N_s and the microwave transient waveform [5.25]

that the ODMR intensity changes with defect density also provides a useful probe of the defect states.

Depinna and *Cavenett* [5.51] have reported the spectral dependence of ODMR, in which they show that two resonances are present but centered at different parts of the luminescence band. They conclude that two unresolved luminescence bands are present in the spectrum. There have, in fact, been other reports of structure in the luminescence spectrum, but none of these has been confirmed further. The ODMR results are complicated by the observation that the two resonances occur at distinctly different recombination times so that the energy shift could be explained by the known time-resolved shift of the luminescence. The question of a second luminescence transition clearly has important implications for the density of states in a-Si:H, and so further work in this area is needed.

b) Optical Absorption

Because the optical band gap energy is one of the most basic parameters of a semiconductor, there have been many measurements of optical absorption in amorphous semiconductors. It is convenient to divide the absorption spectrum up into three regions. First, there is the region above the fundamental gap, corresponding to absorption coefficients α above about 10^3 cm^{-1}. In this region, absorption occurs by transitions between extended states and so no information can be obtained about localized states. It is common to find this absorption data plotted as [5.53]

$$(\alpha\hbar\omega)^{1/2} = \text{const}\,(E_0 - \hbar\omega) \tag{5.10}$$

to obtain the band gap E_0. This equation is justified if the bands are both parabolic and the optical matrix elements are independent of energy [5.54], conditions that are plausible but by no means certain.

The second part of the absorption spectrum is the region around the fundamental edge with absorption coefficients from 10 to 10^3 cm^{-1}. Below this is the third region of extrinsic absorption. Both of these regions potentially involve localized states and so are of interest to the present discussion. An immediate problem with measurements at or below the absorption edge is that it is inconvenient to make samples of a-Si:H of thickness d greater than 10 μm. Direct absorption measurements rapidly become inaccurate when $\alpha d \ll 1$ and so substantial effort has gone into finding better measurement techniques.

Photoacoustic spectroscopy (PAS) and the related photothermal deflection spectroscopy (PDS) methods seem to be the most promising techniques [5.55, 56]. These measure the absorbed power P, rather than transmitted power as in the conventional optical measurements. When αd is small,

$$P \simeq P_0 \alpha d , \tag{5.11}$$

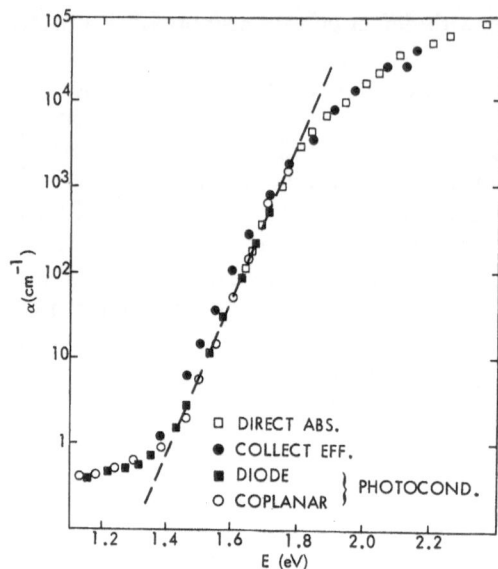

Fig. 5.16. The optical absorption edge measured by a variety of different techniques [5.57]

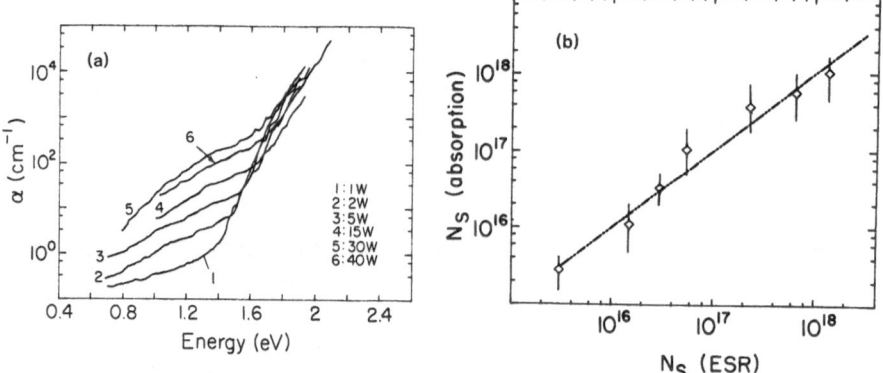

Fig. 5.17. (a) The extrinsic absorption tail as measured by photothermal deflection spectroscopy on various samples deposited at different powers as indicated; (b) plot of the absorption tail strength versus spin density for the same samples. The absorption density of defects is obtained using a unity oscillator strength [5.55]

where P_0 is the incident illumination power. The measurement gives αd directly and the lower limit on αd depends on the detection sensitivity. The photoacoustic method typically detects the pressure increase when a volume of gas in contact with the sample is heated by the light absorption, but other methods are also used. The PDS method uses the change in refractive index of a fluid in contact with the sample to deflect a laser beam. This technique has a sensitivity limit for a-Si:H of $\alpha d \sim 10^{-5}$ and so can conveniently measure α down to 10^{-1} cm^{-1}.

A similar wide range of α can be obtained, in principle, from photoconductivity. The photoresponse, again for small αd, is given by

$$\sigma_{ph} = G\eta\mu\tau e\,\alpha d \ , \tag{5.12}$$

where G is the generation rate, η the generation efficiency, τ the carrier lifetime and μ the mobility. Again, α can be obtained directly, but only if $\eta\mu\tau$ is constant or known. Changes in $\mu\tau$ can be corrected for by measuring the light intensity dependence of σ_{pc} [5.57] or by making the measurements at constant photocurrent (i.e., varying G). Alternatively, measurement of the primary photocurrent alone avoids the uncertainty in the $\mu\tau$ product [5.58]. It is not possible to correct for changes in η without measuring this quantity directly, which has not been done in many cases.

Some examples of the optical absorption obtained by these different techniques are shown in Figs. 5.16, 5.17. There is widespread agreement that the absorption edge is exponential from $\alpha \sim 10$ to 10^3 cm^{-3}. This is the well-known Urbach edge which is a universal feature of all amorphous semiconductors. The slope of this edge is ~ 0.07 eV, but with substantial variation depending on the deposition conditions of the a-Si : H films.

The origin of the Urbach edge has not been satisfactorily resolved and remains controversial. The optical absorption at the fundamental edge is given by

$$a(\hbar\omega) = \int N_v(E)N_c(E + \hbar\omega)M(E, E + \hbar\omega)dE \tag{5.13}$$

and contains information about the density of states at the valence band N_v and conduction band N_c, and the matrix element M of the transition. It is interesting to note that the initial studies of the Urbach edge in amorphous semiconductors (mostly on chalcogenide glasses) dismissed the possibility that α was given simply by the convolution of N_c and N_v and other interpretations were sought [5.53]. Recent measurements on a-Si:H have tended to take exactly the opposite view. For example, *Crandall* [5.58] interpreted the slope of the absorption edge as a direct measurement of N_v by assuming that N_c has a sharper edge.

It is therefore worth reviewing the other suggested mechanisms. One of these is an electric-field broadened exciton transition proposed by *Dow* and *Redfield* [5.59]. The internal field was assumed to arise either from charged defects or from frozen-in phonons. The second mechanism is that of a strong electron-phonon coupling [5.60]. The simplest model of this type gives a Gaussian absorption edge, although a modification gives an exponential and there are alkali-halides in which an exponential edge attributed to strong coupling is observed. This model clearly relates to the discussion of a Stokes shift in the luminescence. It is important to recognize that if there is a Stokes shift in the band edge luminescence, then phonon participation in the absorption is inevitable. We consider that the evidence for a Stokes shift is strong and so it unlikely to be valid to interpret the Urbach edge simply as a joint density of states.

Below the Urbach edge there is a region of extrinsic absorption observed as a broad shoulder extending down to about 0.5 eV. The origin of this absorption has been greatly clarified by recent PDS measurements [5.55]. Figure 5.17 a shows examples of the absorption for samples deposited under different deposition conditions. Specifically the rf power is changed, which is known to change the ESR spin density N_s. Figure 5.17 b shows that in fact there is a direct proportionality between the integrated absorption strength and N_s. *Jackson* and *Amer* [5.55] therefore conclude that the extrinsic absorption is due to transitions involving dangling bond defects. They also show that an oscillator strength of unity gives the correct magnitude of the absorption. These measurements have been extended to doped and compensated a-Si:H and provide a direct measurement of the defect density in those samples as described in Sect. 5.3.3.

Jackson et al. [5.61 a] have also measured the photoconductivity spectrum in the same samples and find that it follows the absorption fairly closely down to ~1 eV but drops off faster below this energy. They conclude that the $\mu\tau$ product is approximately the same for the extrinsic absorption as for the band-edge transitions. Based on the usual assumption that the photocurrent is carried by electrons, this would identify the final state of the extrinsic absorption as the conduction band. Thus, the absorption is deduced to be from singly occupied dangling bonds about 1 eV below the conduction band. From comparisons of phosphorus doped material and undoped material, they conclude that the dangling-bond electron correlation energy is 0.25–0.45 eV [5.61 b]. These values are consistent with DLTS and ESR data.

c) Transient Induced Absorption

Further information can be obtained from induced absorption measurements. In this experiment, electron-hole pairs are created by strong above-gap illumination and the subsequent change in the sub-gap absorption is measured. When these experiments are performed on a picosecond time scale, information about the thermalization of carriers at or above the mobility edge is obtained [5.62]. At longer times, carriers have relaxed into localized states, which are therefore probed by the experiment [5.63]. Examples of the induced absorption spectra are shown in Fig. 5.18. The data are plotted on the assumption of a transition from a sharp level to a parabolic band and a threshold energy is obtained which increases from 0.4 eV at 80 K to 0.6 eV at 275 K. Similar data have been reported by *Olivier* et al. [5.64].

The decay of the induced absorption is shown in Fig. 5.19. The long times (~10^{-3} s) confirm that localized states are involved and *Vardeny* et al. [5.65] interpret the decay in terms of bimolecular recombination and dispersive diffusion. The detailed interpretation of the spectrum remains rather speculative. *O'Connor* and *Tauc* [5.63] suggest that the transition is from

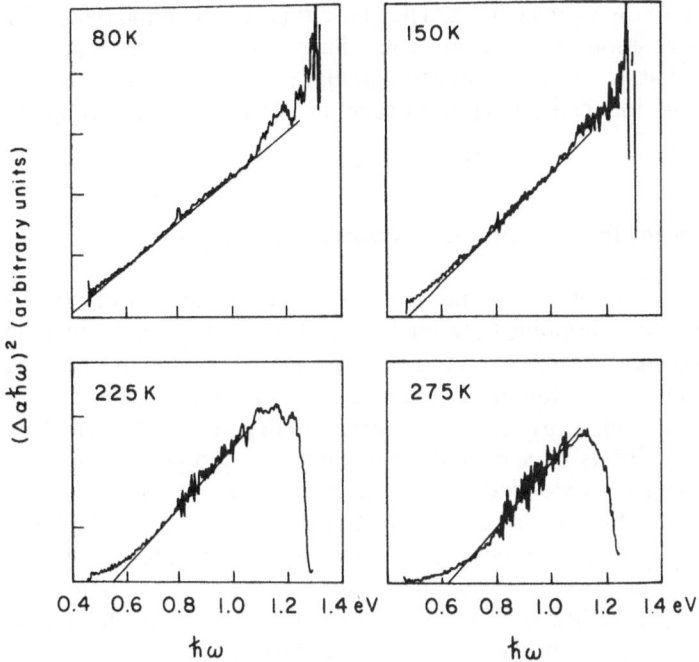

Fig. 5.18. Photoinduced optical absorption spectra at various temperatures plotted as $(\alpha\hbar\omega)^2$ versus $\hbar\omega$ [5.63]

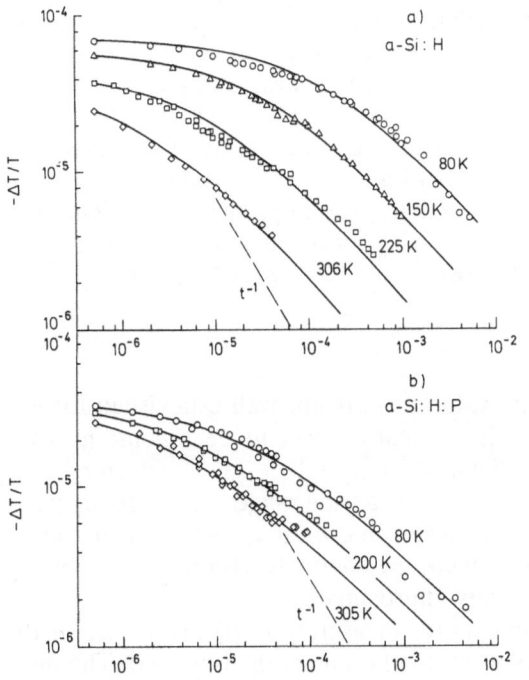

Fig. 5.19. The decay of optically-induced absorption in doped and undoped a-Si:H [5.65]

band-tail holes into the valence band. They base this model on the belief that the equivalent transition for electrons would have a much smaller threshold energy. *Olivier* et al. [5.64] also deduced a similar mechanism. However, the possibility that the induced absorption occurs at defects should perhaps be considered.

5.2.3 Techniques Involving Transport Measurements

Transport in an amorphous semiconductor can occur by the drift of carriers in extended states or by hopping between localized states. DC conductivity in extended states is characterized by a relatively well-defined activation energy which depends on the position of the Fermi energy E_F. Hopping conduction at E_F has a distinguishing $\exp(-B/T^{1/4})$ temperature dependence [5.66]. Hopping far from E_F exhibits an activated conductivity, and in general it is not easy to distinguish from extended state conduction.

Hopping at E_F is observed in unhydrogenated a-Si for which the ESR spin density is of order 10^{19} cm^{-3}, indicative of a large density of states in the gap. The magnitude of the hopping conductivity can be related to the density of states and in one study, $N(E_F) \simeq 3 \times 10^{19}$ cm^{-3} eV^{-1} was obtained, roughly consistent with the spin density [5.67]. No information about the energy distribution is obtained except that it must be approximately constant over a few kT for the calculation of the conductivity to be valid.

In hydrogenated a-Si, the defect density is reduced so that hopping at E_F is no longer observed. The conductivity is activated and in undoped material it is widely believed that transport is by electrons in extended states at the mobility edge. However, transport measurements can still give information about localized states by making use of various processes. The first is the trapping and release of carriers in localized states. This effect cannot be directly measured in the dc conductivity but can be observed in the transport of excess carriers. Thus, photoconductivity, and in particular transient measurements, provides a means of exploring localized states. The release of carriers from a trap occurs with rate

$$p = \nu_0 e^{-E/kT} ,$$
(5.14)

where E is the trap depth. The value of ν_0 is not well established for any amorphous semiconductor and it is usually taken to be a constant 10^{12}–10^{13} s^{-1} based on theoretical estimates. In crystalline semiconductors, ν_0 is known to vary from case to case over quite a wide range and is also temperature dependent. It should be recognized that until ν_0 and its temperature dependence are accurately determined, the energy levels obtained from any experiment using (5.14) will be very approximate.

The second class of experiments for measuring $N(E)$ makes use of the conductivity change that occurs when the Fermi energy is moved. The most direct method is through the field effect in which E_F is moved by the application of an external electric field through an insulating layer. A different group of techniques uses the band bending at a Schottky barrier. Deep level transient spectroscopy (DLTS) [5.68] which uses this structure has proved to be a very powerful method for the study of localized states in crystalline semiconductors. Tunneling experiments are the third type of transport measurement used to obtain information about localized states in a-Si:H.

Transport in a-Si:H is covered in detail in another chapter, as is the field effect technique. In the remainder of this section we concentrate on a review of the information obtained from each of these techniques about the energy distribution of localized states and a comparison of the different results.

a) Transient Photoconductivity

The best known technique of transient photoconductivity is the time-of-flight method of measuring the drift mobility [5.69]. A pulse of carriers is excited close to one electrode of a sample and the current resulting in the transit of the carriers to the other electrode is measured. Under the simplest conditions the measurement gives directly the time taken for the carriers to transit, and from this the mobility is obtained. However, various other conditions are possible. For example, if deep trapping occurs (in the sense that the release time is long compared to the experimental time), then the current decreases exponentially with time and has a specific thickness dependence. However, the most common condition in amorphous semiconductors is dispersion of the charge [5.69]. This occurs when the transit times of individual carriers have a very broad time distribution. This situation has been studied in great detail and is characterized by a current with a power law time dependence

$$I = I_0 t^{-(1-a)} \qquad t < t_T$$
$$I = I_1 t^{-(1+a)} \qquad t > t_T . \tag{5.15}$$

t_T is a characteristic transit time and is obtained from a log-log plot of current pulse, an example of which is shown in Fig. 5.20. a is a parameter describing the dispersion.

The details of the dispersive transport can give information about the localized state distribution and it is this aspect of the experiment that is of interest to us. The original analysis of dispersion used a model of hopping between localized states in which the dispersion came from the wide distribution of hopping times. Subsequently, it was found that dispersion also occurs for transport in extended states with trapping and release from shallow levels. *Pollak* [5.69 c] analyzed the dispersive transport for trap limited transport assuming a continuous distribution of localized states. He showed that the dispersion parameter α was related to $N(E)$ by

$$\alpha = kT \left. \frac{d \ln N(E)}{dE} \right|_{E = kT \ln(v_0 t)} . \tag{5.16}$$

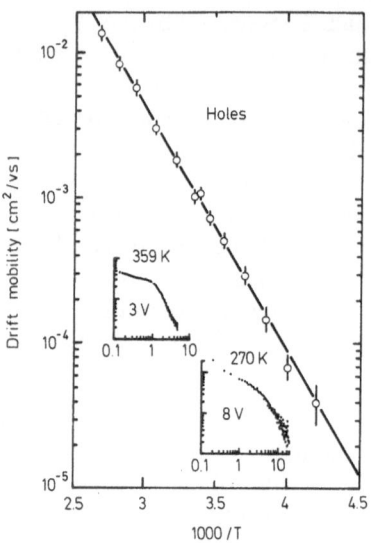

Fig. 5.20. An example of the photocurrent transient data for hole transport in a time-of-flight experiment showing pulse shapes typical for dispersive transport and the temperature dependence of the drift mobility [5.70]

This relation in principle provides a means of obtaining $N(E)$ from the drift mobility measurements. Recently, *Tiedje* et al. [5.70] have made detailed measurements on a-Si:H for just this reason. They point out that if there is an exponential distribution of states

$$N(E) = N_0 e^{-E/kT_c}, \tag{5.17}$$

then α is given by T/T_c, which also follows directly from (5.16). Data for electrons and holes in a-Si:H are shown in Fig. 5.21. In both cases α does increase approximately linearly with T, which leads Tiedje et al. to deduce that the band tails are approximately exponential. From the slope of $\alpha(T)$, the characteristic energy kT_c of the band tails is found to be 0.027 and 0.043

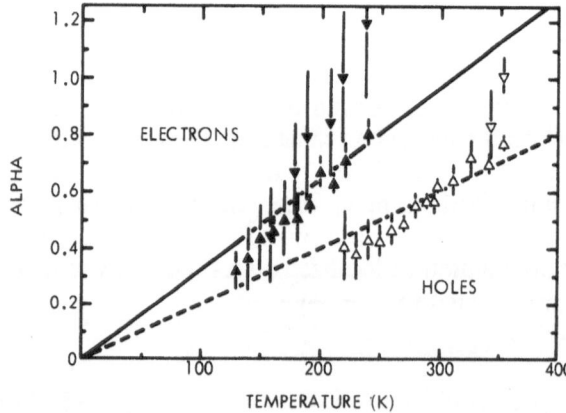

Fig. 5.21. Temperature dependence of the dispersion parameter α for electrons and holes in a-Si:H [5.70]

eV for the conduction and valence band tails, respectively. These values also give a good fit to the temperature dependence of the mobilities of both electrons and holes (Fig. 5.20), with reasonable values of the extended state mobility and v_0.

One concern with this method of obtaining $N(E)$ is that α is not very sensitive to the distribution of states. The data in Fig. 5.21 show that α is not accurately linear in T. From (5.15) it is seen that a Gaussian energy distribution in the band tail results in $\alpha \sim T^2 \ln v_0 t$. Inspection of Fig. 5.21 shows that this expression gives almost as good a fit to the data as exponential distribution does. This result ephasizes that the method requires very accurate values of α to obtain $N(E)$.

In these drift mobility measurements, states of binding energy $kT \ln(v_0 t)$ are investigated at time t. For typical experimental conditions, this energy is 0.2–0.3 eV for electrons and 0.4–0.5 eV for holes (Fig. 5.20). These energies are expected to be in the band tailing due to disorder, and indeed this is the interpretation given to the results. It should be noted that even though the drift mobility changes with temperature, the range of energies investigated does not. To increase the energy range requires extending the time scale of the experiment.

One means of achieving an increased time-scale is to measure the transient photoconductivity in a gap cell [5.71]. This configuration increases the path length from a few microns to a millimeter or more. It is impractical to attempt an observation of the transit of a carrier packet because of the low capacitance of the structure, and instead the gap is illuminated uniformly. The photoconductivity is given by

$$\sigma_{\text{ph}} = ne\mu(t) , \tag{5.18}$$

where n is the number of generated carriers and $\mu(t)$ is the effective mobility, which for dispersive transport is given by

$$\mu = \mu_0 t^{-(1-\alpha)} . \tag{5.19}$$

Provided that the recombination time is sufficiently long so that n is constant after the excitation pulse, then $\mu(t)$ and consequently α can be obtained from the measurement. Figure 5.22 shows examples of this experiment in which the photoconductivity is observed over many decades of time, up to $\sim 10^{-2}$ s.

So far, only a limited number of measurements of this type have been reported on a-Si:H. The results show that α tends to be constant over a wide range of times but varies with doping and between samples deposited differently. The constant α implies that the functional form of $N(E)$ extends over quite a wide energy range. For example, there is no sign of any bands of deep states.

This technique is a useful addition to the drift mobility method but has a number of disadvantages. It is not immediately obvious from the data

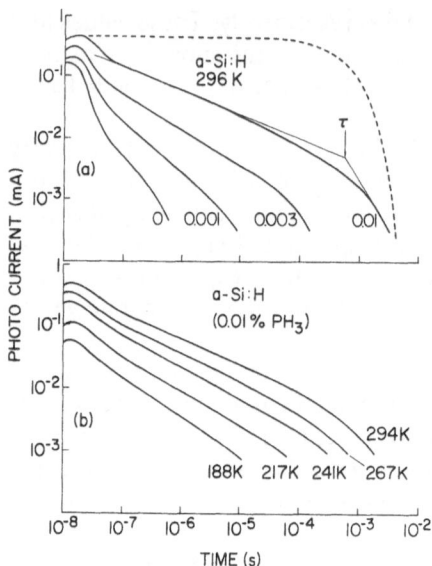

Fig. 5.22. a,b. Photocurrent transients in a gap-cell geometry showing a very long response time. Data in (a) are for different phosphorus doping as shown [5.71]

whether the time dependence of σ_{ph} is due to $\mu(t)$ or to recombination, and this must be carefully checked. There is also no clear distinction between the transport of electrons and holes which must be determined separately. Finally, the transport in the gap cell configuration is close to the surface which raises questions of surface states and band bending; these are also discussed in later sections.

Photoconductivity is the fourth technique we have discussed which measures the transient response of photo-induced carriers. The other three are induced ESR, induced absorption and luminescence. In three of these, the observed signal is proportional to the number of n of excited carriers. The exception is luminescence in which the intensity is proportional to dn/dt. A detailed comparison of the different transient responses would be very helpful in understanding the photo response of a-Si:H.

b) The Field Effect

The field effect is the first of the techniques that use band bending and is probably the most widely tried method for determining the density of gap states in a-Si:H. It is described in more detail in another chapter. The technique is to measure the planar conductance of a sample as a function of an electric field applied to the surface. A common method is to use the sample substrate as the gate insulator across which the field is applied. The observed change in conductance depends on how far the Fermi energy is moved by the field, which in turn depends on the density of states. $N(E)$ is obtained by solving Poisson's equation and several numerical techniques have been reported to achieve this.

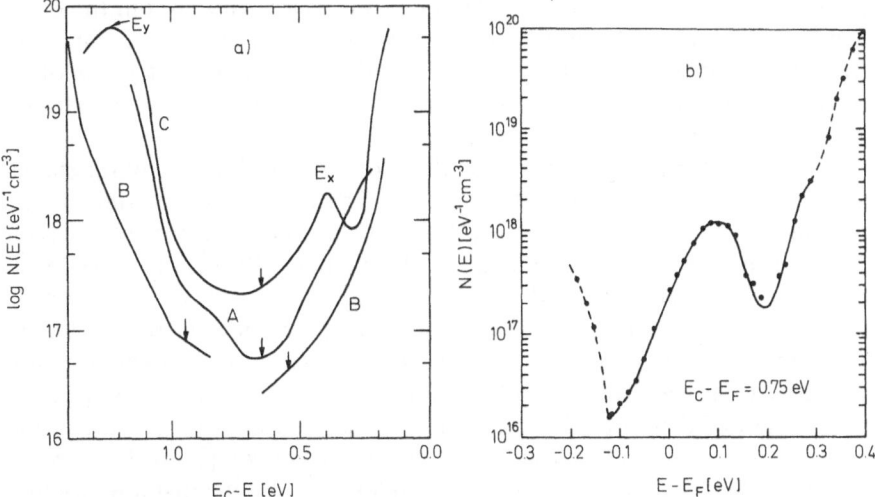

Fig. 5.23. (a) Density-of-states curves for a-Si : H. Curves *A* and *C* are deduced from field effect measurements; curves *B* are from capacitance voltage. **(b)** Another example of the density of states deduced from the field effect [5.72–75, 80]

The earliest measurements of $N(E)$ from the field effect and some later studies are shown in Fig. 5.23 [5.72]. The notable features are the minimum in $N(E)$ of $\sim 10^{17}$ cm^{-3} eV^{-1} near mid-gap, the celebrated E_x and E_y peaks in the upper and lower halves of the gap and a rapidly increasing $N(E)$ near the band tails.

Recently the field effect has been subjected to close scrutiny [5.73]. The criticism has focused on two aspects of the problem. One is the accuracy of the analysis, and the second concerns the different properties of a-Si : H in the bulk and at the surface. *Goodman* and *Fritzsche* have studied the reliability with which $N(E)$ can be obtained from a set of data [5.73]. In this calculation, a particular choice of $N(E)$, intentionally chosen to be that reported by *Spear* et al. [5.72], is used to generate a set of hypothetical "data points." This data is then converted back to a density of states using various reasonable fitting parameters. The results show that although the overall shape is preserved, there can be substantial differences in detail. From this type of analysis, it was concluded in [5.73] that the exact location and shape of the peak E_x is uncertain, as is the shape of E_y.

Weisfield and *Anderson* [5.74] have, in turn, reanalyzed some (measured) data by *Goodman* and *Fritzsche* using a simpler and more approximate method. The overall features are similar but there are differences in detail, particularly near the minimum of $N(E)$. It was also pointed out in [5.73] that their fit to this data is rather insensitive to the structure near the minimum of $N(E)$. A comparison of all the results in Fig. 5.23 is indicative of the accuracy with which the field effect can be analyzed [5.75]. It is clear, therefore, that the details of the spectrum must be interpreted with caution. On the other hand, two features of the data remain. In the best samples the minimum $N(E)$ is no larger than 10^{17} cm^{-3} eV^{-1} and the shape of $N(E)$ cannot be accounted for simply by extrapolating the band tails, indicating that extrinsic states are indeed being observed.

The second criticism of the field effect concerns the location of the states, and this is much harder to resolve. By the nature of the experiment, the current induced by the electric field flows in a channel which is within about 100 Å of the surface of the sample [5.75]. The field effect is thus very sensitive to surface states and if these are present the derived $N(E)$ will not be characteristic of the bulk. An illustration of the problem is the finding that a-Si:H sputtered onto fused quartz produces a very weak effect response indicative of at least 10^{19} cm^{-3} eV^{-1} states at E_F. However, if a thin layer of SiO$_x$ is sputter deposited first, then a large field effect is observed and the apparent density of states drops two orders of magnitude [5.76]. *Weisfield* et al. suggest that the improvement occurs simply because the first few hundred angstroms of deposited film contains defects or impurities. If these are left in the insulating layer rather than the a-Si:H film, then their influence on the field effect results is reduced. There have been several reports of surface states in a-Si:H and these are discussed in Sect. 5.3.3. This problem represents a major drawback to using the field effect to obtain $N(E)$.

c) Capacitance Measurements

Capacitance measurements of the depletion layer has become a popular way of measuring and profiling the dopant concentrations in crystalline semiconductors. The effect uses the property that free carriers are swept out of the depletion layer. The capacitance of the resulting insulating layer is measured. By using a *p-n* or MOS structure and changing the applied voltage, the depletion width is varied. The capacitance C should then be given by

$$C^2 \sim N/V , \qquad (5.20)$$

where N is the dopant concentration (or closely related to it) and V is the total voltage, including the built-in potential. Plots of C^{-2} versus V then yield N and deviations from linearity give the profile.

The main problem in applying this technique to a-Si:H is the complication introduced by having a distribution of deep gap states. Localized states contribute to the capacitance depending on the release rate for thermal excitation to the extended states, as given by (5.14). At sufficiently high frequency ω (typically 10^4 Hz at 300 K), none of the deep states respond and so the geometrical capacitance of the sample is observed. As the frequency ω is reduced, the cutoff energy for detrapping increases and moves into the depletion layer and the capacitance increases. In principle, one can then derive $N(E)$ from measurements of $C(\omega)$. Various groups have attempted to ana-

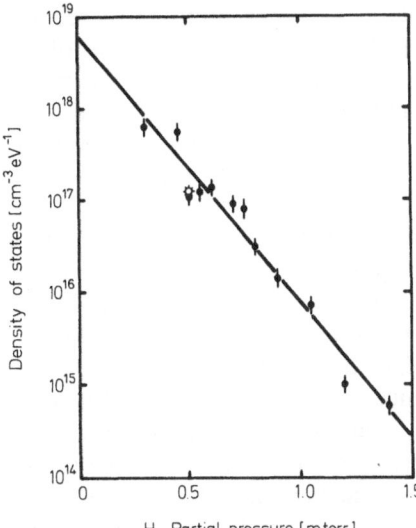

Fig. 5.24. Density of gap states as a function of hydrogen partial pressure in sputtered a-Si:H, derived from $C-V$ measurements, and assuming a constant distribution in energy [5.80]

lyze $C(\omega)$ data in this way with mixed success. For example, *Beichler* et al. [5.77] found that $C(\omega)$ agrees with their predictions for undoped a-Si:H, but that doped material introduces other complications. *Tiedje* et al. [5.78] obtained similar $C(\omega)$ data for undoped samples and from it derived the shape of the depletion potential. This shape was then fitted with an exponential distribution of localized states with a characteristic energy of 0.07 eV.

Capacitance-voltage data have also been used to obtain $N(E)$, based on an extension of the analysis used for crystalline semiconductors [5.79, 80]. Figure 5.23 shows the results of the measurement compared to $N(E)$ obtained from the field effect [5.80]. Figure 5.24 shows the density of states as a function of hydrogen partial pressure in sputtered a-Si:H, also obtained from $C-V$ measurements [5.80]. In these experiments a constant density of states at the Fermi energy is assumed. The density of states is found to be in good agreement with ESR data on one of the samples.

The capacitance techniques have the advantage of measuring bulk rather than surface states, since the depletion layer width is typically a sizeable fraction of the sample thickness. The methods clearly need further development to establish what conditions give reliable results.

d) Deep Level Transient Spectroscopy (DLTS)

DLTS is a powerful technique for studying deep states in semiconductors [5.68]. As for the capacitance measurements described above, this method involves a depletion layer and the proper analysis of the results depends on an accurate understanding of this layer. The method involves expanding the depletion layer of the reverse-biased junction and measuring the transient response as the carriers are thermally released from the traps. Equation

(5.14) is then applied to find the trap energy and the capture cross section. The details of the experiment can vary: the measurement can be of the ·capacitance or of the current, and trap-filling can be performed by optical excitation rather than change of bias. The analysis is complicated in a-Si : H by a continuous distribution of localized states and consequently, loses some of the simplicity it has in crystalline semiconductors which usually have discrete levels.

To overcome this problem, *Cohen* et al. [5.81] used a numerical analysis to evaluate the response of the space charge layer for an arbitrary density of states. They then obtained $N(E)$ by fitting the calculation to the data. Figure 5.25 shows an example of $N(E)$ derived from their DLTS data. They used a capacitance measurement with bias changes to observe electron states, and optical excitation for holes. The capacitance measurements required samples with a relatively large bulk conductivity and so all of their results are on doped samples. Because of the conductivity limitation at low temperatures, the measurement is only sensitive to traps deeper than about 0.4 eV. For these *n*-type samples, they observed a feature in the density of states near mid-gap with a density of up to 10^{18} cm^{-3} eV^{-1} and a deep minimum about 0.5 eV below the conduction band. This $N(E)$ can be contrasted with those shown in Fig. 5.23. They also concluded that the results represent a bulk $N(E)$ by calculating the saturation of the DLTS.

Crandall [5.82] has measured capacitance DLTS on undoped a-Si : H. The problem of low bulk conductivity is avoided by performing the measurements at low frequency and with long transients. An electron trap is observed

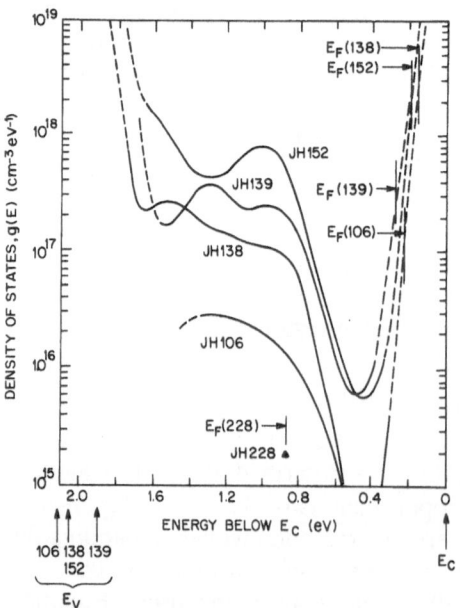

Fig. 5.25. Density-of-states measurements from DLTS experiments in phosphorus doped a-Si : H. Samples with the largest phosphorus content have the greatest density of states in the center of the gap [5.81]

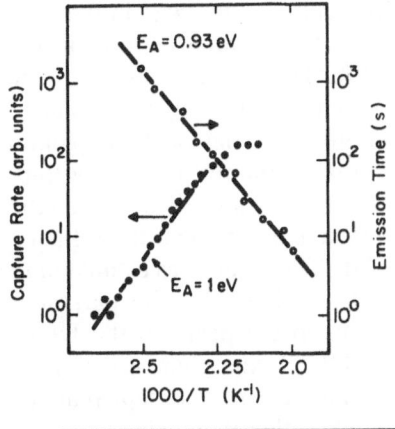

Fig. 5.26. Activated capture and emission rates for a deep trap observed in DLTS [5.82]

Fig. 5.27. Current transient DLTS spectra using optical excitation of carriers and different reverse bias voltages, showing shallow trapping states [5.83]

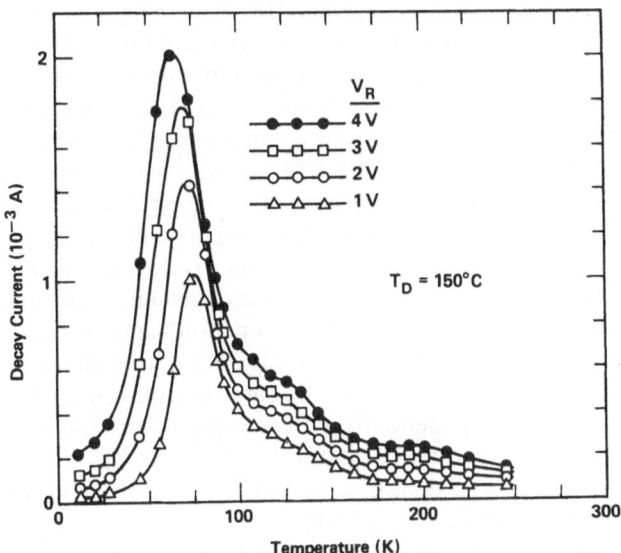

with a DLTS peak at ~ 450 K. This peak is evidently not the same as that observed by Cohen et al. because the longer measurement time should cause a shift of the peak to low temperature rather than high temperature. The trap observed by Crandall has the unusual property of a capture rate that is activated with about the same energy (~ 1.0 eV) as the emission rate, as shown in Fig. 5.26. Crandall found that the electron trap is associated with air contamination during deposition. He also related the activated, high energy capture and release of electrons to the Staebler-Wronski effects discussed in Sect. 5.3.1 b.

Current transient DLTS has also been studied in undoped a-Si : H [5.83]. This measurement has the advantage of not being limited by the bulk conductivity and so low temperature measurements can be made. Data in the temperature range 10 to 250 K are shown in Fig. 5.27. Three features in the spectra can be discerned at 50, 120 and 200 K. These results illustrate one problem with DLTS in an amorphous semiconductor. Usually it is assumed that thermally excited carriers are immediately swept out of the depletion layer so that the response time characterizes that of the release from a trap. In a-Si:H, the effective carrier mobility is low and activated. Thus, when the temperature is such that the drift time equals the measurement time, then a DLTS signal will be observed from this process. This effect was estimated to occur at about 50 K for electrons and 200 K for holes for the particular experimental conditions used in Fig. 5.27, and so the peaks in the DLTS spectra have been interpreted in this way. The DLTS experiment is then little different from a time-of-flight drift mobility measurement except that the carriers are initially distributed across the sample rather than being in a well-defined position. The remaining feature of the current transient DLTS in Fig. 5.27 is determined by the release time and is attributed to an electron trap.

DLTS measurements are usually made by keeping a constant release time and varying the temperature. The alternative of fixing the temperature and varying the time has also been reported [5.84]. This method has the advantage that one does not need to know the temperature dependence of the material properties (for example, the capture cross section) and the disadvantage of needing observations over a very wide time range to cover a reasonable energy range (5.14). Figure 5.28 shows some data which indicates a band of localized states about 0.8 eV below the conduction band of n-type a-Si:H, in general agreement with the data of Fig. 5.25. Recent measurements [5.85] using this technique report that the capture cross section for electrons is very small and varies exponentially with energy. In consequence, the peaks in Fig. 5.28 are reanalyzed and found to be shallower by 0.2–0.3 eV. However, the small cross sections do not agree with the measurements of the emission rate prefactor by *Cohen* et al. [5.81].

It seems likely that DLTS will prove to be the best technique for the electrical characterization of deep states. Its primary advantage over the field effect is that surface states are less influential and their effect can be determined. Clearly much work needs to be done in measuring $N(E)$ from DLTS in well-characterized samples. There is also an evident need for further study of the emission rate prefactor to obtain a reliable energy scale, and possibly better understanding of the saturation properties to ensure that the correct state densities are being measured.

e) Tunneling Spectroscopy

Tunneling spectroscopy is known to give information about localized gap states in semiconductors. This technique has also been applied to a-Si:H [5.86]. A thin oxide is grown on the sample by keeping it about 150 °C in air for a few days and a Cr contact is evaporated on top. There is a steep onset of

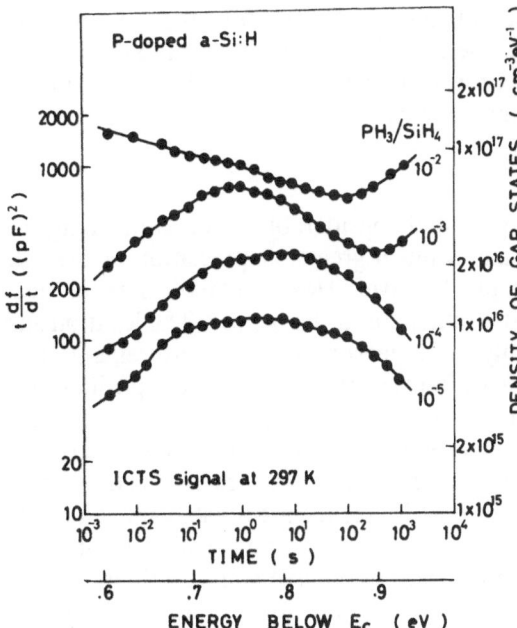

Fig. 5.28. Density-of-states measurements using isothermal DLTS for different doping levels [5.84]

conductance due to tunneling into the valence or conduction band, depending on the polarity. In addition a small peak near the conductance minimum is from tunneling to localized states in the gap. Analysis of this feature leads to the density of states shown in Fig. 5.29 for phosphorus doped a-Si:H. Measurements on undoped material have a similar feature which is interpreted as a band of states about 0.45 eV below the conduction band edge.

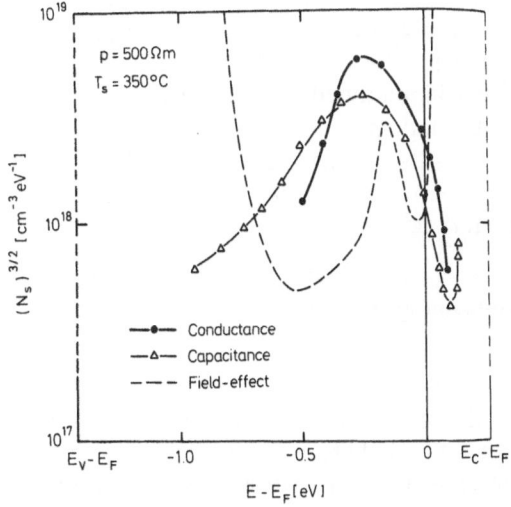

Fig. 5.29. Density of surface states deduced from tunneling conductance measurements compared to other experimental results [5.86]

From the frequency dependence of the results, and for other reasons too, *Balberg* and *Carlson* [5.86] concluded that in undoped a-Si : H, bulk states of density 10^{18}–10^{19} cm^{-3} eV^{-1} are observed, whereas in doped a-Si : H, surface states are responsible for the tunneling current.

f) Summary

Transport measurements have provided considerable detail concerning the distribution of band-tail states and a much greater appreciation of the role and properties of the surface depletion layer. However, the electrical measurements of deep states present a rather confusing picture. The location and density of states vary substantially from measurement to measurement. This in itself is not necessarily surprising because the defects and impurities may be characteristically different in the samples prepared in different laboratories. However, this contrasts with the ESR and luminescence results which seem to give remarkably uniform results. In the face of such diversity it is impossible to state with confidence which results are reliable and which are not. We note particularly that some measurements (i.e., field effect, tunneling) find a broad feature in the upper half of the gap and a minimum at mid gap, whereas DLTS finds exactly the reverse. Possible explanations are that surface states as opposed to bulk states are being measured: that the states are, in fact, identical but there are substantial errors in determining the energy levels; or simply that a different set of levels is being observed.

5.3 Material Properties

The previous section described the different spectroscopic methods that have been applied to the study of localized states in a-Si : H. In this section we discuss different aspects of the localized gap states and compare the information obtained by each of the experimental techniques. We shall be concerned particularly with native gap states in undoped a-Si : H including surface states, and the states introduced by doping and compensation. One of the main problems in comparing results is the uncertainty over whether equivalent samples have been used. For this reason we shall concentrate on information obtained from well-characterized a-Si : H.

5.3.1 Undoped Hydrogenated Amorphous Silicon

a) Band-Tail States

A tail of localized states at the band edges is a central feature of most models of amorphous semiconductors. In a-Si : H there is clear evidence for band tailing and reasonable agreement as to the distribution of states within the tail (at least for the conduction band). At present the most reliable information on the shape of the band tails seems to be from drift mobility data [5.70]. The dispersive transport is described in terms of an exponential distribution of traps $\exp(-W/W_0)$, and the characteristic energy W_0 is 0.027 eV for the conduction band and 0.043 eV for the valence band of low defect density glow discharge material. The reported room temperature drift mobilities vary by about two orders of magnitude between different groups, and this is perhaps an indication of the substantial dependence of the tail shape on sample preparation conditions.

Neither the field effect nor DLTS gives much information about the tail state distribution because they are unable to probe the high density of states near the mobility edges. Nevertheless, a simple extrapolation from the density of states at the foot of the band tails (Fig. 5.25) yields values of W_0 within about a factor of two of those given by the drift mobility. Capacitance-frequency measurements have also been analyzed in terms of an exponential band tail and again, roughly comparable values of W_0 are obtained for the conduction band.

The 1.4 eV luminescence band is described in terms of radiative tunneling between band-tail states. These experiments, therefore, also provide direct evidence for the existence of tails as well as information as to the nature of the tail states. Thus, as described in Sect. 5.2.2, band-tail electrons are relatively weakly bound with a Bohr radius of about 10 Å. Holes, on the other hand, are more strongly localized, and according to the luminescence models, have a substantial distortion energy which contributes to the Stokes shift. None of the electrical measurements give any definitive information about the distortion energy of band-tail states. However, the larger width of the valence band tail is at least consistent with greater localization and/or lattice distortion.

The slope of the conduction band tail can be deduced from the temperature dependence of the luminescence peak energy and intensity dependence. These data, shown in Figs. 5.13, 5.14, are related to the shape of the band tail by adopting a model similar to that used to describe dispersive transport. It is assumed that a demarcation energy exists separating shallow states which will be thermally excited to the conduction band, from deep states which will recombine. Both measurements can then be explained by an exponential band tail, with W_0 of 28–35 meV, in fair agreement with the drift mobility data.

Despite the evidence for localized band tails, there is as yet no clear experimental evidence as to the position and nature of the mobility edge in a-Si:H. Conductivity experiments are generally interpreted in terms of transport at a well-defined mobility edge, but this model is based more on theoretical arguments than on clear experimental information.

Since the band tails arise from disorder in the Si–H network, it is to be expected that any modification of the network will be reflected in the shape of the band tails. There is indeed ample evidence for this process. One

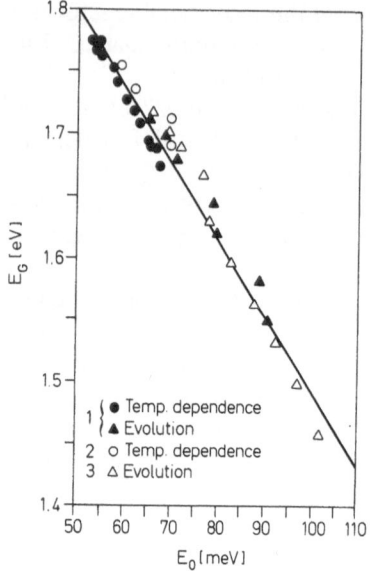

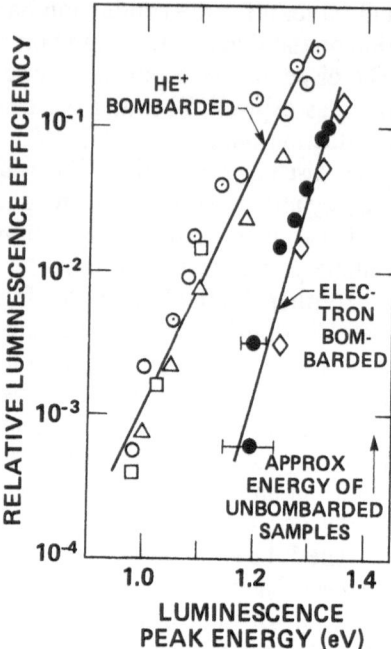

Fig. 5.30. Optical gap E_g as a function of the slope of the Urbach tail E_0. (○, ●) correspond to different measurement temperatures; (△, ▲) to different stages of hydrogen evolution [5.87]

Fig. 5.31. Relative luminescence intensity versus luminescence peak energy for samples bombarded with He$^+$ ions or electrons and then annealed [5.49]

method of changing the disorder potential is by annealing and hydrogen evolution. *Cody* et al. [5.87] found that the optical band gap, and more significantly, the slope of the Urbach edge, changes with hydrogen evolution as shown in Fig. 5.30. They interpreted the results in terms of a changing slope of the band tails. Although we have argued in Sect. 5.2.2b that the shape of the Urbach edge is probably complicated by phonon interactions, the changes of slope probably do reflect the changing distribution of band-tail states.

Another means of introducing disorder is by electron or ion bombardment. Measurements of the luminescence peak energy after bombardment and subsequent annealing are shown in Fig. 5.31 [5.49]. Since the luminescence is a transition between the band tails, its energy is a measure of the width of the tails. It is observed that after bombardment, the luminescence peak drops in energy but is restored to its original value by annealing. Of interest is that the shift of the peak is much larger after bombardment by He$^+$ ions compared to electrons, and this has been interpreted in terms of the more extensive secondary damage caused by the heavier particle. Analysis of the luminescence data does not provide a good quantitative estimate of the broadening of the tails, nor does it determine which band tail is more affected.

There are various other potential sources of disorder. For example, the microstructure which is known to occur predominately in samples deposited in the presence of Ar or other inert gasses [5.88] is a likely source of disorder broadening, although there is as yet no detailed information. Doping and impurities can also introduce additional disorder and this is discussed in Sect. 5.3.2.

b) Defect States

The defect that has received the most attention in a-Si:H is the paramagnetic state observed in ESR at $g = 2.0055$. When this defect was first observed, it was associated with a threefold coordinated silicon atom with a dangling bond. The basis for this interpretation was that the dangling bond is the simplest paramagnetic defect in a random network and therefore the most plausible. The evidence for this identification of the ESR is now much stronger and has been discussed in detail in Sect. 5.2.1.

A great deal of evidence has been accumulated about the properties of the dangling bond and much of this is contained in the description of the different spectroscopic techniques discussed in Sect. 5.2. The importance of dangling bonds as nonradiative recombination centers limiting carrier lifetimes is seen by the quenching of luminescence and photoconductivity when the density of these defects is high. The understanding and control of the dangling bonds are consequently of importance in device applications of a-Si:H.

Several measurements have attempted to identify the energy levels of the dangling bonds, but there is not yet a consensus. Some information can be obtained directly from the ESR data. The observation that the defects are paramagnetic in undoped material but diamagnetic in doped material leads directly to the conclusion that the energy level diagram is qualitatively as shown in Fig. 5.32. There must be two energy levels corresponding to the singly and doubly occupied defect (the lower and upper Hubbard bands) and

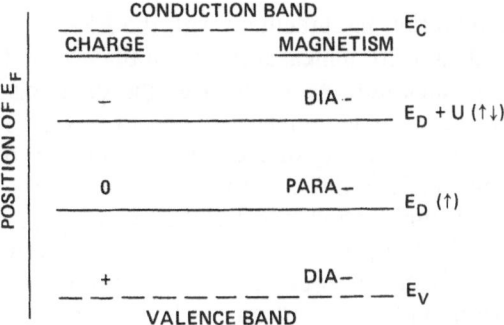

Fig. 5.32. Schematic energy level diagram of dangling bonds showing how the charge and magnetic state depend on the position of the Fermi energy

three charge states ($+$, 0, and $-$). The energy levels span the position of E_F in undoped material, which is close to mid gap, and are described by a positive correlation energy. We note that the extent to which the defect bands overlap has not been determined yet.

The luminescence band observed at 0.9 eV has been associated with transitions from the dangling-bond level. The evidence for this is that the 0.9 eV transition occurs predominately in samples known to have a large dangling-bond density. The transition is interpreted as being from the doubly occupied defect to the valence band tail. When account is taken of the Stokes shift, the doubly occupied dangling-bond level is estimated to be about 0.5 eV below the conduction band.

Field effect data provide mixed information. The earliest data, and some subsequent experiments, show a broad feature in the upper half of the gap which might be associated with the dangling-bond level, although the alternative explanation as surface states (which might also be dangling bonds) is a possibility, as discussed in Sect. 5.2.3 b.

The evidence of DLTS and PDS seems clearer. A broad and sometimes double-peaked structure is observed near mid-gap in DLTS measurements (Fig. 5.25). The density of states of this structure increases with doping level and parallels the known changes in dangling bond density (see Sect. 5.3.2). In PDS an absorption band below the gap is observed whose intensity is directly proportional to the $g = 2.0055$ ESR spin density. From this it is deduced that the absorption band involves one of the dangling bond levels. It is interpreted as being from the singly occupied level to the conduction band because the transition is observed to excite photoconductivity. From the shape of the absorption band in doped and undoped a-Si:H, the singly occupied dangling-bond level is deduced to be ~ 1.2 eV below the conduction band, with a correlation energy of 0.25–0.45 eV. These values are in general agreement with ESR data. The PDS data, in conjunction with luminescence, provide some confirmation that the DLTS spectrum of Fig. 5.25 is attributable to dangling bonds, because similar densities of defects are obtained at equivalent doping levels in the two experiments.

Recently, some more direct evidence that the DLTS feature near midgap is due to dangling bonds has been obtained by changing the depletion length of a *p-n* junction [5.88]. The dangling-bond ESR is observed to increase as E_F moves towards midgap in *n*-type material, in agreement with the ESR data discussed in Sect. 5.2.1. More significant, from measurements of the bias and temperature dependence of the voltage-induced resonance, the dangling-bond state is shown to have the same carrier release times as the DLTS feature. It is worth pointing out that the energy of the bands attributed to dangling bonds in the DLTS (Fig. 5.25) is difficult to reconcile with the requirement that the upper Hubbard band is above the dark Fermi energy, as discussed in connection with Fig. 5.32. E_F in undoped a-Si:H is usually 0.6–0.8 eV below the conduction band. However, these values of E_F obtained from dc conductivity are themselves questionable because of surface-related effects. Accurate and reliable measurements of E_F are needed to help resolve the questions concerning the dangling-bond binding energies.

Dangling bonds are clearly an important trapping and recombination center. This is seen by the quenching of both luminescence and photoconductivity as the defect density is raised, and by a range limitation in time-of-flight drift mobility measurements. At low temperatures, trapping occurs by tunneling of electrons from band-tail states, as has been deduced from luminescence data, and at high temperatures by a process of diffusion and capture. Although the luminescence at 0.9 eV has been associated with the recombination at the dangling bonds, this is a process of very low efficiency so that the recombination is evidently mostly nonradiative, but the mechanism is as yet unknown.

It is now known that the dangling bond density is very sensitive to the details of how the material is prepared. The reason that hydrogen or other gases are used in the deposition is because this results in a dramatic reduction in the dangling-bond density. However, even in the hydrogenated material, the density can vary by several orders of magnitude. Figure 5.33 shows how some of the glow discharge deposition parameters, substrate temperature and gas dilution, and rf power influence the ESR spin density [5.89]. Other variables of the deposition also have an influence. By judicious choice of the deposition conditions, the spin density can be reduced to a level of $\sim 10^{15}$ cm^{-3}. The existence of surface dangling bonds as discussed in Sect. 5.3.1c makes the determination of the minimum bulk level ambiguous, and it may be well below 10^{15} cm^{-3}.

The dangling-bond density is also influenced by post-deposition treatment. If samples deposited at a substrate temperature of 150 °C or below are

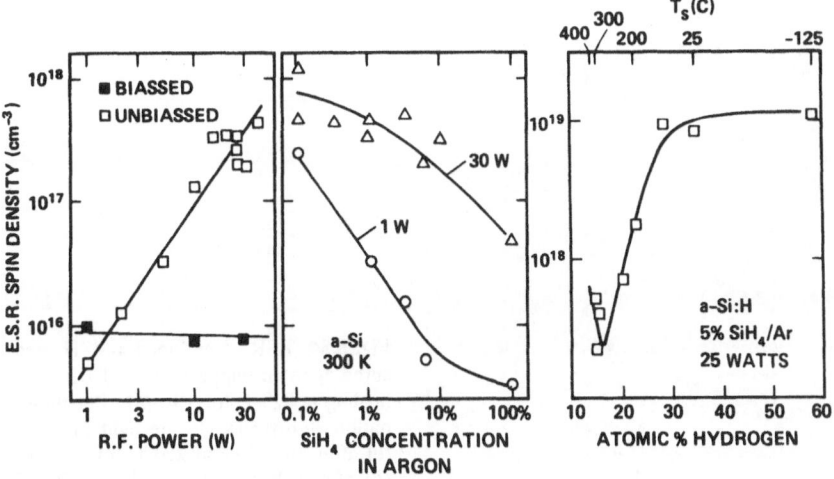

Fig. 5.33. The dangling-bond spin density shown as a function of the deposition parameters for glow discharge deposited a-Si : H [5.89]

annealed to 200–300 °C, then the spin density decreases [5.90]. At higher annealing temperatures, hydrogen is evolved and the dangling-bond density increases. Hydrogen can then be reintroduced into the material by exposure to a plasma [5.91]. Although there are no measurements of the ESR after this treatment, the recovery of the luminescence is a clear signal that the dangling-bond density decreases.

Bombardment by ions or electrons also generates defects as observed by the dangling bond ESR and the quenching of luminescence. Again, annealing to 200–300 °C removes the ESR signal. This experiment is interesting because the annealing does not return the luminescence to the high value usually associated with the low dangling-bond density and so it was deduced that nonradiative centers must remain which are diamagnetic. There is no further evidence as to the structure of these states, particularly whether they are qualitatively distinct defects or perhaps band-tail-like states that have been distorted by the bombardment. However, this is apparently the only case where there is some evidence of deep native defects other than dangling bonds.

Prolonged illumination with above band gap light is another means of creating defects in a-Si : H. This effect, first reported by *Staebler* and *Wronski* [5.92] from conductivity and photoconductivity measurements, is not yet completely understood and may involve both a bulk and a surface effect. However, there is now considerable evidence for a change in the bulk density of states. *Dersch* et al. [5.93] have observed directly an increase in the $g = 2.0055$ ESR spin density after illumination, as shown in Fig. 5.34. The

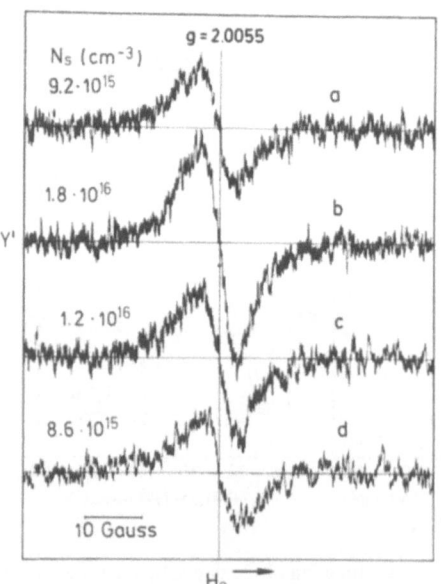

Fig. 5.34. ESR of undoped a-Si : H measured at room temperature. *(a)* After annealing at 220 °C for 1 h; *(b)* after illumination with focused white light for 15 h; *(c)* after annealing at 220 °C for 30 min.; *(d)* after subsequent annealing at 220 °C for an additional 30 min. [5.93]

luminescence is also observed to change, with intensity transferred from the 1.4 eV peak to the 0.9 eV peak which has been attributed to dangling-bond recombination. In addition, changes in the luminescence decay parallel those which occur when the defect density is increased [5.94]. DLTS measurements also find an increase in the density of deep traps after prolonged illumination, although the details differ widely. The Marburg group (*Stuke:* private communication) found an increase in mid-gap states, consistent with the increase in dangling-bond density. *Cohen* et al. [5.95] observed an increase in hole traps, whereas *Crandall* [5.82] related the changes on illumination to electron traps with a DLTS signature, very different from that of dangling bonds. An increase in the dangling-bond density (or another deep trap) also readily accounts for the observed decrease in photoconductivity, and pinning of the Fermi energy might possibly explain the changes in the dark conductivity. However, some of the effects have been associated with surface properties, as discussed in Sect. 5.3.1 c.

The creation of defects by light is a very inefficient process. For example, 15 hours of fairly strong illumination generates only about 10^{16} spins cm^{-3} [5.93]. The efficiency is also not linear in the light intensity, although the detailed dependence is not documented [5.94]. There is some debate as to how illumination can create dangling bonds. *Adler* [5.96] argues that the energy of the light is insufficient to break Si–Si bonds and so some other defect must be assumed, despite the strong evidence that the defects are dangling bonds. The same argument could be applied to the creation of dangling bonds by evolution of hydrogen at temperatures of only 400 °C. Here the explanation is that much of the bond energy is recovered by reconstruction of the remaining atoms or by the formation of H_2 molecules. The same processes could account for the creation of dangling bonds by illumination if this occurred close to SiH structures, for example, on the surface of small voids. In fact, the subsequent annealing of the dangling bonds at typically 150 °C almost certainly involves the motion of hydrogen to saturate the defect. This mechanism is used to explain the annealing of native dangling bonds [5.49] and the motion of hydrogen at these temperatures is observed in NMR [5.97] and hydrogen evolution [5.90].

A parallel can be drawn between the photoproduction of defects in a-Si:H or in chalcogenide glasses and recombination enhanced defect motion in crystals. In each case the recombination of electron-hole pairs must transfer energy to highly localized lattice vibrations. In all such cases in crystals, nonradiative processes involving strong electron-lattice coupling provide the transfer of energy. A similar model has been proposed in chalcogenides, and probably the effect in a-Si:H has the same origin. Motion of a defect occurs by breaking bonds and their subsequent reconstruction in a slightly different configuration, so that the energy required to move a defect is less than the defect creation energy. Our description of the defect creation in a-Si:H is very similar and involves a Si–H bond breaking and the hydrogen atom reconstructing into some different configuration.

One of the striking features of a-Si:H is that one single type of defect, the dangling bond, appears to dominate. This contrasts with crystalline silicon, for example, which has a large number of distinct native defects. The only case in which the presence of another defect is indicated is after bombardment, which might be expected to have a fairly drastic effect on the structure. There is, of course, a natural distinction between defects in crystalline and amorphous materials. Whenever an atom is out of a lattice position in a crystal, a defect results. An amorphous network does not have a lattice and so the concept of a defect is not so simple. There are two extremes which are fairly well defined. One of these is the band-tail states. These are attributed to fairly small distortions in bond lengths and bond angles of the fully coordinated network that arise from the disorder. The distortions are distributed continuously with the largest having the lowest probability. Provided the tails do not extend very far, as seems to be the case in a-Si:H, the states can be characterized unambiguously as part of the conduction band or valence band. The other extreme is the coordination defect of which the dangling bond is the prime example. The different coordination makes this a qualitatively different state from the rest of the network (e.g., it either has a spin or a charge). The twofold coordinated Si atom is another example of a well-characterized defect, although there is no direct evidence that this defect exists in significant numbers in a-Si:H.

In between these two extremes is a less well-defined area. Consider, for example, a vacancy resulting in removing one Si atom from the network. Inevitably there will be some reconstruction in which the four dangling bonds pair off. One possibility is that the reconstruction is minimal and that a distinct deep defect results. Alternatively, the network might allow sufficient relaxation so that the reconstructed bonds are not substantially more distorted than those of the rest of the network. Any localized states would consequently be an indistinguishable part of the band tails. If this degree of relaxation occurred generally, then any particular defect structure would tend to relax to give a combination of band-tail states and possibly a single coordination defect. This represents a major simplification of the defect state structure compared to that in a crystal, but it is a model which is consistent with the majority of the experimental data.

c) Surface States

As the density of deep defect states is reduced, the surfaces of a semiconductor play an increasingly important role. In a-Si:H, depletion layer widths of order 1 μm can be achieved, which is similar to the typical sample widths. Good control of the surface is then essential for an understanding of the electrical properties. Two types of surface state are found to be significant in a-Si:H. One of these comprises a fixed charge that can be introduced on the surface, giving strong band bending. One source is adsorbed gas molecules, particularly H_2O vapor or NH_3. When a fresh a-Si:H surface is exposed to

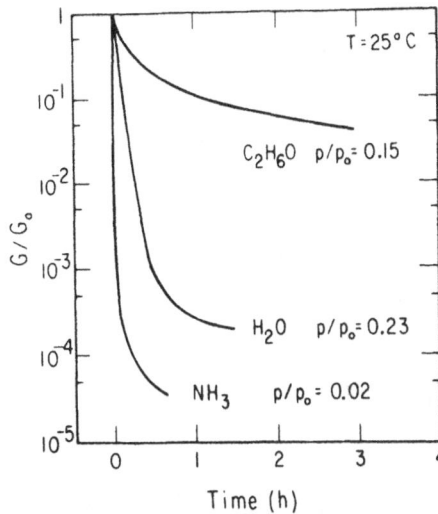

Fig. 5.35. Relative change in conductance of a 0.36 μ-thick a-Si:H film due to exposure by the adsorbates indicated. The changes are reversed by annealing in vacuum at 160°C [5.98]

either of these, the conductivity can change by several orders of magnitude owing to the field effect in the surface layer, as shown in Fig. 5.35 [5.98]. Annealing to 150°C drives off the adsorbed gas so that the process is reversible. A similar effect has been seen by allowing charge to diffuse through an insulator to the a-Si:H interface [5.99]. The similarity of the changes observed with surface charges and the effects of prolonged illumination have led to the suggestion that the latter is, in fact, a surface effect. As described in the previous section, there seems little doubt that changes in the bulk defect density do occur with illumination. However, it may be that some of the effects reported could be attributed to surface charge. In any event, it is evident that conductivity measurements on exposed films will tend to be dominated by the surface effects.

The second type of surface state is a localized level within the band gap originating from an a-Si:H defect at the surface or interface. Such states have been of particular concern because of their influence on the field effect results. In Sect. 5.2.3 we decribe measurements showing that the field effect was very sensitive to the way the interface to the insulator was prepared, and this was attributed to surface states. Tunneling measurements have also been interpreted in terms of surface states.

ESR and PDS provide some direct evidence as to the density and properties of surfaces states. From thickness dependence measurements in undoped samples as shown in Fig. 5.36, surface-related states can be identified [5.100 a]. These states are paramagnetic and have $g = 2.0055$, showing that they are dangling bonds, and a density of $\sim 10^{12}$ cm^{-2}, depending on the sample preparation conditions. In a-Si:H with a columnar growth structure, it is found that a dangling-bond ESR signal grows as oxidation proceeds slowly down the columns [5.100 b]. It is argued that the defects occur at the a-Si:H/oxide interface which covers the surface of the columns and that the

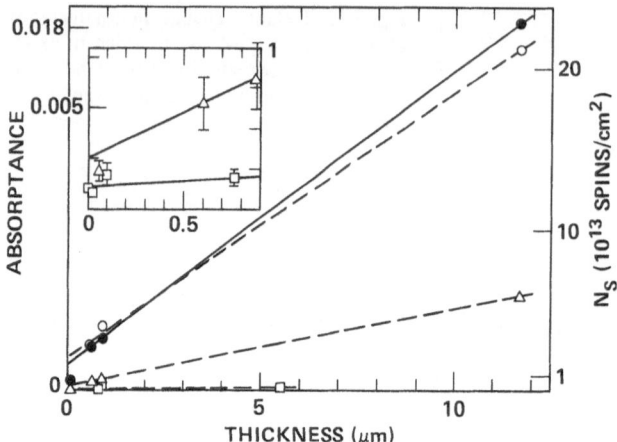

Fig. 5.36. Dangling-bond spin density and extrinsic absorption as a function of sample thickness [open (solid) symbols are PDS(ESR) measurements. (○) 140 °C sample 1; (△) 140 °C sample 2; (□) 230 °C sample]

surface spin density is ~3 × 10^{11} cm^{-2}. It is also observed that crushing a-Si:H films leads to an increase of the dangling-bond spin density, which also could be explained by defects at the a-Si:H/oxide interface.

Apart from their role in the field effect experiment, little is known of the influence of these surface states. Surface recombination is observed in luminescence experiments and also in solar cells, but it is not known whether the same states are involved. Surface dangling bonds would be expected to pin the Fermi energy near midgap, therefore having different properties from the fixed charges which tends to pin the Fermi energy near one or other band edge.

5.3.2 Doping

The Fermi level E_F in a-Si:H can be shifted by more than 1 eV by the addition of Column I, III or V elements from the periodic table. Column IV, VI or VII atoms, on the other hand, can be and are routinely incorporated at high concentrations with little direct effect on E_F, but often with important effects on the band gap energy and tail widths. The use of doping to reveal intrinsic properties of a-Si:H have been discussed in Sect. 5.2. The states introduced into or removed from the gap and their origin are the subjects of this section.

a) Doping Mechanism and Efficiency

Spear and *LeComber* [5.5] first demonstrated that a-Si:H can be doped n or p-type by the addition of PH$_3$ or B$_2$H$_6$ to the SiH$_4$ during deposition. This was

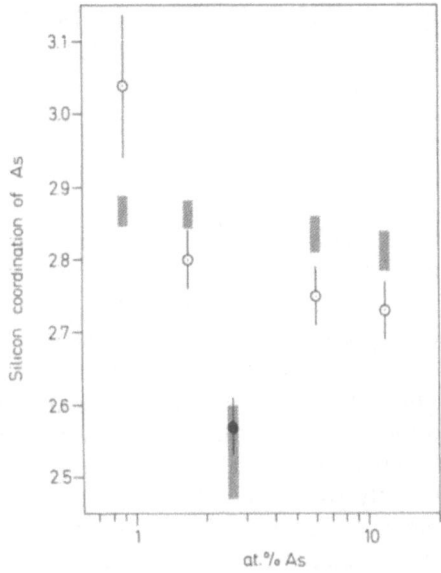

Fig. 5.37. The coordination of arsenic in a-Si : H as a function of arsenic concentration (○). Note that the coordination at low As concentrations is increasing [5.101]

surprising in the context of the pseudo-equilibrium chemical picture of a-Si held at the time. In glasses quenched from the melt, impurity atoms are generally accommodated in the molten or quenched phase in their lowest energy bonding configurations, e.g., 3-fold coordination for phosphorus and boron. Evidently in a-Si : H, some of the dopants are instead in fourfold coordination. As we shall see later, the fraction of four-coordinated dopant, known as the doping efficiency, is not unity, and in fact is rather small in heavily doped material.

The most direct experimental probe of the dopant coordination is extended x-ray absorption fine structure (EXAFS). *Knights* et al. [5.101] studied the EXAFS of a-Si : H samples with levels of arsenic varying from 100% to < 1%. Figure 5.37 shows their results for the average As coordination. The upturn at low incorporation levels is argued to be indicative of the change from alloying to doping. (The reduction in As–Si coordination below three is a result of the presence of As–H bonds.) Unfortunately, the sensitivity of the experiments is proportional to the As concentration and becomes too weak in the regime of greatest interest. Also, assumptions concerning the distribution of hydrogen bonded to Si and As, necessary for the quantitative interpretation, are not experimentally demonstrated.

In crystalline Si, the most direct test of substitutional doping comes from hyperfine studies (using ESR or ENDOR) of the neutral impurity at low temperatures [5.102]. In a-Si no paramagnetic signal has been observed from P or B. This has been interpreted to follow from the relatively large density of states deeper in the gap which compensate the dopant, i.e., E_F is always deeper than the dopant energy levels. Thus, in a-Si no clear evidence exists which proves the substitutional nature of doping. However, because it is plausible, simple conceptually and in agreement with the measured effects of doping, substitutional doping has become the accepted model.

Measurements of conductivity, thermo power and mobility all indicate that E_F shifts towards the valence (conduction) band edge as p-type (n-type) dopant incorporation increases. These measurements set upper bounds for the dopant binding energy levels of $E_B \sim E_v + 0.3$ eV and $E_p \sim E_c - 0.2$ eV, corresponding to the maximum observed shifts of E_F in doped a-Si:H. Above $\sim 1\%$ concentration, the conductivity drops again [5.103] concomitantly with the decrease in the dominant dopant coordination number indicated by EXAFS. Early estimates of doping efficiencies at low doping levels (E_F near midgap) [5.5] gave estimates of $\eta \sim 10$–30%. These were based on field effect measurements which have subsequently been thought to overestimate the midgap density of states and therefore overestimate η. LESR and luminescence data (Fig. 5.38 below) tend to suggest that the doping efficiency is much less than unity, particularly when E_F is close to the band edges. DLTS measurements, in principle, could also give the doping efficiency but no measurements have been published yet which have been correlated with dopant concentration over a wide range.

Doping efficiencies less than unity can arise, for example, from the presence of 3-fold incorporation or autocompensating dopant-induced defect formation. Evidence for the latter effect will be presented in the next section. Doping has also been achieved by ion implantation [5.104] (e.g., B, P, As) and indiffusion (e.g., Li) [5.105]. The doping efficiencies, as measured by shifts in E_F for a given concentration of impurity, is approximately an order of magnitude less than incorporation during film growth. This is not surprising for the implantation doping because the heavy damage production leads to a large density of compensating deep states.

b) Defect Creation by Doping

There is a great deal of evidence from luminescence, ESR, PDS and DLTS for defect creation associated with dopant incorporation. It is clearly important for technological applications of a-Si:H to determine if defect creation is an intrinsic result of doping or if deposition techniques can be found so that doping can be achieved without defect creation.

The first clear identification of the defect states came, as for undoped material, from correlated studies of ESR and luminescence. In moderately doped samples grown under conditions for which the undoped sample has a negligible spin density, there is also no ESR in the dark. However, light induced ESR is observed in both n and p-type samples with the neutral dangling-bond characteristics [5.35]. The line appears in n-type (p-type) material along with the band-tail electron (hole) ESR line. These data are shown in Fig. 5.2. In each case the induced dangling-bond density is $\sim 10^{17}$ cm^{-3} compared with a density of $\sim 10^{15}$ cm^{-3} in undoped a-Si:H. It is

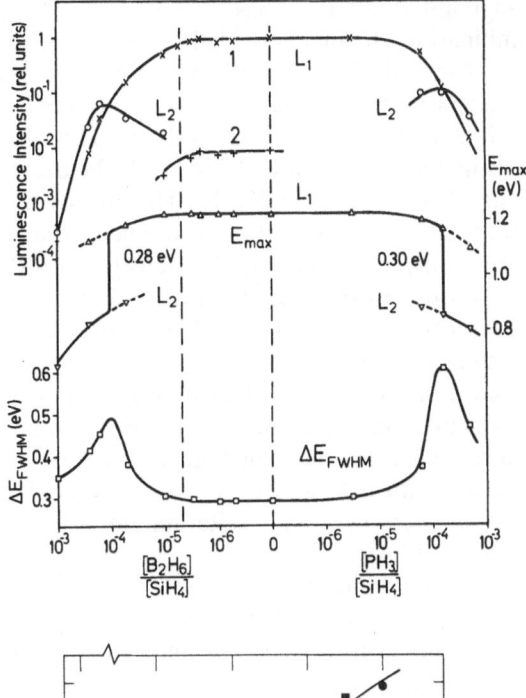

Fig. 5.38. Quenching of band edge luminescence (L_1) and introduction of defect luminescence (L_2) with doping [5.31]

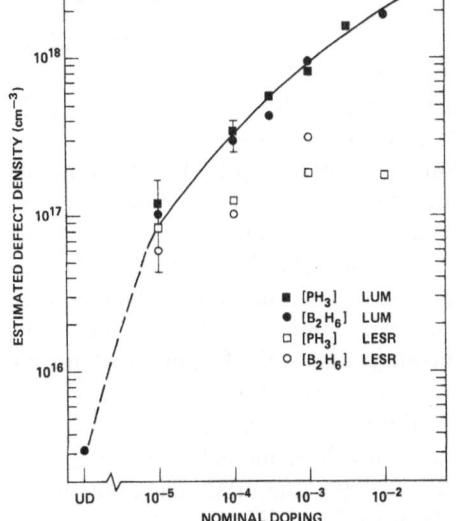

Fig. 5.39. Defect densities versus doping concentration as estimated by LESR and luminescence measurements. Note that these data agree with the PDS data shown in Fig. 5.43 [5.31]

thus clear that doping introduces additional dangling-bond-like defect states which are charged in equilibrium. The fact that the defects are charged even at very low doping levels indicates that E_F has moved past the energy level of the charged defect denoted by E_0 and $E_0 + U$ in Fig. 5.32.

The increase in defect density with doping is also evident from luminescence data. In undoped samples the inverse correlation between band-edge luminescence efficiency and dangling-bond density has been shown to be a result of the competitive recombination channel provided by the dangling bond. Figure 5.38 demonstrates that with doping, both n and p-type, the band edge luminescence is quenched [5.31]. Similarly, the defect luminescence band at 0.8–0.9 eV found in heavily defective undoped a-Si : H is also observed in doped a-Si : H. By using the relationship of Fig. 5.12, the defect density introduced by doping has been estimated [5.31]. The results are shown in Fig. 5.39 together with the LESR spin densities. The smaller values for the LESR values are, in fact, lower bounds on the defect densities because steady-state signal depends on the excitation intensity and recombination times and saturation is not reached [5.36, 41]. Another manifestation of the defects introduced by doping is observed in the transient decay of luminescence. Just as in undoped a-Si : H samples of increasing dangling-bond density, so for increasing doping concentrations the initial fast (non-radiatively-driven) decay increases, whereas the long time recombination remains the same [5.41]. Further confirmation comes from the similar dependence of luminescence on temperature and excitation intensity in doped and defective undoped samples [5.48].

From the presence of the same defect luminescence band at ∼ 0.9 eV, the same binding energy is assigned to doping-related defects as to those in undoped material. Furthermore, the similar ESR signatures indicate that the defects are structurally similar to those in undoped a-Si : H. Two possibilities discussed for the defect structure are (a) isolated dangling bonds, and (b) dangling bonds neighboring threefold coordinated dopant atoms. The absence of a dopant-related hyperfine broadening for the ESR line would imply, for case (b), a strong localization of the defect wave function.

Results of photothermal deflection spectroscopy (PDS) measurements [5.55] of extrinsic, subband gap absorption are in close quantitative agreement with the results of luminescence and ESR. In Fig. 5.17 it was shown that the integrated extrinsic absorption in undoped a-Si : H was directly proportional to the dangling-bond density. PDS measurements on doped a-Si : H show a similar subgap absorption band as in defective undoped material. Examples of absorption spectra are shown in Fig. 5.40. The defect densities are then derived from the integrated absorption strength, and Fig. 5.41 demonstrates the excellent agreement with the estimates from luminescence in the same samples. The PDS estimate assumes the same matrix elements for band-tail defect transitions in undoped and doped material. The agreement, therefore, is further evidence of the identity of the defects.

The DLTS data shown in Fig. 5.25 also shows that there is an increase in midgap state density with phosphorus doping. These states have been associated with dangling bonds as described in Sect. 5.3.1 b. The density of states introduced is also in reasonable quantitative agreement with the PDS and the other results. Thus, the evidence for dangling bond incorporation into doped a-Si H is very strong. The reasons for the increased defect density are discussed in the next section.

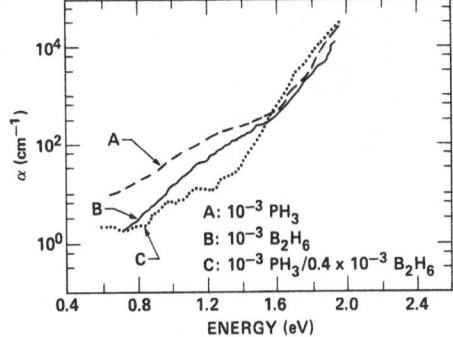

Fig. 5.40. Typical PDS spectra for doped and compensated a-Si:H. Gas phase doping level is 10^{-3}. Shoulder near 1.2 eV decreases with compensation [5.55]

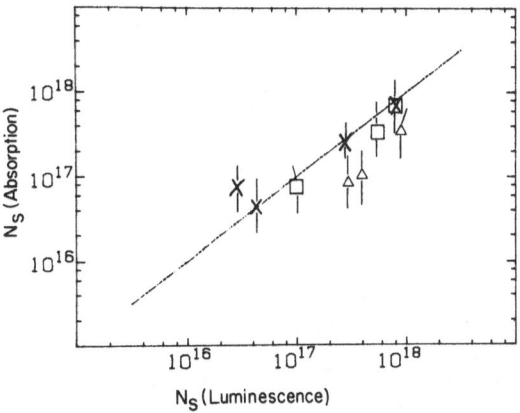

Fig. 5.41. Defect densities obtained by PDS plotted versus the estimates from luminescence data showing agreement between both experiments [($\square$) phosphorus doped, ($\triangle$) boron doped, ($\times$) compensated] [5.55]

Knowledge of the defect density allows the doping efficiency to be estimated. Each fourfold coordinated phosphorus atom donates one electron which can either occupy a dangling-bond site, a band-tail state or remain on the phosphorus. Each of these can potentially be observed by ESR or LESR. In this way, the total number of active donors can be found. The results are shown in Fig. 5.42. It is evident that the doping efficiency is far from being constant and instead drops steadily with increased doping. It is interesting that both boron and phosphorus doping behave similarly.

Finally, there is direct evidence from photoemission measurements [5.106] in very heavily phosphorus-doped (8%) a-Si:H for the presence of a shoulder in the density of states ~ 0.4 eV above E_v (Fig. 5.43). The origin of these states is unclear. However, it is certainly to be expected that alloying should increase the distribution of disorder potentials and the width of the band tails. Strong electron-phonon coupling could increase the binding energy of such states and create a distinct shoulder.

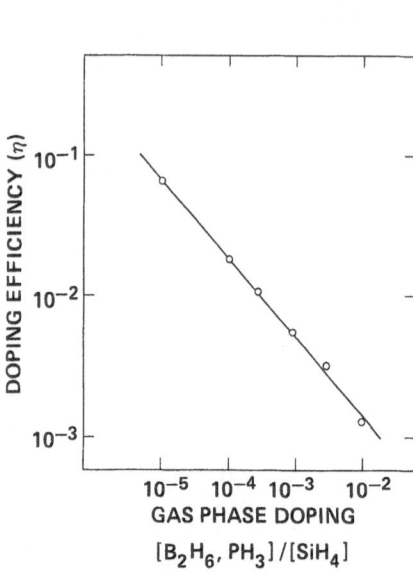

Fig. 5.42. The doping efficiency for both boron and phosphorus obtained from the defect densities estimated in the luminescence and ESR measurements of Fig. 5.39. There is a pronounced decrease in efficiency at high doping levels

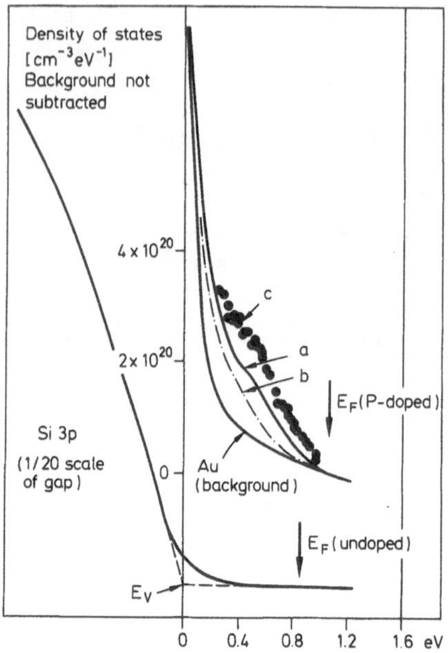

Fig. 5.43. Detail of valence band edge obtained from photoemission for three samples: *(a)* undoped, deposited at 250 °C; *(b)* annealed to 350 °C; *(c)* deposited at 250 °C with heavy phosphorus doping [5.106]

c) Compensation

Compensation is, in principle, a powerful technique. It allows one to vary E_F and the impurity concentrations independently. Co-doping of a-Si:H with PH_3 and B_2H_6 has indeed been shown to return the Fermi energy to midgap [5.31, 32, 107]. However, the mechanism of compensation has not been conclusively demonstrated. Incorporation, as in crystalline silicon, may occur as electrically compensating isolated donors and acceptors. In glow discharge deposition there is evidence [5.32] for enhanced incorporation efficiency of the minority impurity in the gas phase. This is indicative, at least at high doping levels ($\geq 10^{-3}$), of some degree of chemical pairing and therefore creation of neutral centers in the sample. Until experiments such as nuclear magnetic resonance are performed to ascertain the local dopant configurations, it seems most plausible to assume that compensation is dominated by isolated ionized impurities.

As discussed above, increased numbers of dangling-bond-like defects are correlated with increased dopant levels. Studies of compensated samples

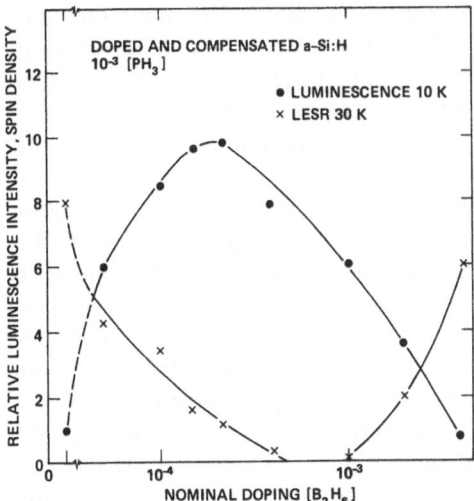

Fig. 5.44. Luminescence and LESR data for samples doped with 10^{-3} [PH$_3$] and compensated with various amounts of boron showing the reduction in defect density and corresponding increase in luminescence with compensation [5.108]

have been carried out in order to determine if the induced gap states are chemically related to the impurities, or if their presence is associated solely with the position of the Fermi energy [5.32]. Figure 5.44 shows results of LESR and luminescence experiments on a set of samples having fixed phosphorus concentration and varying boron concentration. Conductivity activation energies and signs of thermopower measurements on the same samples show that E_F moves through midgap very nearly at the nominal compensation level. Figure 5.44 shows that the dangling-bond light-induced spin density (which for all samples is undetectable in the dark) disappears near compensation even though twice as many impurities are present than for the singly doped samples. Also observed in LESR is a resonance which is similar to the valence band-tail hole signature. Thus, the compensated material differs from undoped material.

Luminescence measurements (Fig. 5.44) also reflect the decrease of nonradiative deep gap states. The luminescence efficiency of singly doped a-Si:H increases dramatically with the addition of complementary dopant; however, not to the level of undoped material. Furthermore, the 0.9 eV luminescence band also disappears with compensation. It is thus clear that co-doping removes dangling bonds. The luminescence transition in compensated a-Si:H has been identified to arise from the recombination of band-tail carriers. However, there is a continuous shift of the peak to low energy with co-doping which has been associated with band-tail spin states.

PDS measurements have been performed on the same samples. In Fig. 5.41, the crosses show agreement with luminescence data and indicate again that the dangling-bond density is reduced by compensation. Figure 5.40 shows PDS spectra for samples doped with P, B or both at the 10^{-3} level. The curves demonstrate that compensation both decreases the density of deep states and increases a shallower subgap absorption. From luminescence decay and temperature dependence [5.32] and from drift mobility measure-.ments [5.107], it is found that the electron tail states are not noticeably affected by doping or codoping. The data are therefore taken to indicate the growth of a band of states deep in the valence band tail to which the photo-induced holes rapidly thermalize before radiatively recombining with band-tail electrons. Two missing pieces of this picture remain. First, LESR should occur in pairs of lines if the excited states are isolated, positive U states, but the band-tail electron LESR has not been observed. Second, a competitive nonradiative channel has not been identified. It is possible that the distribution of deep states giving rise to the low energy luminescence, as in the case of the 0.9 eV defect luminescence band, is dominated by sites where nonradiative recombination is more probable than radiative recombination.

If compensation is strictly a chemical pairing during deposition, then it is possible that defect creation in singly-doped material is chemically-induced. However, at low gas phase concentrations, chemical pairing is quite improbable. If, then, on the other hand, compensation is dominantly electrical, the implication of the data is that doping-induced defect creation is related only to the position of the Fermi energy. Furthermore, the strongly sublinear increase of the defect density with doping concentration shown in Fig. 5.39 implies that the highest probability for defect creation occurs at the lowest doping levels, i.e., when $|E_d - E_F|$ is the greatest (here E_d is the energy of the ionized donor or acceptor).

An autocompensation mechanism analogous to that occuring in crystalline semiconductors has been proposed to explain the results [5.31, 108]. In this model deep compensating defects are created along with dopant incorporation to reduce the net electrical and structural energy of the system. The electronic energy gains from defect creation are greatest when $|E_d - E_F|$ is greatest and so defect creation is most probable. The microscopic details of these defects have not been directly demonstrated. No hyperfine broadening of the doping-induced dangling-bond state has been observed. Therefore, if, as is most likely, localization of the dangling-bond electron is strong, we cannot yet determine if the self-compensating dopant-defect pairs are distant pairs or nearest neighbors.

In summary, then, the most likely picture is that at low doping levels the doping efficiency (i.e., tetrahedral dopant incorporation) is high and defect creation is high. In material with low native defect density, E_F quickly jumps from midgap to the energy levels of the charged state of the induced defects. At high dopant concentration, the doping efficiency declines as threefold incorporation becomes more probable.

5.3.3 Impurities

A host of impurities have been incorporated into a-Si, either intentionally or inadvertently, usually at levels much higher than those in crystalline silicon.

The effects of impurities can be roughly categorized as follows: (a) alloying, (b) doping and (c) deep gap state formation. Alloying here includes modification and removal from the gap of otherwise deep states, and band and band-tail state alteration. Because of the disordered nature of a-Si, almost all impurities are probably involved in more than one modality. In Sect. 5.3.1 we have already discussed that impurities such as boron and phosphorus can introduce 4-fold coordinated, shallow, electrically active dopant states as well as 3-fold coordinated, "electrically inactive" states lying outside the gap. We have also shown that impurity incorporation can induce defect formation in the matrix. The role of impurities is thus very complex, depending on species, concentration, Fermi level, etc. In this section we shall, therefore, discuss only a small subset of possible impurity effects, treating only the nature of localized states associated with the impurities.

a) Hydrogen and Halogens

The role of hydrogen in purging states from the a-Si gap is almost entirely accepted. Ultraviolet photoemission (UPS) [5.109] measurements yield densities of states for a-Si and a-Si : H as shown in Fig. 5.45. It is evident that states predominantly at the top of the valence band are removed from the gap and replaced by deeper Si–H bonding states. No hydrogen-related levels are added in the bandgap. Similar UPS results have been obtained for fluorine alloying [5.110].

Although UPS measurements have not been reported for a-Si : H samples after thermally driving off the hydrogen, it is plausible to assume that the

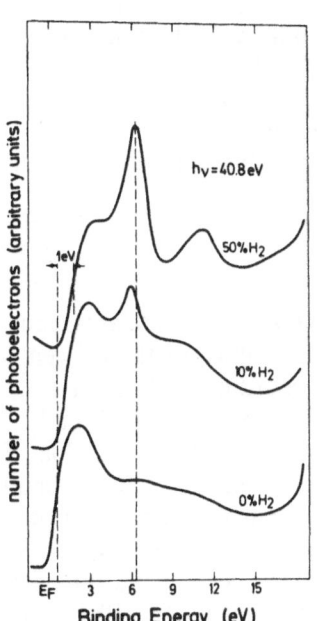

Fig. 5.45. Valence band density of states deduced from photoemission at 40.8 eV for a-Si samples reactivity sputtered in H_2 atmospheres. [H_2] indicated are gas phase concentrations [5.109]

UPS spectrum of a-Si would result. Then one could infer that the major effect of removing hydrogen is to introduce valence band-tail states. Dangling-bond defects at much lower levels (and therefore unobservable in UPS) are also to be expected. There have been many studies of the effects of dehydrogenation and rehydrogenation of a-Si : H. (Again, fewer but similar results have been found for a-Si : F.) Details of the hydrogen-related effects depend on such sample characteristics as microstructure (e.g., columnar biphasic structure or homogeneous alloying) and hydrogen bonding order (e.g., $\equiv$Si–H, $=$Si–H$_2$, etc.). The literature is too large to cover fully. However, a large degree of generalization about the nature of localized states can be achieved. Most hydrogen-related effects can be consistently explained by local transition between a-Si : H and a-Si-like densities of states. For example, when hydrogen is thermally evolved from a-Si : H, the absorption edge shifts to lower energy [5.18]. The band-edge luminescence peak also shifts to lower energies closely tracking the absorption edge. The band-edge luminescence efficiency is quenched and the defect luminescence increases as the dangling-bond spin density increases. The changes in conductivity and photoconductivity [5.107] are also in accord with an increase of valence band-tail states and deep defect states. It has been noted that a much greater quantity of hydrogen is evolved than spins created (or vice versa for deposition of a-Si in the presence of hydrogen). It is argued that during growth or H-evolution, most of the dangling bonds left behind reconstruct into weak Si–Si bonds [5.14, 18].

Annealing at low temperatures ($< 250\,^\circ$C), whether in as-deposited defective material or radiation damaged material, is explained by hydrogen motion (via bond switching) leading to increased dangling-bond passivation (as shown by *Engemann* et al. [5.111], oxygen can play a similar role in defect passivation). The shift of the luminescence peak to higher energies and narrowing with increasing annealing temperature are quantitatively consistent with a reduction in the disorder potential in the band tails. Finally, rehydrogenation by in-diffusion of atomic hydrogen [5.112] seems to reverse the evolution effects.

b) Oxygen, Nitrogen, Carbon and Germanium

After hydrogen, the next most studied, and probably most abundant impurity is oxygen. Infrared absorption studies show that most of the oxygen is bonded into the a-Si matrix predominantly as two-coordinated Si–O–Si [5.15, 113]. Transmission measurements show that alloying increases and broadens the absorption edge. At low concentrations, evidence for introduction of localized gap states comes from luminescence measurements. A broad band peaking at ~ 1.1 eV has been interpreted to arise from a transition between conduction band-tail electrons and holes trapped at deep, negatively charged, oxygen-related defects (similar results have been found for post-deposition oxidation of a-Si having columnar microstructure) [5.100]. *Grif-*

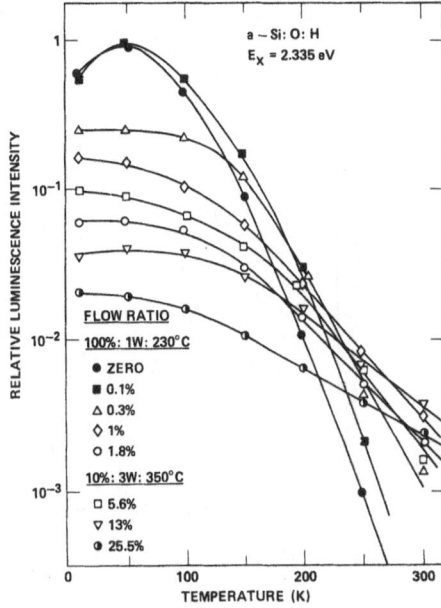

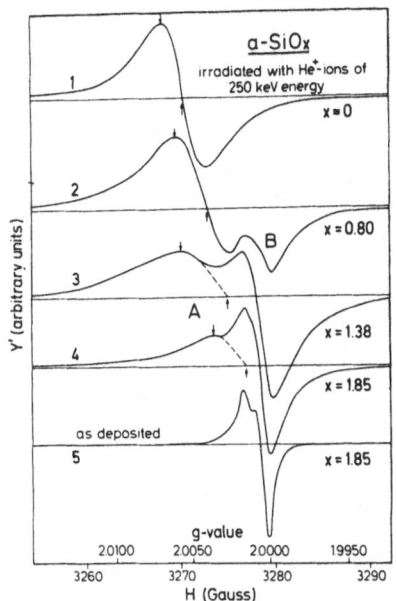

Fig. 5.46 Band-edge luminescence intensity versus temperature for varying oxygen concentrations in a-Si : O : H alloys [5.114]

Fig. 5.47. ESR spectra of a-Si : O : H alloys after He$^+$ irradiation. The spectra have been theoretically related to superpositions of dangling bond centers with 0–3 oxygen neighbors [5.115]

fith et al. [5.114] have found that oxygen incorporation into otherwise pure a-Si : H modifies the photoconductivity temperature dependence, thus adding or modifying existing gap states. No estimates of number or energy are obtained. Also, *Crandall* observed an increased density of states with oxygen content from DLTS [5.82]. At higher oxygen concentrations, only a single luminescence band is observed [5.15]. From Fig. 5.46 it can be seen that the low temperature luminescence efficiency decreases with increasing oxygen content [5.115]. This effect which is common to all a-Si alloys (except a-Si : H) is indicative of an increase in nonradiative, defect density. The Marburg group has observed [5.116] ESR (Fig. 5.47) from a-Si : O : H alloys after defect creation by He$^+$ bombardment. The spectra (Fig. 5.47) have been attributed to a sum of curves corresponding to silicon dangling bonds with 0–3 oxygen back-bonded neighbors. Presumably the native defects which quench the luminescence are similar dangling-bond sites.

For alloying in glow discharge deposited samples, the Xerox group found that optical properties all evolve smoothly from the a-Si : H values (Fig. 5.48). The absorption edge increases and broadens, the luminescence increases in energy (but by much less) and broadens. The results are quantitatively explained by an increasing electron-phonon coupling (and resultant

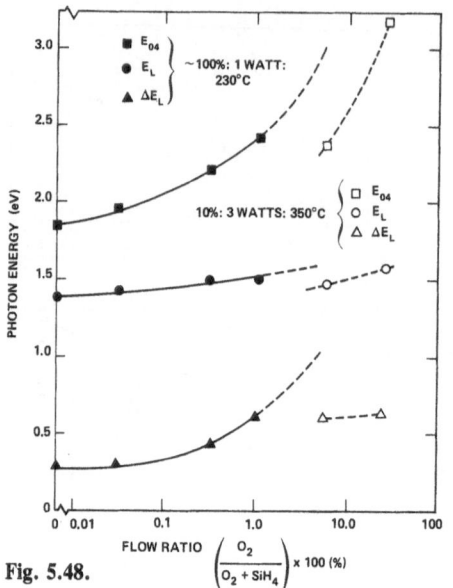

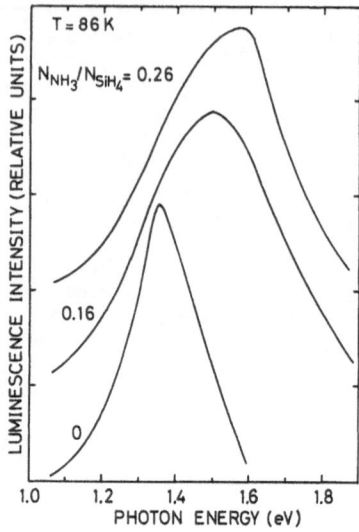

Fig. 5.48.   Fig. 5.49.

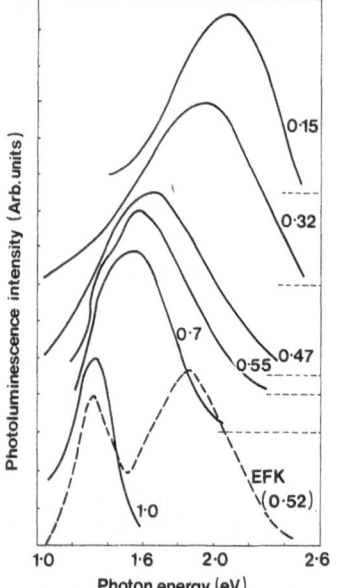

Fig. 5.48. Variations of luminescence peak energy (E_L), width (ΔE_L) and optical absorption (E_{04}) with gas phase oxygen concentration in a-Si : O : H alloys [5.115]

Fig. 5.49. Luminescence spectra of a-Si : N : H alloys [5.117]

Fig. 5.50. Luminescence spectra of a-Si : C : H alloys [5.120]. Dashed line is from the work of *Engemann* et al. [5.121]

Stokes shift) arising from the increased flexibility (lower average coordination) of the a-Si/a-SiO$_2$ alloy. Moreover, from the results of time-resolved luminescence decay measurements and temperature dependence studies (Fig. 5.46), it has been inferred that oxygen incorporation increases the width of the conduction band.

Just as in crystalline silicon, nitrogen in a-Si:H is not a shallow donor level. No changes in luminescence have been observed for low concentrations of nitrogen. (As for the case of oxygen impurities, *Griffith* et al. [5.114] found that nitrogen modifies the gap state distribution, but the details of the distribution have not been ascertained.) At higher nitrogen levels, the band-edge luminescence peak has been found to increase in energy as seen in Fig. 5.49 [5.117]. An ESR signal of increasing strength evolves as a single line from the dangling bond line of a-Si:H.

It is of interest to note [5.114] that much as in the case of boron/phosphorus co-doping, there is evidence for cooperative chemical effects between nitrogen and oxygen and introduction of valence band-tail hole traps.

Group IV elements are isoelectronic substitutional impurities. It was hoped that, if there were no topological changes, there migh be a simple alloy variation of the band gap from a-Ge:H to a-C:H [5.118]. Several groups [5.119–122] have studied the luminescence of group IV alloys and have found a single broad luminescence band arising from band to band recombination which follows the band gap. The work of *Sussmann* and *Ogden* is shown in Fig. 5.50 [5.120]. In all samples the low temperature luminescence efficiency is strongly reduced from that of a-Si:H by the presence of deep states associated with dangling bonds on Si, C or Ge [5.123].

Oxygen, nitrogen and carbon are the major impurities in a-Si:H. The evidence indicates that although the majority of impurity atoms are inactive, a low density of deep states is associated with each type of impurity. These states have proved to be difficult to study. Investigations of the alloys are useful, but the extrapolation of the results to low concentrations is of limited validity. A much better approach is to improve the purity of a-Si:H so controlled impurity concentrations can be studied.

5.4 Concluding Remarks

Through the spectroscopic techniques described in this chapter, a good understanding of localized states in a-Si:H is developing. Inevitably, as the broader issues are resolved with reasonable certainty, the investigations center around the finer questions for which the interpretation of the results is less sure. There are many examples of this in a-Si:H. The band tails are believed to be roughly exponential and of limited width in energy. The extent of localization, the magnitude of the correlation energy and of lattice distortion have each been estimated. However, the nature of the mobility edge and the details of the transport processes remain experimentally unresolved. Similarly, the existence of dangling-bond defects is widely accepted, as is the fact that a very low density of deep states can be achieved in a-Si:H. The energy distribution of the residual deep states is, on the other hand, very controversial.

Substitutional doping and the shift of the Fermi energy is well documented and so is the introduction of defects by the doping process. However, the details of the doping process are still unclear, and the question of what states are introduced by other impurities is a very important current problem. Regarding the role of localized states in the electronic properties, the low temperature recombination seems well understood, including the predominant role of dangling bonds as nonradiative centers. There is less agreement about the room temperature recombination and not much is known about the nonradiative mechanisms. The pace of progress in this field is such that we do not expect to have to wait long for an answer to many of these questions.

References

5.1   P. W. Anderson: Phys. Rev. **109**, 1492 (1958)
5.2   N. F. Mott: Adv. Phys. **16**, 49 (1967)
5.3   D. Weaire, H. F. Thorpe: Phys. Rev. B **4**, 2508 (1971)
5.4a   W. E. Spear (ed.): Proc. 5th Intern. Conf. on Amorphous and Liquid Semiconductors (CICL 1974), p. 1
5.4b   W. E. Spear, P. G. LeComber: Phil. Mag. **33**, 935 (1976)
5.5   J. C. Knights, R. Lujan: Appl. Phys. Lett. **35**, 244 (1979)
5.6   J. D. Joannopoulos, G. Lucovsky (eds.): *The Physics of Hydrogenated Amorphous Silicon I,* Topics Appl. Phys., Vol. 55 (Springer, Berlin, Heidelberg, New York 1983)
5.7   See, for example, G. D. Watkins, J. W. Corbett: Phys. Rev. **138**, A 543 (1965)
5.8   M. H. Brodsky, R. S. Title: Phys. Rev. Lett. **23**, 581 (1969)
5.9   Z. Z. Ditina, L. P. Strakhov, H. H. Helms: Fiz. i Tekh. Poluprov. **2**, 1199 (1968) [Engl. Trans.: Sov. Phys.-Semicond. **2**, 1006 (1969)]
5.10   K. L. Brower, W. Beezhold: J. Appl. Phys. **43**, 3499 (1972)
5.11   See, for example, G. E. Pake: *Paramagnetic Resonance* (Benjamin, New York 1962)
5.12   Y.-H. Lee, J. W. Corbett: Phys. Rev. B **8**, 2810 (1973)
5.13   J. C. Phillips: Comments Solid State Phys. **3**, 67 (1970)
5.14   H. Fritzsche: Proc. 7th Intern. Conf. on Amorphous and Liquid Semiconductors ed. by W. E. Spear (CICL 1977), p. 3
5.15   R. A. Street, J. C. Knights, D. K. Biegelsen: Phys. Rev. B **18**, 1880 (1978)
5.16   C. C. Tsai, H. Fritzsche, M. H. Tanielian, P. J. Gaczi, P. D. Persans, M. A. Vesaghi: Proc. 7th Intern. Conf. on Amorphous and Liquid Semiconductors ed. by W. E. Spear (CICL 1977), p. 339
5.17   U. Voget-Grote, W. Kömmerle, R. Fischer, J. Stuke: Phil. Mag. B **41**, 127 (1980)
5.18   D. K. Biegelsen, R. A. Street, C. C. Tsai, J. C. Knights: Phys. Rev. B **20**, 4839 (1979)
5.19   U. Voget-Grote, J. Stuke, H. Wagner: AIP Conf. Proc. **31**, 29 (1976)
5.20   R. A. Street, D. K. Biegelsen, J. Stuke: Phil. Mag. B **40**, 451 (1979)
5.21   H. Dersch, J. Stuke, J. Beichler: Phys. Stat. Sol. B **105**, 265 (1981)
5.22   S. Hasegawa, S. T. Yazaki: Thin Solid Films **55**, 15 (1978)
5.23   A. Abragam: *Principles of Nuclear Magnetism* (Clarendon Press, Oxford 1973), p. 128
5.24   C. A. J. Ammerlaan, A. Van der Wiel: J. Mag. Res. **21**, 387 (1976)
5.25   R. A. Street, D. K. Biegelsen, J. C. Zesch: Phys. Rev. B **25**, 4334 (1982)
5.26   P. J. Caplan, E. H. Poindexter, B. E. Deal, R. R. Razouk: J. Appl. Phys. **50**, 5847 (1979)

5.27 J. Singh, A. Madhukar: Appl. Phys. Lett. **38**, 884 (1981)
5.28 N. M. Johnson, D. K. Biegelsen, M. D. Moyer: In *Physics of MOS Insulators,* ed. by G. Lucovsky, S. T. Pantelides, F. L. Galeener (Pergamon, New York 1980), p. 311
5.29 H. Dersch, J. Stuke, J. Beichler: Phys. Stat. Sol. B **107**, 307 (1981)
5.30 S. Hasegawa, J. Kasajima, T. Shimizu: Phil. Mag. B **43**, 149 (1981)
5.31 R. A. Street, D. K. Biegelsen, J. C. Knights: Phys. Rev. B **24**, 969 (1981); R. Fischer, W. Rehm, J. Stuke, U. Voget-Grote: J. Non-Cryst. Solids **35/36**, 687 (1980)
5.32 A. Friederich, D. Kaplan: J. Phys. Soc. Jpn. Suppl. A. **49**, 1233 (1980)
5.33 S. Hasegawa, T. Kasajima, T. Shimizu: Solid State Commun. **29**, 13 (1979)
5.34 N. Ishii, H. Kumeda, T. Shimizu: J. Appl. Phys. Jpn. **20**, L 673 (1981); L 920
5.35 R. A. Street, D. K. Biegelsen: Solid State Commun. **33**, 1159 (1980)
5.36 D. K. Biegelsen, R. A. Street, W. Jackson: Physica **117/118** B, 899 (1983)
5.37 R. A. Street: Adv. Phys. **30**, 593 (1981)
5.38 R. A. Street: Phil. Mag. **37**, 35 (1978)
5.39 R. A. Street, J. C. Knights, D. K. Biegelsen: Phys. Rev. B **18**, 1880 (1978)
5.40 D. Engemann, R. Fischer: In *Amorphous and Liquid Semiconductors,* ed. by J. Stuke, W. Brenig (Taylor and Francis, London 1974), p. 947; R. A. Street, C. Tsang, J. C. Knights: Inst. Phys. Conf. Ser. **43**, 1139 (1979)
5.41 C. Tsang, R. A. Street: Phys. Rev. B **19**, 3027 (1979)
5.42 D. G. Thomas, J. J. Hopfield, W. M. Augustyniak: Phys. Rev. **140**, 202 (1965)
5.43 C. Tsang, R. A. Street: Phil. Mag. B **37**, 601 (1978)
5.44 R. W. Collins, M. A. Paesler, W. Paul: Solid State Commun. **34**, 833 (1980); I. G. Austin, T. S. Nashashibi, T. M. Searle, P. G. LeComber, W. E. Spear: J. Non-Cryst. Solids, **32**, 373 (1979)
5.45 R. A. Street: Phys. Rev. B **23**, 861 (1980)
5.46 R. A. Street, T. M. Searle, I. G. Austin: In *Amorphous and Liquid Semiconductors,* ed. by J. Stuke, W. Brenig (Taylor and Francis, London 1974), p. 953
5.47 R. A. Street: J. Phys. (Paris) C 4, **42**, 575 (1981)
5.48 D. Engemann, R. Fischer: Phys. Stat. Sol. (b) **79**, 195 (1977); R. A. Street: Phys. Rev. B **21**, 5775 (1980)
5.49 R. A. Street, D. K. Biegelsen, J. Stuke: Phil. Mag. B **40**, 451 (1979)
5.50 W. Rehm, D. Engemann, R. Fischer, J. Stuke: Proc. 13th Intern. Conf. on Physics of Semiconductors ed. by F. G. Fumi (1976), p. 525
5.51 B. C. Cavenett: Adv. Physics, **30**, 475 (1981); S. Depinna, B. C. Cavenett: Phil. Mag. (to be published)
5.52 R. A. Street: Phys. Rev. B **26**, 3588 (1982)
5.53 See, for example, N. F. Mott, E. A. Davis: *Electronic Processes in Non-Crystalline Solids* (Oxford 1979)
5.54 J. Tauc: In *The Optical Properties of Solids,* ed. by F. Abeles (North-Holland, Amsterdam 1970), p. 277
5.55 W. B. Jackson, N. M. Amer: Phys. Rev. B **25**, 5559 (1982)
5.56 S. Yamasaki, H. Okushi, A. Matsuda, H. Oheda, N. Hata, K. Tanaka: J. Appl. Phys. Jpn. **20**, L 665 (1981)
5.57 G. Modell, D. A. Anderson, W. Paul: Phys. Rev. B **22**, 1918 (1980); B. Abeles, C. R. Wronski, T. Tiedje, G. D. Cody: Solid State Commun. **36**, 537 (1980)
5.58 R. S. Crandall: Phys. Rev. Lett. **44**, 749 (1980)
5.59 J. D. Dow, D. Redfield: Phys. Rev. B **1**, 3358 (1970)
5.60 J. Toyozawa: Prog. Theor. Phys. **22**, 455 (1959)
5.61 W. B. Jackson, R. J. Nemanich, N. Amer: Phys. Rev. B **27**, 4861 (1983) W. B. Jackson: Solid State Commun. **22**, 133 (1982)
5.62 Z. Vardeny, J. Tauc: Phys. Rev. Lett. **46**, 1223 (1981)
5.63 P. O'Connor, J. Tauc: Solid State Commun. **36**, 947 (1980)
5.64 M. Olivier, J. C. Peuzin, A. Chenevas-Paule: J. Non-Cryst. Solids **35/36**, 693 (1983)
5.65 Z. Vardeny, P. O'Connor, S. Ray, J. Tauc: Phys. Rev. Lett. **44**, 1267 (1980)

5.66 N. F. Mott: J. Non-Cryst. Solids **1**, 1 (1968)
5.67 M. L. Knotek: Solid State Commun. **17**, 1431 (1975)
5.68 D. V. Lang: In *Thermally Stimulated Relaxation in Solids,* ed by P. Bräunlich, Topics Appl. Phys. Vol. 37 (Springer, Berlin, Heidelberg, New York 1979), Chap. 3
      J. Bourgoin, M. Lannoo: *Point Detects in Semiconductors* II, Springer Ser. Solid-State Sci., Vol. 22 (Springer, Berlin 1983)
5.69 H. Scher, E. W. Montroll: Phys. Rev. B **12**, 2455 (1975);
      G. Pfister, H. Scher: Adv. in Phys. **27**, 747 (1978);
      M. Pollak: Phil. Mag. **36**, 1157 (1977)
5.70 T. Tiedje, A. Rose: Solid State Commun. **37**, 49 (1981);
      T. Tiedje, A. Rose, J. M. Cebulka: AIP Conf. Proc., **73**, 197 (1981)
5.71 J. M. Hvam, M. H. Brodsky: Phys. Rev. Lett. **46**, 371 (1981);
      R. A. Street: Solid State Commun. **39**, 263 (1981)
5.72 W. E. Spear, P. G. LeComber: Phil. Mag. **33**, 935 (1976);
      A. Madan, P. G. LeComber, W. E. Spear: J. Non-Cryst. Solids **20**, 39, 239 (1976)
5.73 N. Goodman, H. Fritzsche: Phys. Mag. B **42**, 149 (1980)
5.74 R. L. Weisfield, D. A. Anderson: Phil. Mag. B **44**, 83 (1981)
5.75 For a review see H. Fritzsche: Solar Energy Mat. **3**, 447 (1980)
5.76 R. L. Weisfield, P. Viktorovitch, D. A. Anderson, W. Paul: Appl. Phys. Lett. **39**, 263 (1981)
5.77 J. Beichler, W. Fuhs, H. Mell, H. M. Welsch: J. Non-Cryst. Solids **35/36**, 587 (1980);
      P. Viktorovich, G. Moddel: J. Appl. Phys. **51**, 4847 (1980)
5.78 T. Tiedje, C. R. Wronski, B. Abeles, J. M. Cebulka: Solar Cells **2**, 301 (1980)
5.79 A. J. Snell, K. D. Mackenzie, P. G. LeComber, W. E. Spear: Phil. Mag. B **40**, 1 (1979);
      W. E. Spear, P. G. LeComber, A. J. Snell: Phil. Mag. B **38**, 303 (1978);
      R. A. Abram, P. Doherty: J. Phil. Mag. B **45**, 167 (1982)
5.80 T. Tiedje, T. D. Moustakas, J. M. Cebulka: Phys. Rev. B **23**, 5634 (1981);
      M. Hirose, T. Suzuki, G. H. Dohler: Appl. Phys. Lett. **34**, 234 (1979)
5.81 J. D. Cohen, D. V. Lang: Phys. B **25**, 5321 (1982);
      D. V. Lang, J. D. Cohen, J. P. Harbison: Phys. Rev. B **25**, 5285 (1982)
5.82 R. S. Crandall: Phys. Rev. B **24**, 7457 (1981)
5.83 M. J. Thompson, N. M. Johnson, R. A. Street: J. Phys. (Paris) C 4, **42**, 617 (1981)
5.84 H. Okushi, Y. Tokumaru, S. Yamasaki, H. Oheda, K. Tanaka: J. Phys. (Paris) C 4, **42**, 613 (1981)
5.85 H. Okushi, Y. Tokumaru, S. Yamasaki, H. Oheda, K. Tanaka: Phys. Rev. B **25**, 4313 (1982)
5.86 I. Balberg, D. E. Carlson: Phys. Rev. Lett. **43**, 58 (1979)
5.87 G. D. Cody, T. Tiedje, B. Abeles, B. Brooks, Y. Goldstein: Phys. Rev. Lett. **47**, 1480 (1981)
5.88 J. C. Knights, R. Lujan: Appl. Phys. Lett. **35**, 244 (1979)
5.89 J. C. Knights: Jpn. J. Appl. Phys. **18** (Suppl. 18-1) 101 (1979)
5.90 D. K. Biegelsen, R. A. Street, C. C. Tsai, J. C. Knights: Phys. Rev. B **20**, 4839 (1979);
      D. Engemann, R. Fischer: In *The Physics of Semiconductors,* ed. by M. H. Pilkuhn (Teubner, Stuttgart 1974), p. 1042
5.91 J. I. Pankove: Phys. Lett. **32**, 812 (1978)
5.92 D. L. Staebler, C. R. Wronski: Appl. Phys. Lett. **31**, 292 (1977)
5.93 H. Dersch, J. Stuke, J. Beichler: Appl. Phys. Lett. **38**, 456 (1981)
5.94 I. Hirabayashi, K. Morigaki, S. Nitta: J. Appl. Phys. Jpn. **19**, L 357 (1980);
      J. Shah, A. E. Digiovanni: Solid State Commun. **37**, 153 (1981)
5.95 J. D. Cohen, D. V. Lang, J. P. Harbison, A. M. Sergent: J. Phys. (Paris) C 4, **42**, 371 (1981)
5.96 D. Adler: J. Phys. (Paris) C 4, **42**, 3 (1981)
5.97 J. Reimer, R. W. Vaughan, J. C. Knights: Phys. Rev. B **23**, 2567 (1981)
5.98 M. Tanelian, H. Fritzsche, C. C. Tsai, E. Symbalisty: Appl. Phys. Lett. **33**, 353 (1978)

5.99   I. Solomon, T. Dietl, D. Kaplan: J. Phys. (Paris) **39**, 1241 (1978)

5.100a W. B. Jackson, D. K. Biegelsen, R. J. Nemanich, J. C. Knights: Appl. Phys. Lett. **42**, 105 (1983)

5.100b R. A. Street, J. C. Knights: Phil. Mag. B **43**, 1091 (1981)

5.101 J. C. Knights, T. M. Hayes, J. C. Mikkelsen Jr.: Phys. Rev. Lett. **39**, 712 (1977)

5.102 G. Feher: Phys. Rev. **114**, 1219 (1959)

5.103 J. C. Knights: Phil. Mag. **34**, 663 (1976)

5.104 P. G. LeComber, W. E. Spear, G. Muller, S. Kalbitzer: J. Non-Cryst. Solids **35/36**, 327 (1980)

5.105 W. Beyer, R. Fischer: Appl. Phys. Lett. **31**, 850 (1977)

5.106 B. von Roedern, G. Moddel: Solid State Commun. **35**, 467 (1980)

5.107 D. Allen, P. G. LeComber, W. E. Spear: Proc. 7th Intern. Conf. on Amorphous and Liquid Semiconductors, ed by W. E. Spear (CICL 1977), p. 323

5.108 D. K. Biegelsen, R. A. Street, J. C. Knights: AIP Conf. Proc. **73**, 166 (1981)

5.109 B. von Roedern, L. Ley, M. Cardona: Phys. Rev. Lett. **39**, 1576 (1977)

5.110 L. Ley, H. Richter, R. Kärcher, R. L. Johnson, J. Reicherdt: Proc. 9th Intern. Conf. on Amorphous and Liquid Semiconductors, p. 753

5.111 D. Engemann, R. Fisher, F. W. Richter, H. Wagner: Proc. 6th Intern. Conf. on Amorphous and Liquid Semiconductors, ed. by B. T. 2 Kolomiets (NAUKA, Moscow 1976), p. 217

5.112 J. I. Pankove: Appl. Phys. Lett. **32**, 812 (1978)

5.113 M. A. Paesler, D. A. Anderson, E. C. Freeman, G. Moddel, W. Paul: Phys. Rev. Lett. **41**, 1492 (1978);
M. A. Paesler, W. Paul: Phil. Mag. B **41**, 393 (1980)

5.114 R. W. Griffith, F. J. Kampas, P. E. Vanier, M. D. Hirsch: J. Non-Cryst. Solids **35/36**, 391 (1980)

5.115 J. C. Knights, R. A. Street, G. Lucovsky: J. Non-Cryst. Solids **35/36**, 179 (1980);
R. A. Street, J. C. Knights: Phil. Mag. B **43**, 1091 (1981)

5.116 E. Holzenkämpfer, F. W. Richter, J. Stuke, U. Voget-Grote, J. Non-Cryst. Solids **32**, 327 (1979)

5.117 H. Kurata, M. Hirose, Y. Osaka: Jpn. J. Appl. Phys. **20**, L811 (1980)

5.118 A. Onton, H. Wiecker, D. Chevallier, C. R. Gvarnieri: Proc. 7th Intern. Conf. on Amorphous and Liquid Semiconductors ed. by W. E. Spear (CICL 1977)

5.119 H. Munekata, S. Murasato, H. Kunimoto: Appl. Phys. Lett. **37**, 536 (1980)

5.120 R. S. Sussmann, R. Ogden: Phil. Mag. B **44**, 137 (1981)

5.121 D. Engemann, R. Fischer, J. Knecht: Appl. Phys. Lett. **32**, 567 (1978)

5.122 D. Hauschildt, R. Fischer, W. Fuhs: Phys. Stat. Sol. (b) **102**, 563 (1980)

5.123 T. Shimizu, M. Kumeda, Y. Kiriyama: AIP Conf. Proc. **73**, 171 (1981)

6. Time-Resolved Charge Transport in Hydrogenated Amorphous Silicon

By T. Tiedje

With 25 Figures

In this chapter, we review recent work on time-resolved charge transport in hydrogenated amorphous silicon (a-Si : H). Most of the emphasis is on transient photoconductivity, an area in which substantial progress has been made recently. Perhaps the most important conclusion of this work is that a multiple-trapping model can explain a large body of time-dependent charge-transport phenomena in a-Si : H. As a result, the multiple-trapping model will be a unifying theme.

6.1 Background

We begin with a discussion of some recent dc transport results that bear on a problem of fundamental importance in the time-dependent problem, namely, the energy position of the mobility edges. Section 6.2 contains a discussion of the experimental problems associated with transient photoconductivity measurements. In Sect. 6.3, the multiple-trapping model of dispersive transport is explained first in terms of physical concepts and then with exact solutions so that the spatial distribution of the charge density can be illustrated. In Sect. 6.4, experimental results from time-of-flight experiments, photoconductivity decay and photoinduced absorption are described and compared with the predictions of the multiple trapping model.

Several more general reviews of charge transport in a-Si : H have been published recently. *Fritzsche* [6.1] gave a comprehensive description of the status of work on glow discharge a-Si : H in 1980; *Moustakas* [6.2], and *Paul* and *Anderson* [6.3] emphasized sputtered material; *Nagels* [6.4] concentrated on fundamental aspects of charge transport in amorphous semiconductors; *LeComber* and *Spear* [6.5] focused on doped material. In addition, the book edited by *Mort* and *Pai* [6.6] is a valuable reference on transient photoconductivity.

6.1.1 Mobility Gap

The mobility edges separate the localized gap states from the extended transport states. Although the focus of this chapter is on time-resolved experiments, dc experiments provide the most reliable information about the position of the mobility edges. In practice, only the separation between the mobility edges or mobility gap can be measured by these techniques. The most convincing data comes from the activation energy of the conductivity when the Fermi level is at midgap, but even this technique has problems when one asks for better than 0.1 eV accuracy. The problem is that the prefactor in the conductivity and its temperature dependence are not well understood [6.7–10].

Some of the best data on the mobility gap by this technique has been obtained by *Beyer* et al. [6.11] who studied the temperature dependence of the conductivity and thermoelectric power in a series of compensated a-Si : H films with different Fermi level positions. In one of the films the thermoelectric power was zero over 150 K, an indication that the Fermi level is at midgap. The conductivity activation energy of the sample was 0.87 eV, which implies a mobility gap of 1.74 eV. The optical gap for this material, defined in the usual way [6.12] from the extrapolation of $(\alpha h\nu)^{1/2}$, was 1.58 eV. The relevant optical gap for comparison with the thermal activation gap is the extrapolated zero temperature gap (temperature coefficient 4.3×10^{-4} eV/K [6.13]) or 1.71 eV.

This close agreement between the optical gap (1.71 eV) and the mobility gap (1.74 eV) suggests that the localized state distributions near the mobility edges are relatively narrow ($\lesssim 0.1$ eV) and appears to contradict suggestions in the literature [6.9, 14, 15] that the band-tail states have distributions whose widths are given roughly by the activation energies of the electron (0.15–0.19 eV) and hole (0.35–0.4 eV) drift mobilities.

6.1.2 Transport Path

Another important question for time-resolved experiments is the importance of hopping between deep gap states as a transport mechanism. This effect has been used in the interpretation of certain transport experiments on a-Si : H diode structures [6.16, 17]. An approach that gives insight into the transport mechanism is the effect of a magnetic field on the transport properties. Various magnetic field effects have been studied in a-Si : H, namely, the Hall effect [6.18, 19], photoelectromagnetic (PEM) effect [6.20] and the magnetic field dependence of photoconductivity [6.21–23] and dark conductivity [6.24]. The magnetic field dependence of the conductivity demonstrates that if hopping between deep gap states does occur, it does not make a significant contribution to the overall transport process in undoped and lightly doped material [6.24]. The argument is as follows. The magnetic field affects the transition rate of electrons between localized states because it couples to the electron spin, and the spin affects the allowed transitions through the exclusion principle. A reduction in the transition rate between localized states will reduce the conductivity in a hopping system, but have exactly the opposite effect on the photoconductivity in a conventional band conductor. Here a reduction in the transition rates will reduce recombination and enhance the photoconductivity.

A reversal in the sign of the magnetic field effect is, in fact, observed [6.24] in going from photoconductivity in lightly doped material to dark conductivity in heavily doped ($[PH_3]/[SiH_4] \gtrsim 10^{-2}$ or $[B_2H_6]/[SiH_4] \gtrsim 10^{-3}$) material. Presumably hopping near the Fermi level dominates in the heavily doped material. Also, the effect of the magnetic field on the dark conductivity falls off very rapidly at low doping levels [6.24], as illustrated in Fig. 6.1. From this data, one can conclude that in undoped and lightly doped material, transport in states close to the band edge dominates, regardless of the Fermi level position.

In the remainder of this chapter we will use these assumptions, namely, that charge transport in undoped and lightly doped a-Si:H requires thermal activation of the carriers to the band edge and that the localized state distributions near the mobility edges are relatively narrow ($\lesssim 0.1$ eV).

6.2 Transient Photoconductivity Measurements

6.2.1 Time-of-Flight Technique

The time-of-flight experiment has been discussed extensively in the literature [6.25, 26]. The objective of the experiment is the measurement of the length of time required for a photo-injected packet of electrons to drift from one side of the sample to the other. The electron (or hole) drift mobility is then defined by the transit time t_T through the relation $\mu_D = L^2/(t_T V)$, where L is the sample thickness and V the applied voltage.

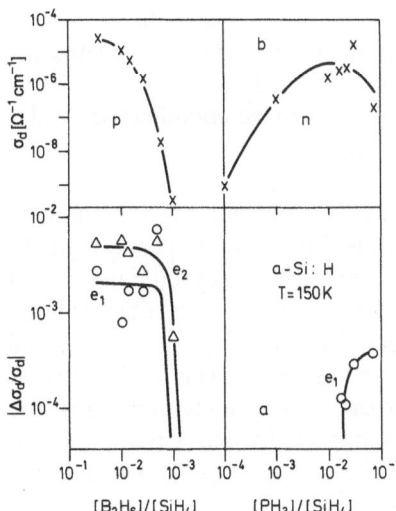

Fig. 6.1. Dark conductivity and magneto-conductance as a function of dopant concentration [6.24]. e_1 and e_2 refer to two different components of the magnetic field response in the 0–10 kG range. Charge transport occurs by hopping near the Fermi level in the region of nonzero magneto-conductance

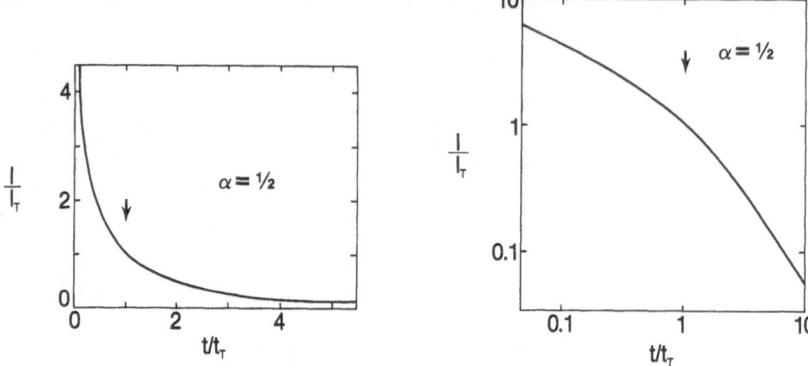

Fig. 6.2. Typical dispersive time-of-flight current transient for an amorphous semiconductor. The current decay was calculated from (6.22) and is plotted on both a linear and logarithmic scale. The arrows mark the transit time

In an ideal crystalline material, the drifting electrons generate a constant displacement current while they are in motion. The current terminates when the electrons reach the back contact and this termination is broadened by the diffusive spreading of the charge packet during its transit across the sample.

In amorphous semiconductors, the charge packet spreads out or disperses to a much greater extent than one would expect from diffusion alone and the current measured in the external circuit starts to decay immediately with a form analogous to Fig. 6.2. In most situations the dispersion can be modeled by a drift mobility that decreases as a power law in time ($t^{\alpha-1}$, for $0 < \alpha < 1$) [6.27]. It is a straighforward matter to show that this time dependence for the mobility leads to a transit time that is a nonlinear function of sample thickness and electric field E, according to the relation $t_T \propto (L/E)^{1/\alpha}$. As a result, when the drift mobility is defined in terms of the transit time in the usual way, it depends on sample thickness and field, in contrast to the normal situation where the mobility in a given material has a single value. Although the dispersion makes carrier transport measurements more complicated, it also contains interesting information about the electronic structure of the material.

The basic condition that must be met for a time-of-flight experiment to be feasible is that the carrier transit time t_T must be short compared with the time required to screen out an injected space charge through redistribution of the background charge inside the material. This requirement is satisfied if the transit time is short compared to the dielectric relaxation time $t_d = \varrho\varepsilon$, and the contacts are blocking. The experiment is still feasible even if t_d is short, as long as the density of trapped charge that can be thermally emitted in the time t_T is not great enough to perturb the applied field. This situation would exist, for example, in a very clean material in which a depletion region can extend across the entire thickness of the sample.

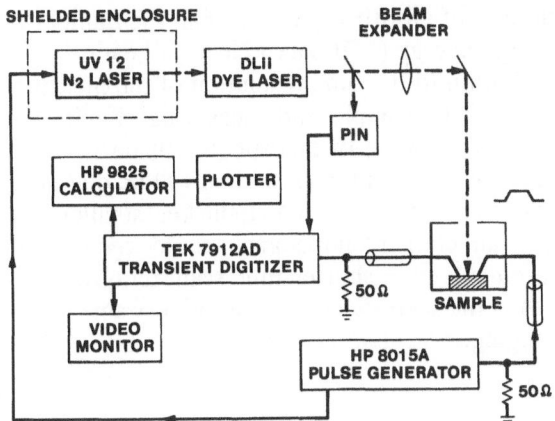

Fig. 6.3. Typical experimental apparatus used in the time-of-flight experiment [6.97 a]. The calculator is used for signal averaging and log conversion

In a typical experiment on amorphous silicon, electron-hole pairs are generated by a 10 ns flash of light from a N_2 or dye laser incident on a semi-transparent Schottky barrier contact [6.28]. One experimental configuration is shown in Fig. 6.3. One sign of carrier is pulled across the thin film sample by a bias voltage applied between the Schottky contact and a collecting contact on the opposite surface. The bias is applied in the form of a pulse so that the applied field is uniformly distributed across the sample and not concentrated at the contacts [6.29]. Even so, the field inside the sample can still be quite nonuniform because of the space charge field near the Schottky contact. In good electronic quality material with a band-bending of 0.5 V at the contact, the space charge layer can easily extend 0.5 μm into the film [6.30–33]. Wide space charge layers are desirable for a good solar cell [6.34] but can distort the photocurrent decay in a time-of-flight experiment.

The space charge due to the field of the photogenerated carriers themselves is another potential source of distortion. These space charge effects, namely, current saturation at high injection intensities and a current cusp near the space-charge free transit time, have been studied extensively in crystalline semiconductors [6.35]. Both transient [6.36] and steady-state space-charge limited currents [6.37] have been observed in amorphous silicon. Space charge effects can be neglected as long as the photogenerated charge is small compared to CV, where C is the sample capacitance and V the applied voltage. In this limit the photocurrent should scale linearly with the density of injected charge.

In practice, however, it is possible for nonlinearities to set in below the space charge limit because of trap saturation effects [6.38], although such phenomena have not been reported. Because of potential ambiguities from these and other effects, it is desirable to have a measure of the total injected charge density that does not depend on integrating the measured current transient. From this point of view, optical carrier injection may be preferable to electron beam injection (which is sometimes used [6.25]) since it is probably easier to measure the number of absorbed photons than the number of free electrons produced by a high energy (~ 10 keV) electron beam.

The main experimental limitation to the time resolution of the measurement is the RC response time of the sample and measuring circuit. For example, a 1 µm thick, 2 mm^2 area sample of amorphous silicon in series with a 50 Ω load, has a 10 ns RC time. This response time will be degraded further by the sheet resistance of the contacts. Although thicker, smaller area samples measured at lower characteristic impedance could reduce the response time to 1 ns, measurements on a shorter time scale require structures with lower capacitance than the sandwich structure, or very small samples and extensive signal averaging.

6.2.2 Photoconductivity Decay

Better time resolution can be obtained in measurements of photoconductivity decay with coplanar electrodes. If the photocurrent can be monitored on a time scale which is sufficiently short that no significant recombination can occur, then the photoconductivity decay is a measure of the time-dependent transport properties. The amplitude of the photocurrent per injected electron-hole pair is then a measure of the sum of the electron and hole drift mobilities. Very fast measurements are possible because the coplanar electrodes can have very small capacitance. In principle, fast photoconductivity measurements are the most direct probe of the charge transport in the vicinity of the mobility edge. In the first few picoseconds or fractions of a picosecond, the photoexcited carriers generated in band states thermalize down to the mobility edge. Initially the thermalized carriers interact primarily with the high density of shallow localized states close to the mobility edge.

Auston and his collaborators have developed an elegant electro-optic correlation technique for measuring photoconductivity on a ~ 10 ps time scale [6.39]. This technique is a variation of the pump-probe technique used in time-resolved picosecond optical experiments [6.40]. The photoconductor of interest is excited by one optical pulse and then electrically biased by a pulse from a fast photoconductive switch driven by a delayed optical pulse. A suitable microstrip transmission line pattern that transmits the bias pulse to the sample is shown in Fig. 6.4. The bias pulse drives a quantity of charge through the photoconductor which is proportional to the convolution of the photoconductive response of the sample with the bias pulse. The charge, which is proportional to the sample's photoconductivity, is monitored as a dc current through the sample induced by a high repetition rate mode-locked laser. The 10 ps response time limit of the measurement is defined by the relaxation rate of the fast photoconductive switch (radiation damaged silicon) and the circuit time constant [6.39]. Direct electrical measurements of the photoconductivity decay are possible down to 40 ps with a sampling oscilloscope.

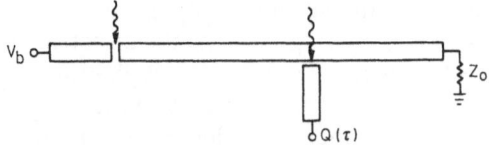

Fig. 6.4. An electrode pattern used in picosecond photoconductivity measurements [6.39]. The bias voltage V_b is applied to the sample through a fast photoconductive switch located at the gap in the electrode pattern on the left. The sample is located at the gap on the right-hand side

A similar sample geometry is used in measurements of the photoconductivity into the millisecond regime, where the decay of the carrier density by recombination becomes the dominant factor. For these experiments, a 10 ns N_2 pumped dye laser with a 0–100 Hz repetition rate is an appropriate excitation source. Larger electrodes are acceptable for measurements in the nanosecond regime. For example, in the nanosecond work that has been reported, the active part of the sample is a 1 mm gap between two parallel finger electrodes that are 1 cm long [6.41–43], while for the picosecond experiments, the active part of the sample is a 50 μm wide gap between two 0.5 mm wide striplines meeting end on [6.44, 45].

In both the picosecond work of *Johnson* et al. [6.44, 45] and the nanosecond work of *Hvam* and *Brodsky* [6.41, 42], the excitation pulse generates of order 10^{17} electron-hole pairs per cm^3. This density is substantially higher than the maximum density tolerable in a time-of-flight experiment. In the latter experiment, the space charge of the separated electrons and holes limits the maximum density to $\sim 10^{14}$ cm^{-3} averaged over a 1 μm thick film [6.29].

The high carrier densities used in the photoconductivity decay experiments introduces two potential problems. First a-Si:H excited with 10^{17} electrons cm^{-3}, each electron with mobility 1 cm^2/V s, has a dielectric relaxation time of about 70 ps. Thus, the picosecond experiments are on the edge of a space-charge relaxed condition, while in the nanosecond experiments the sample will be fully relaxed. In this case, reliable measurements of the photocurrent decay require contacts that remain ohmic during the entire photocurrent decay process. Unless the contacts are ohmic to both electrons and holes, a condition that is difficult to test for experimentally, space charge layers could build up near the contacts. The photoconductivity decay at long times would then be complicated by the build-up and relaxation of space charge layers near the contacts.

6.3 Multiple-Trapping Model of Dispersive Transport

Theoretical work on dispersive transport in amorphous semiconductors has focused on two different mechanisms, usually referred to as hopping [6.26,

27, 46] and multiple trapping [6.47–50]. In the pure hopping model, the transport occurs by tunneling between neighboring sites and the dispersion is due to the random distribution of site separations. In the multiple-trapping model, an electron in a localized site must first be thermally excited above the mobility edge before it can move to another site. The dispersion arises from the distribution of energy levels for the localized states. In both cases a small fluctuation in the relevant parameter of the system, namely, the site separation in the hopping model or the trap depth in the trapping model, can lead to enormous fluctuations in the transition rate between sites. Dispersive transport occurs when the average electron experiences a single localization event whose characteristic time is comparable to the mean transit time defined by the electric field and sample geometry. (In the following we restrict the discussion to electrons and assume it can be trivially generalized to holes.)

Probably the most significant conceptual difference between the hopping and trapping models of charge transport is that in the hopping model, the capture and release rates are symmetric whereas in the multiple-trapping model, the two rates are asymmetric. In the trapping model, all of the localized states can have the same capture rates independent of their release rates. In the hopping model, on the other hand, if a site is difficult to get out of it is also difficult to get into. This feature of hopping transport has led several authors to the conclusion that at long times, dispersion always disappears in pure hopping systems [6.51–53].

Another difference between the two models is in their temperature dependences. In the hopping model all the sites have the same energy (within kT) and both the electron drift mobility and its dispersion are independent of temperature. On the other hand, since thermal excitation is involved in the multiple-trapping process, this model always leads to strong temperature dependences for both the drift mobility and the dispersion. Various combinations of the two transport mechanisms ("trap-controlled hopping") have been proposed as an explanation of certain observations such as a temperature-dependent drift mobility but a temperature-independent dispersion, for example [6.26]. In this review we will interpret the experimental transport data on amorphous silicon in terms of the multiple-trapping model. In amorphous silicon, both the dispersion and the drift mobility are temperature dependent and the multiple-trapping model seems to be the simplest model that explains the data.

Although the strict multiple-trapping model assumes a sharp mobility edge and ignores the possibility of tunneling even between weakly localized states near the mobility edge, the details of the transport mechanism near the mobility edge are not likely to be very important in the interpretation of room temperature experiments on a nanosecond or longer time scale. On this time scale the charge carriers spend most of their time trapped in localized states a few tenths of an electron volt away from the mobility edge. For the model to give an accurate description of these experiments, it may be sufficient that the mobility drop off more rapidly towards midgap than the density of states. Photoconductivity measurements on a picosecond time scale will be more sensitive to the transport mechanism in the vicinity of the mobility edge. These experiments are discussed in Sect. 6.4.3.

6.3.1 Physics of the Multiple-Trapping Process

a) Electron Thermalization in an Exponential Distribution of Traps

In view of the numerous complicating factors involved in charge transport experiments on amorphous silicon discussed in Sect. 6.2, exact solutions of particular models are not as crucial to the interpretation of experiments as an understanding of the underlying physical mechanisms. This statement must be reemphasized when the models involved are obviously highly idealized, as is the multiple-trapping model. In this section we develop a physical picture of the multiple-trapping version of the dispersive transport process [6.54, 55]. We present approximate expressions for the experimentally measurable parameters when the trapping states are distributed exponentially in energy. Then we develop an exact solution of the model for the special case where the dispersion parameter $\alpha = 1/2$ to illustrate the time evolution of the shape of a charge packet drifting in an electric field.

Perhaps the most outstanding feature in the transient photocurrent in amorphous semiconductors is the nearly universal observation of power-law behavior in the photocurrent decay as a function of time [6.26], similar to that illustrated in Fig. 6.2. *Scher* and *Montroll* [6.27] showed that this behavior is a consequence of a broad distribution of event times in the carrier transport process. A suitably broad distribution has the form

$$\psi(t) \propto \begin{array}{ll} t^{-1-\alpha} ; & t \geq \omega_0^{-1} \\ \omega_0^{-1-\alpha} ; & t < \omega_0^{-1} . \end{array} \tag{6.1}$$

The $\psi(t)$ is a probability distribution function for the time t that an electron must wait before moving to another site. This distribution function has no characteristic long time cut-off. The parameter α $(0 < \alpha < 1)$ which characterizes this distribution function is determined by the disorder in the material. The short time cut-off ω_0^{-1} is of the order of a phonon frequency. Shortly after this event time distribution function was first proposed as the origin of dispersive transport, *Silver* and *Cohen* [6.47] and others [6.49, 56] pointed out that one source of such a power law distribution of event times is thermal emission from an exponential distribution of localized states.

For an exponential distribution of localized states of the form $N_t(\varepsilon) = N_{t0}\exp(-\varepsilon/kT_c)$ and a thermal emission rate given by $\omega(\varepsilon) = \omega_0\exp(-\varepsilon/kT)$, the distribution of thermal emission times can be related to the probability distribution for localized state energies $N_t(\varepsilon)/\int N_t(\varepsilon)d\varepsilon$ as follows:

$$\psi(t) = \frac{N_t(\varepsilon)}{\int N_t(\varepsilon)d\varepsilon} \frac{d\varepsilon}{dt}$$

$$= a\omega_0 (\omega_0 t)^{-1-a} ; \quad t > \omega_0^{-1} , \tag{6.2}$$

where $a = T/T_c$ and ω_0 is an energy independent attempt rate. In this analysis the conduction band mobility edge is the zero of energy and positive energy ε points towards midgap.

Detailed balance requires that the attempt rate $\omega(\varepsilon)$, for escape from a localized state, be related to its capture rate by a Boltzmann factor [6.57]. This relationship means that in the case of a diffusive capture process, $\omega_0 = 4\pi N_c Da$ where D is the free electron diffusion constant, a is the capture radius of the localized state and N_c [cm^{-3}] is the effective density of states at the conduction band mobility edge [6.58]. The diffusive capture limit applies when the free electron mean free path between scattering events is comparable to or shorter than the capture radius of the localized state. Otherwise the ballistic expression [6.57] $\omega_0 = \pi N_c v_{th} a^2$ applies, where πa^2 is the capture cross section and v_{th} is the thermal velocity of a free electron. These expressions for the attempt rate will be used later in the comparison of the model with the experimental data.

There is no reason why the localized states should all have the same capture cross section. However, one expects the band-tail states close to the mobility edge all to have the same origin, say local potential fluctuations in the lattice, for example. In this case the energy dependence of the capture cross section is probably weak, at least compared to the exponential energy dependence of the thermal emission rate [6.59]. Furthermore, the present state of experiments on amorphous silicon is not yet sufficiently advanced for the energy dependence of these cross-sections to be measured experimentally. Although we have assumed a constant cross section, an exponential energy dependence can easily be absorbed into the same formalism. *Arkhipov* and *Rudenko* [6.60] have included an energy dependence of this form in their analysis of dispersive transport in vitreous As$_2$S$_3$: Sb$_2$S$_3$.

The photocurrent generated by a flash of light decays with time as the photoelectrons accumulate in deep traps in their attempt to reach thermal equilibrium. The nature of the photocurrent decay is determined by the details of the thermalization process. In the conventional picture, immediately after the excitation light flash ($\sim 10^{-12}$ s), the photo-excited electrons thermalize to the bottom of the conduction band and in a time interval of order ω_0^{-1}, almost all of the electrons are captured by localized states. The instant after the electrons are trapped, their distribution parallels the density of states as illustrated in Fig. 6.5, since by assumption, every localized state has an equal chance of capturing an electron. Somewhat later the electrons in shallow traps are thermally re-emitted to the band and are then retrapped. As this process continues, more and more of the electrons accumulate in the deep states because these states have a low probability of re-emission.

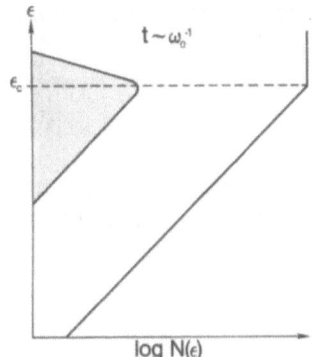

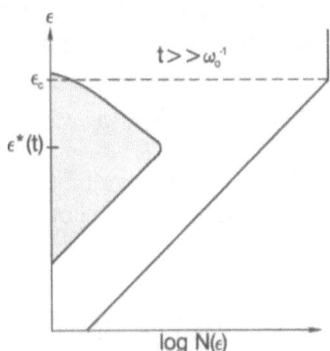

Fig. 6.5. Distribution of photoexcited electrons one trapping time after they have thermalized to the bottom of the conduction band. The states below the mobility edge ε_c are localized

Fig. 6.6. Distribution of electrons many trapping times after photo-excitation. The electrons in localized states above $\varepsilon^*(t)$ have thermalized into a Boltzmann distribution

After a large number of trapping-retrapping cycles, the population of the electrons in the shallow states reaches a quasi-equilibrium and assumes a Boltzmann distribution. Meanwhile the deep states have by definition been accumulating electrons, without re-emitting them. Thus, the electron distribution in the deep states parallels the density of states and always maintains the same form. The net result is that the density of electrons in deep traps falls off exponentially toward midgap because of the decreasing density of states, and the density of electrons in shallow states falls off exponentially in the other direction, toward the conduction band edge, because of the Boltzmann factor. It follows that the electron density must peak somewhere in the middle. The distribution of trapped electrons is illustrated in Fig. 6.6. The maximum density is at $\varepsilon^*(t) = kT \ln \omega_0 t$, which is the threshold energy level from which an electron can just be thermally emitted in the time t.

b) Current Decay at Short Times ($t \leqslant t_T$)

The idea that $\varepsilon^*(t)$ separates the thermalized states from the nonthermalized states has been used in the analysis of the frequency response of the capacitance of amorphous silicon Schottky barrier diodes [6.29, 61, 62]. Also, *Arkhipov* and *Rudenko* [6.60] used the concept of $\varepsilon^*(t)$ as a thermalization threshold to obtain an approximate solution to the multiple trapping equations for a continuous distribution of taps. *Orenstein* and *Kastner* [6.55] and *Tiedje* and *Rose* [6.54] took this concept one step further and showed that the electron drift mobility can be calculated as if all the trapped electrons were concentrated at $\varepsilon^*(t)$, and that this level sinks deeper with time. This recipe means that the electron drift mobility can be calculated from the standard expression for the drift mobility with a single trap level:

$$\mu_D = \mu\, n_{\text{free}}/(n_{\text{free}} + n_{\text{trap}}) \,, \tag{6.3}$$

where μ is the free carrier mobility. When this equation is expressed in terms of the parameters of the exponential trap distribution, we find that for $T_c \geqslant T$,

$$\mu_D(t) \cong \left(1 - \frac{T}{T_c}\right) \frac{T}{T_c} \frac{N_{c0}\exp\left[-\varepsilon^*(t)/kT\right]}{N_{t0}\exp\left[-\varepsilon^*(t)/kT_c\right]} \,. \tag{6.4}$$

The factors of T and T_c in front take into account the relative widths of the peak in the electron density near $\varepsilon^*(t)$ and the thermal tail in the free electron density above the conduction band mobility edge. Also $n_{\text{trap}} \gg n_{\text{free}}$ is assumed. $N_{c0}[\text{cm}^{-3}\,\text{eV}^{-1}]$ is the free electron density of states just above the conduction band mobility edge and $N_{t0}[\text{cm}^{-3}\,\text{eV}^{-1}]$ is the density of electron traps just below the conduction band mobility edge.

Substitution of the logarithmic time dependence of the thermalization threshold $\varepsilon^*(t)$ into (6.4) produces the power law decay in the drift mobility and photocurrent that is expected, based on the waiting time distribution function in (6.2). The time dependence of the photocurrent is

$$I(t) \cong \frac{e\mu E g}{L}\, \alpha(1-\alpha)(\omega_0 t)^{\alpha-1} \,, \tag{6.5}$$

where E is the electric field, L is the sample thickness and g is the total number of electrons injected at $t = 0$. In (6.5) and in the remainder of this article, we assume that $N_{c0} = N_{t0}$. Since the current decay in (6.5) reflects the time dependence of the electron drift velocity, it can be used to define an integral equation for the average transit time t_T for the electrons. The equation is

$$\int_0^{t_T} E\mu_D(t)dt = L \,. \tag{6.6}$$

The solution of (6.6) with $\mu_D(t) = I(t)L/eEg$ is

$$t_T = \omega_0^{-1}\left(\frac{\omega_0}{1-\alpha}\right)^{1/\alpha} \left(\frac{L}{\mu E}\right)^{1/\alpha} \,. \tag{6.7}$$

This transit time expression has the nonlinear dependence on sample thickness and inverse field that is characteristic of dispersive transport [6.26, 27].

c) Current Decay at Long Times ($t \geq t_T$)

After the transit time ($t \geq t_T$), an electron that is thermally emitted from a deep state no longer has enough time to thermalize all the way down to $\varepsilon^*(t)$ before being collected. It is only able to thermalize to $\varepsilon^*(t_T)$. In this regime the thermal emission rate from the deep states below $\varepsilon^*(t)$ is the limiting factor in the current decay rather than the retrapping of the electrons elsewhere in the sample. Incidentally, this is the condition that must be met for a current DLTS experiment to give a measure of the density of states [6.32, 33, 63]. In this limit every factor of "e" in time, a slice of trapped electrons kT wide, is thermally emitted off the top of the distribution of trapped electrons. This description of the current decay process can be expressed mathematically as follows:

$$I(t > t_T) \cong \frac{\Delta Q}{\Delta t} = ef_T \frac{kTN_{t0}}{t} \exp\left[-\frac{\varepsilon^*(t)}{kT_c}\right]. \tag{6.8}$$

The occupation factor f_T for the deep states at the transit time is given by

$$f_T = \frac{g}{LN_{t0}} \frac{1}{kT_c}\left(1 - \frac{T}{T_c}\right) \exp\left[\frac{\varepsilon^*(t_T)}{kT_c}\right]. \tag{6.9}$$

To evaluate the current decay explicitly, we substitute the logarithmic time dependence of $\varepsilon^*(t)$ into (6.8, 9) as before. The result has the anticipated t^{-a-1} time dependence [6.26, 27]

$$I(t > t_T) = \frac{e\mu Eg}{L} a(1-a)(\omega_0 t_T)^{2a}(\omega_0 t)^{-a-1}, \tag{6.10}$$

where the transit time t_T is defined in (6.7). The distribution of trapped electrons for $t > t_T$ is illustrated in Fig. 6.7. Note that the electrons thermalize down to $\varepsilon^*(t_T)$ and have a distribution that follows the trap distribution below $\varepsilon^*(t)$.

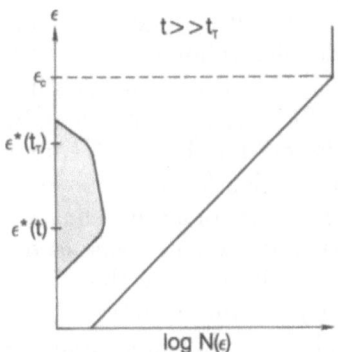

Fig. 6.7. Distribution of trapped electrons many transit times after the excitation ($t \gg t_T$). The electron distribution is thermalized down to $\varepsilon^*(t_T)$ but thermal emission can take place as far down as $\varepsilon(t)$

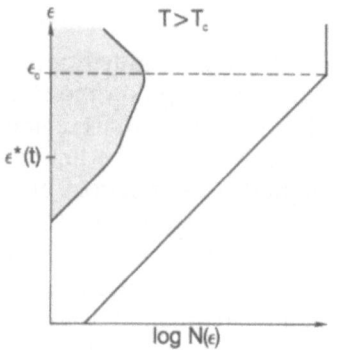

Fig. 6.8. Distribution of electrons at high temperatures. Although the charge has a Boltzmann distribution down to $\varepsilon^*(t)$, the peak in the electron distribution remains at the mobility edge because of the rapid variation in the density of states

d) High Temperature Regime ($T > T_c$)

The photocurrent has a power law decay at long time ($t > \omega_0^{-1}$) and low temperatures ($T < T_c$). At high temperatures ($T > T_c$), the bulk of the electron density remains in the vicinity of the mobility edge at all times because the rapid fall-off in the density of traps more than compensates for the increase in the occupation function towards midgap. The trapped electron distribution analogous to Fig. 6.6 but for $T > T_c$ is shown in Fig. 6.8. In this case the fraction of the injected charge in deep traps below $\varepsilon^*(t)$ decreases with time and the transport process becomes less dispersive [6.54] with increasing time. The temperature dependence of the drift mobility also changes at T_c. Below this temperature it is thermally activated and above T_c it has a power law behavior.

6.3.1 Spatial Distribution of Charge

a) Dispersive Regime ($0 < \alpha < 1$)

The analysis of the multiple-trapping model presented in the preceding section gives a clear picture of the physical mechanisms involved in the transport process, but it gives no information about the spatial distribution of the charge density. For this information a more detailed solution is required. Several solutions of the multiple-trapping equations have been published [6.48–50]. *Noolandi* [6.48] and *Schmidlin* [6.49] solved the problem by Laplace transform techniques for a series of discrete trap levels. *Arkhipov* and *Rudenko* [6.60] developed an approximate solution for an exponential distribution of traps by formulating the problem in terms of a wave equation with time-dependent coefficients. Their approximation was the same as the approximation used in the preceding section, namely, that the states with $\varepsilon < \varepsilon^*(t)$ are thermalized and those with $\varepsilon > \varepsilon^*(t)$ are frozen. In the following, we write down the multiple-trapping equations for an exponential distribution of traps and illustrate the nature of the spatial distribution of the charge for the special case of $\alpha = \frac{1}{2}$.

In *Noolandi's* [6.48] notation, the multiple-trapping equations for the free electron density in the i-th trapping level $n_i(x, t)$ are

$$\frac{\partial n}{\partial t} + \mu E \frac{\partial n}{\partial x} = \sum_i (r_i n_i - \omega_i n) + g\delta(x)\delta(t) \tag{6.11}$$

and

$$\frac{\partial n_i}{\partial t} = \omega_i n - r_i n_i \;, \tag{6.12}$$

where μ is the free carrier mobility and r_i and ω_i are the thermal emission rate and capture rate, respectively, of the i-th trap level. The emission and capture rates are related by the detailed balance requirement $\omega_i/r_i = \exp(\varepsilon_i/kT)$. The delta functions in (6.11) define the initial condition for the time-of-flight experiment. In these equations we have assumed that the space charge field of the injected charge can be ignored compared with the applied field E. The injected charge density is assumed to be low enough that the electron traps are not saturated. Finally, we have neglected diffusion because the spreading in the charge packet as a result of the distribution of trapping times is expected to dominate diffusion. *Ferreira* [6.64] showed that this assumption is rigorously correct at long times. Diffusion cannot be neglected as $t \to 0$ and $\alpha \to 1$.

Since we are interested in the transient response of the system to an impulse at $t = 0$, the equations can be solved conveniently by Laplace transforms. The Laplace transform of $n(x, t)$ is defined as

$$\tilde{n}(x, s) = \int_0^\infty n(x, t)e^{-st}\, dt \;. \tag{6.13}$$

Similarly, the transforms of (6.11, 12) are

$$s\tilde{n} + \mu E \frac{\partial \tilde{n}}{\partial x} = s\tilde{n} \sum_i \frac{\omega_i}{s + r_i} + g\delta(x) \tag{6.14}$$

and

$$\tilde{n}_i = \frac{\omega_i}{s + r_i} \tilde{n} \;. \tag{6.15}$$

We have eliminated $\tilde{n}_i$ from (6.14) with the help of (6.15). The summation over the discrete distribution of traps can be converted into an integral over the density of states as follows:

$$\sum_i \frac{\omega_i}{s + r_i} = \int_0^\infty \frac{\omega_0 e^{-\varepsilon/kT_c}}{s + \omega_0 e^{-\varepsilon/kT}} \frac{d\varepsilon}{kT} , \tag{6.16}$$

with the same notation as in Sect. 6.3.1. After a change of variables the integral reduces to

$$\left(\frac{s}{\omega_0}\right)^{a-1} \int_{s/\omega_0}^\infty \frac{u^{-a} du}{1 + u} . \tag{6.17}$$

The Laplace variable s is a frequency parameter that characterizes the time scale. As long as we are only interested in times much longer than one capture time ω_0^{-1}, the lower limit of integration in (6.17) can be taken to be zero. This approximation is valid provided $a < 1$. It is consistent with our limited understanding of the transport processes when the carriers are close to the mobility edge, since these processes dominate at short times $(t \sim \omega_0^{-1})$.

The integral in (6.17) can now be evlauated analytically and the following equation for the free carrier density results;

$$\mu E \frac{\partial \tilde{n}}{\partial x} + \left[s + \omega_0 \left(\frac{s}{\omega_0}\right)^a \frac{\pi}{\sin a\pi} \right] \tilde{n} = g\delta(x) . \tag{6.18}$$

The solution of this linear first-order differential equation for $\tilde{n}$ is

$$\tilde{n}(x, s) = \frac{q}{\mu E} \exp\left\{ -\frac{x}{\mu E} \left[s + \omega_0 \left(\frac{s}{\omega_0}\right)^a \frac{\pi}{\sin a\pi} \right] \right\} , \tag{6.19}$$

subject to the boundary condition that $\tilde{n}(x, s) = 0$ for $x < 0$. The solution in the time domain is the inverse transform of $\tilde{n}(x, s)$ or $L^{-1}[\tilde{n}(x, s)]$. The mathematical problem associated with the inverse transform of (6.19) can be reduced to $L^{-1}[\exp(-bs^a)]$ by the shifting theorem for Laplace transforms [6.65].

b) Special Case ($a = \frac{1}{2}$)

In general, the inverse transform must be evaluated numerically. However, for the special case of $a = \frac{1}{2}$, an analytic solution exists. From a table of Laplace transforms [6.65],

$$n(x, t) = \frac{1}{2\sqrt{\pi}} \frac{g \cdot \hat{x}}{\ell} \frac{\hat{x}}{t^{3/2}} \exp\left(-\frac{\pi^2}{4} \frac{\hat{x}^2}{t}\right) : x < \mu E t \tag{6.20}$$

$$0 : x > \mu E t .$$

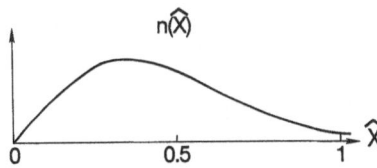

Fig. 6.9. Free electron density as a function of position for $t = t_T/4$ and $\alpha = \frac{1}{2}$. The electrons in shallow traps $[\varepsilon > \varepsilon^*(t)]$ have the same distribution

In this expression, $l = \mu E/\omega_0$ is the mean drift distance between trapping events and $\hat{x} = x/l$ and $\hat{t} = (t - x/\mu E)\omega_0$ are distance and time normalized to the mean trapping distance and time, respectively. The $x/\mu E$ term in $\hat{t}$ ensures that the tiny fraction of the electrons that manage to escape being trapped altogether, do not move faster than μE. The $x/\mu E$ term can be neglected over most of the region where the electron density is nonzero. The spatial distribution of the free electron density is shown in Fig. 6.9.

The total current at any given time is equal to $e\mu E/L$ multiplied by the total number of free carriers. Thus,

$$I(t) = \frac{e\mu E}{L} \int_0^L n(x, t)\, dx .$$ (6.21)

Once again, for $\alpha = \frac{1}{2}$, this current can be evaluated analytically:

$$I(t) = \frac{e\mu E}{L}\, \frac{g}{\pi^{5/2}\hat{t}^{1/2}}\left[1 - \exp\left(-\frac{\pi^2}{4}\frac{\hat{L}^2}{\hat{t}}\right)\right] ,$$ (6.22)

where $\hat{L} = L/l$. This current decay is illustrated in Fig. 6.2. It is easy to verify that this expression for the photocurrent decays as $t^{-1/2}$ at short times and as $t^{-3/2}$ at long times, as expected for $\alpha = \frac{1}{2}$. The extrapolated intersection point of these two branches of the photocurrent is usually defined as the transit time in experiments. According to this definition, the transit time from (6.22) is

$$t_T = \frac{\pi^2}{4}\,\omega_0\left(\frac{L}{\mu E}\right)^2 .$$ (6.23)

Note that this exact expression for the transit time has the same field, thickness and ω_0 dependence as the approximate value calculated from the physical argument in the preceding section (6.7). However, the approximate expression for the transit time in (6.7) is a factor of 1.6 too big for $\alpha = \frac{1}{2}$.

The charge in deep traps below $\varepsilon^*(t)$ has a different spatial distribution from the free charge. By definition, the thermal emission probability for an electron in a deep trap is negligible. This means for the deep traps the release rate can be neglected in the expression for the density of trapped electrons (6.15). The density of electrons in deep traps is

$$\tilde{n}_t(\varepsilon > \varepsilon^*, x, s) = \frac{\omega_0}{s\alpha} e^{-\varepsilon/kT_c} \tilde{n}(x, s) , \tag{6.24}$$

or the density is just the time integral of the free electron density:

$$n_t(\varepsilon > \varepsilon^*, x, t) = \frac{\omega_0}{\alpha} e^{-\varepsilon/kT_c} \int_0^t n(x, u)du . \tag{6.25}$$

For $\alpha = \frac{1}{2}$, the integral is an error function;

$$n_t(\varepsilon > \varepsilon^*, x, t) = e^{-\varepsilon/kT_c} \frac{g}{\pi l} \ \mathrm{erfc}\left(\frac{\pi}{2} \frac{\hat{x}}{t^{1/2}}\right) . \tag{6.26}$$

In (6.24–26) we have assumed that the effective density of states at the conduction band edge $N_c[\mathrm{cm}^{-3}]$ equals kTN_{t0} where $N_{t0}[\mathrm{cm}^{-3}\ \mathrm{eV}^{-1}]$ is the density of states for the traps at the conduction band mobility edge. The spatial distribution of electrons in deep traps is plotted in Fig. 6.10. Note that the density of charge in deep traps in Fig. 6.10 has its maximum value at $x = 0$, while the density of free charge illustrated in Fig. 6.9 peaks away from the origin.

The distribution of the total charge density is also of interest. For $\alpha < 1$ and long times ($t > \omega_0^{-1}$), most of the charge is in traps. Therefore, the free carrier density can be neglected and the total electron density is the sum of (6.15) over all the traps, to a good approximation. This summation leads to the same integral that was evaluated above in (6.16), so that the total charge density $\tilde{n}_T(x, s)$ follows immediately ($0 < \alpha < 1$):

$$\tilde{n}_T(x, s) = \left(\frac{\omega_0}{s}\right)^{1-\alpha} \tilde{n}(x, s) . \tag{6.27}$$

For $\alpha = \frac{1}{2}$, the total charge density is a Gaussian:

$$n_T(x, t) = \frac{1}{\sqrt{\pi}} \frac{g}{lt^{1/2}} \ \exp\left(-\frac{\pi^2}{4} \frac{\hat{x}^2}{t}\right) , \tag{6.28}$$

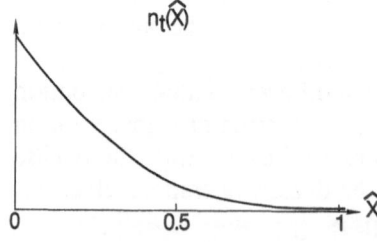

$n_t(\hat{x})$

Fig. 6.10. Spatial distribution of electrons in deep traps below $\varepsilon^*(t)$ for $t = t_T/4$ and $\alpha = \frac{1}{2}$

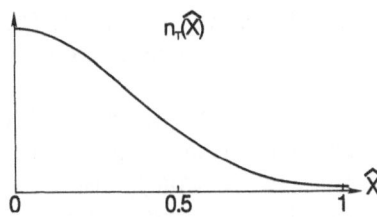

Fig. 6.11. Total electron density in both deep and shallow traps for $t = t_T/4$ and $\alpha = \frac{1}{2}$

where the symbols have the same meaning as in (6.20). This distribution is illustrated in Fig. 6.11. Referring back to the expression for the transit time in (6.23), we note that at the transit time the charge density at the back contact is down by $1/e$ from its value at $x = 0$. It follows that at the transit time the injected charge is spread out rather uniformly across the sample. This is a general property of dispersive transport.

The Gaussian behavior of the total charge density when $\alpha = \frac{1}{2}$ illustrates an interesting relationship (pointed out by *Butcher* and *Clark* [6.66]) between diffusion and drift when the waiting time distribution function $\psi(t)$ has the $t^{-1-\alpha}$ form. The charge distribution for diffusion alone $(E = 0)$ has the same shape as the charge distribution resulting from drift alone without diffusion, provided α is replaced by $\alpha/2$. Thus, dispersive drift transport with $\alpha = \frac{1}{2}$ should resemble diffusion in the absence of dispersion $(\alpha = 1)$; hence, the Gaussian charge distribution for $\alpha = \frac{1}{2}$. *Butcher* and *Clark* [6.66] have also shown that the electric field driven charge density always has its maximum at the origin $(x = 0)$ when $\alpha \leq \frac{1}{2}$. However, when $\alpha > \frac{1}{2}$, the maximum is at nonzero x. The maximum becomes progressively sharper and moves more

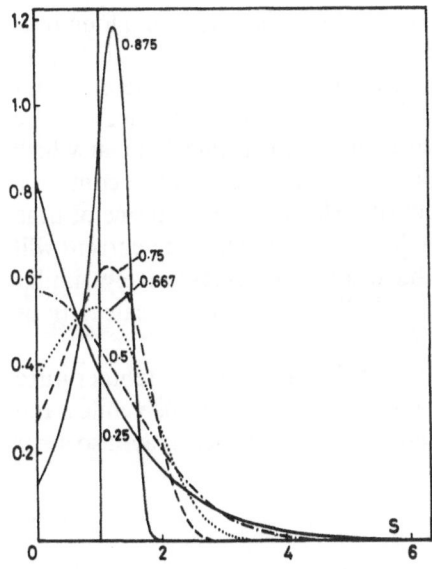

Fig. 6.12. Electron density at $t = t_T$ for various values of α as indicated [6.66]. In each case, the electron density starts off as a delta function at the left-hand boundary $(S = 0)$ and drifts to the right under the influence of an electric field. The horizontal scale is in units of the sample thickness

nearly linearly with time as α approaches the nondispersive $\alpha = 1$ limit. The shape of the charge distribution for a series of different α values is shown in Fig. 6.12.

The reason for the change in behavior at $\alpha = \frac{1}{2}$ is that in the weakly dispersive $\alpha > \frac{1}{2}$ limit, more than half of the total charge is thermalized in the shallow states and its distribution dominates, while in the strongly dispersive $\alpha < \frac{1}{2}$ case, more than half of the charge is in the deep states. The deep charge always peaks at $x = 0$ for any α, as in Fig. 6.10, while the thermalized charge has the same distribution as the free carriers and always peaks at $x > 0$, as illustrated for the special case of $\alpha = \frac{1}{2}$ in Fig. 6.9.

c) High Temperature Limit ($T > T_c$)

Up to now we have focused on the dispersive low temperature regime where $\alpha < 1$. However, in the multiple-trapping model, there is no physical reason why α cannot be larger than unity. The $\alpha > 1$ problem can be solved by the same procedure as the $\alpha < 1$ problem, with one difference. The lower limit of integration in the integral in (6.17) can no longer be approximated by zero because the integral diverges as the lower limit approaches zero when $\alpha > 1$. Most of the contribution to the integral comes from near the lower limit when $(s/\omega_0) \ll 1$; we approximate the denominator of the integrand by unity and the integral becomes trivial. Then the Laplace transform of the free carrier density, analogous to (6.19), is ($\alpha > 1$)

$$\tilde{n}(x, s) = \frac{g}{\mu E} \exp\left(-\frac{sx}{\mu E} \frac{\alpha}{\alpha - 1}\right). \tag{6.29}$$

The inverse transform of this expression is a delta function moving with a constant drift velocity $\mu E(\alpha - 1)/\alpha$. This drift velocity is valid in the long time limit for $\alpha > 1$. The same expression can be derived by the physical arguments used in the previous section and the assumption that all the electrons are thermalized in the traps [6.54]. The requirement that all the electrons be thermalized gives a criterion for the onset of the long-time limit in which (6.29) is valid. In the long-time limit, the deep states below $\varepsilon^*(t)$ contain a negigible fraction of the total electron density which means in terms of time that t should satisfy $t \gg \omega_0^{-1} \exp[1/(\alpha - 1)]$. At shorter times the current will decrease with time for $t < t_T$ and exhibit dispersion. Qualitatively similar behavior has been observed by *Silver* et al. [67] in Monte Carlo calculations with $\alpha = 1.2$.

Clearly, expressions such as (6.29) that predict a power law temperature dependence for the drift mobility should not be taken too seriously unless the temperature dependence of the free-carrier mobility, which may also be a power law, is included.

6.4 Experimental Results

In this section we discuss recent time-resolved transport experiments on a-Si:H. In any discussion of experiments on this material, one must must keep in mind the fact that a-Si:H is not a single material such as crystalline silicon, but rather a whole family of materials with widely different electronic properties. The optical gap, for example, can vary from 1.5 to 1.9 eV [6.68], depending on deposition temperature, hydrogen content, defect density, etc. At the same time, many general features are common to all members of the a-Si:H family.

In the following we will focus primarily on the low defect density glow discharge material that is believed to be optimal or close to optimal for solar cells. These materials are deposited at 200–300 °C; they have an optical gap near 1.7 eV, H contents in the 6–15% range and a density something like 96% of the density of crystalline silicon [6.69].

6.4.1 Time of Flight

a) Electrons

Time-of-flight measurements of the electron drift mobility in a-Si:H were pioneered by *LeComber* and *Spear* [6.14, 70] who studied glow discharge material prepared at 200 °C in an inductive discharge. They found a room temperature drift mobility of 0.1 cm²/V s with a 0.19 eV thermal activation energy. Later, *Moore* [6.71] found similar values for both the electron drift mobility and its activation energy in dc glow discharge material deposited at 330 °C. He used an unconventional technique in which the bias field was applied in the form of a ramp. Even though the shape of the photocurrent transients was nonideal, Moore was able to interpret his data in terms of a nondispersive, field-independent drift mobility [6.71].

More recent work has demonstrated that the electron drift mobility can be significantly higher in glow discharge material (~ 1 cm²/V s at room temperature) and significantly lower in some reactively sputtered material (≲ 0.01 cm²/V s) [6.36, 72]. Also, *Nielsen* and *Dalal* [6.73] have found a correlation between sample thickness and the magnitude of the drift mobility. This phenomenon is not a consequence of dispersive transport, since the observed trend is for the drift mobility to increase with thickness, while (6.7) predicts a decrease.

A possible explanation is that the film quality improves with thickness because of the gettering effect of the deposition process [6.74]. Another possibility is that the surface layer where the electrons are injected may trap electrons preferentially compared to the bulk. The surface layer would then introduce a delay, which would have the effect of reducing the apparent drift mobility in thin samples [6.75]. This hypothesis is supported by the fact that the rise time of the photocurrent is frequently anomalously long, as if some charge has to be released from surface traps [6.76]. Also, the electrons are normally injected in the space-charge region near a blocking contact where the material tends to be *p*-type because of the band-bending. The electron lifetime is known to be short in *p*-type material [6.5, 77]. Surface trapping effects may explain the nonideal current transients that are observed in some relatively thin films ($\lesssim 1$ μm) [6.76, 78]. The transients are nonideal in the sense that they exhibit neither Gaussian nor conventional power law dispersive behavior.

Datta and *Silver* [6.78] took the nonideal current transients and the increase in drift mobility with thickness as evidence that the time-of-flight experiment actually measures a space-charge layer relaxation phenomena, triggered by the light flash. In this interpretation the real drift mobility is significantly larger than 1 cm^2/V s.

The electron drift moblity in reactively sputtered material is generally lower (0.006–0.13 cm^2/V s) and more strongly temperature dependent than in glow discharge material [6.36, 79, 80]. The dispersion in the electron transport in sputtered material depends on how the material is prepared, although which preparation parameters are the crucial ones is not yet clear. Examples of both dispersive electron transport with a field-dependent drift mobility and nondispersive transport have been published [6.3, 36, 81].

The greater degree of variability in the transport properties of the sputtered material compared to the glow discharge material is consistent with the greater range of structural properties achievable in sputtering. For example, the hydrogen content of the films can be controlled independently of the other deposition parameters. This characteristic of the sputtered material has been exploited to demonstrate by drift mobility measurements as a function of hydrogen content, that the conduction band edge is insensitive to the hydrogen content of the material [6.79] for H content in the 14–20% range. One might expect the hydrogen to affect the conduction band edge through the admixture of anti-bonding Si-H orbitals [6.82, 83], or indirectly through its effect on the disorder [6.13, 68].

b) Holes

In the first time-of-flight work on a-Si : H, hole transport was not observed, presumably because of a short deep-trapping lifetime for holes in the slightly *n*-type undoped material [6.70]. However, *Moore* [6.71] discovered that the addition of a small amount of diborane ($[B_2H_6]/[SiH_4] \sim 10^{-5}$–$10^{-3}$) increases the hole lifetime sufficiently for measurements of the hole transit time to be possible. He found a room temperature value of 5–6 × 10^{-4} cm^2/V s and an activation energy of 0.35 eV in 330 °C substrate temperature, dc glow discharge material [6.71]. At the same time, the boron reduces the electron lifetime so that electron and hole transit times normally cannot be observed in the same sample. These observations are consistent with photoconductivity experiments which show that the majority carrier lifetime increases with doping [6.5].

More recent work on material made at RCA (diffusion length ~ 1 μm [6.84]) suggests that in the very best material it may be possible to observe both electron and hole transport in the same sample [6.85].

Published values for the hole drift mobility range from $\sim 10^{-4}$ to 5×10^{-3} cm^2/V s with activation energies from 0.28–0.43 eV [6.69, 86–88]. In all of this work, the hole photocurrent transients were found to be strongly dispersive even though the hole mobility was only weakly dependent on field. *Moore* [6.71], for example, found the current transients to be so dispersive that he was unable to identify the transit time without resorting to a ramp bias field. Nevertheless, he was able to interpret his results in terms of a field-independent drift mobility. Although the time-of-flight experiment with a ramp bias has not been interpreted in terms of a dispersive transport theory, one would still expect to see the effects of a field-dependent drift mobility. *Allan* [6.86] has also reported a field-independent hole drift mobility at room temperature, even though the current transients have the power law decays characteristic of dispersive transport.

From the slope of the hole photocurrent decay, the dispersion parameter α is variously estimated as 0.41–0.59 [6.86, 88]. According to dispersive transport theory, the drift mobility should scale with sample thickness L as $L^{1 - 1/\alpha}$; as a result, a strong thickness dependence is anticipated. However, *Mort* et al. [6. 87] found a hole drift mobility of 8×10^{-4} cm^2/V s at room temperature in a very thick film (32 μm), a value which is not that different from values of 0.15–2×10^{-4} cm^2/V s reported on much thinner (0.8–2 μm) samples [6.86]. For α in the 0.41–0.59 range, one would expect the drift mobility in the thick film to be 10–50 times smaller than in a 2 μm film. Although the apparent discrepancy may simply be due to a difference in the samples, more extensive data on the shape of the hole photocurrent transients as a function of field and sample thickness is required. The nature of the hole transport process is particularly important in the development of a comprehensive picture of the electronic properties of a-Si:H.

There are two schools of thought about the hole trapping states. One theory is that the holes self-trap into polaronic states with a binding energy that is comparable to the thermal activation energy of the hole drift mobility. In this interpretation the polaron binding energy is responsible for the Stokes shift (~ 0.4 eV) in the photoluminescence [6.89]. In the alternate picture which we adopt here, lattice relaxation is a small effect, and the Stokes shift in the luminescence is due to nonradiative capture of photoexcited carriers by localized states [6.90–92]. The activation energy of the drift mobility is then associated with a kinetic limit to the thermalization depth of holes in a relatively narrow localized state distribution.

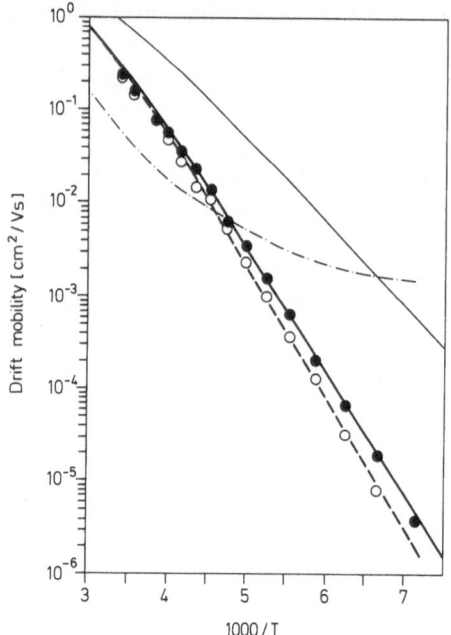

Fig. 6.13. Temperature dependence of the electron drift mobility for three different samples. The thin solid line at the top refers to an rf glow discharge film 3.9 μm thick measured at 4 V [6.88]. The open and solid data points are for a 4 μm thick dc glow discharge sample measured at 8 V and 15 V, respectively [6.97 a]. The theoretical fits are discussed in the text. The dash-dot line is taken from [6.14] by *Lecomber* and *Spear*

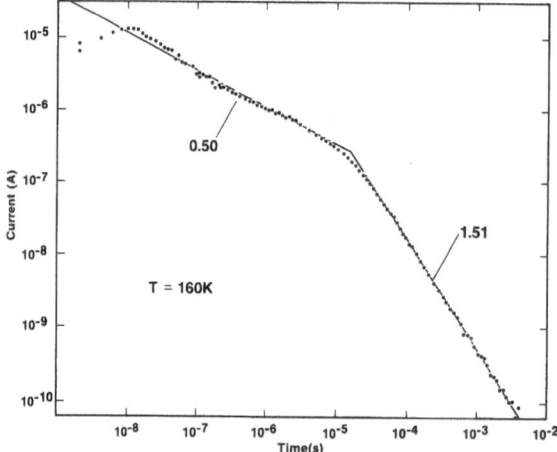

Fig. 6.14. Experimental electron photocurrent transient for the dc glow discharge film in Fig. 6.13. Note the extensive range of the power law behavior [6.97 a]

c) Temperature Dependence

In their original work on the electron drift mobility, *LeComber* and *Spear* [6.14] found that the thermal activation energy of the electron drift mobility decreased from 0.19 eV at room temperature to 0.09 eV below 240 K. A smaller reduction in the activation energy (from 0.19 eV to 0.15 eV) was later observed by *Moore* [6.71] at about the same temperature. This effect was

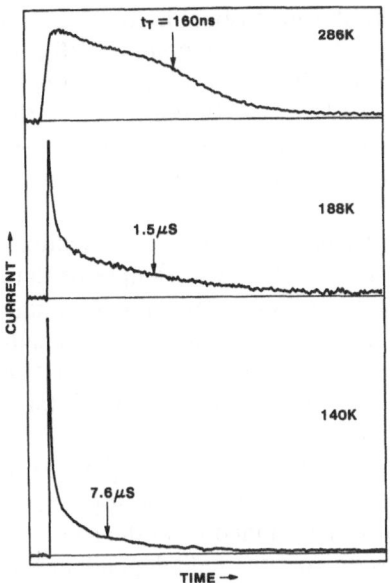

Fig. 6.15. Experimental electron photocurrent decay in a time-of-flight experiment for three different temperatures [6.97 a]

originally interpreted as evidence for a transition from trap-controlled band transport at high temperatures to hopping in band-tail states at low temperatures [6.14, 95]. Perhaps this interpretation should be reviewed in light of subsequent developments in the theory of dispersive transport [6.27] and experiments which show that the dispersion increases at low temperatures [6.88]. Without assurances that the temperature dependence was measured at constant field, one might attribute a lower activation energy at low temperatures to the combined effects of a progressively increasing dispersion and a progressively higher measurement field at low temperatures. More recent measurements made at constant field on material deposited at 300 °C from an rf or dc capacitive discharge, show a higher electron drift mobility and no indication of a decrease in the activation energy at low temperatures [6.88]. This data is illustrated in Fig. 6.13, along with the earlier data of *LeComber* and *Spear* [6.14]. The more recent data is consistent with a multiple-trapping transport model over the entire temperature range (140–400 K).

The heavy lines in Fig. 6.13 through the drift mobility data for the dc glow discharge sample [6.97 a] are theoretical fits based on the approximate solution to the multiple-trapping model outlined in the previous section. An attempt frequency ω_0 of $8 \times 10^{12} \, \text{s}^{-1}$, a free-carrier mobility of 13 cm^2/V s and a density-of-states parameter $T_c = 350$ K have been used in the fit. The thin line at the top is a similar fit to data on an rf glow discharge film with $\omega_0 = 4 \times 10^{11} \, \text{s}^{-1}$, $\mu = 13$ cm^2/V s and $T_c = 312$ K [6.88]. The two sets of data are not meant to be representative of dc and rf glow discharge films, but rather illustrative of the differences that can be observed in the transport properties with different deposition conditions. In any event, the free-carrier mobility and attempt frequencies used in the fits are not unreasonable for an amorphous semiconductor [6.96]. For example, a free-carrier mobility of 10 cm²/ V s corresponds to a momentum randomization length of a few interatomic distances. Similarly, an attempt frequency of $10^{12}\,\mathrm{s}^{-1}$ can be converted by the detailed balance argument described in Sect. 6.3.2a into a localized state capture cross section of atomic dimensions for an effective band edge degeneracy of $10^{19}\,\mathrm{cm}^{-3}$.

Below room temperature, the electron photocurrent decay exhibits power law behavior over many orders of magnitude. An example of data where the power law behavior persists over five orders of magnitude in time is shown in Fig. 6.14 [6.97a]. In the multiple-trapping model, this result means that the localized state distribution is accurately exponential from 0.14 to 0.33 eV below the mobility edge. For $T_\mathrm{c} = 312$ K, the density of states changes by over three orders of magnitude in this energy range.

There is other evidence that the conduction band tail is sharp (i.e., $T_\mathrm{c} \simeq$ 300 K). For example, the dispersion in the electron transport at room temperature is close to zero ($\alpha \sim 1$) whether it is determined from the shape of the transient photocurrent decay (Fig. 6.15) or from the field dependence of the drift mobility [6.88]. The dispersion increases strongly at low temperatures, as the experimental photocurrent transients in Fig. 6.15 illustrate. This temperature dependence is in good agreement with the multiple-trapping expression $\alpha = T/T_\mathrm{c}$ with a T_c of 312 K for the conduction band edge, as illustrated in Fig. 6.16. The possible nonzero intercept in the experimental data for α in Fig. 6.16 can be explained in the context of the multiple-trapping model of dispersive transport by a capture cross section of the localized states that increases with binding energy [6.97b].

DLTS experiments [6.98a] are also consistent with a sharp conduction band edge [6.32, 33]. In a DLTS experiment, it is possible to measure the density of states at a given energy and the thermal emission energy required to excite an electron from these states into the conduction band. From the measured density of states 0.3 eV below the conduction band mobility edge

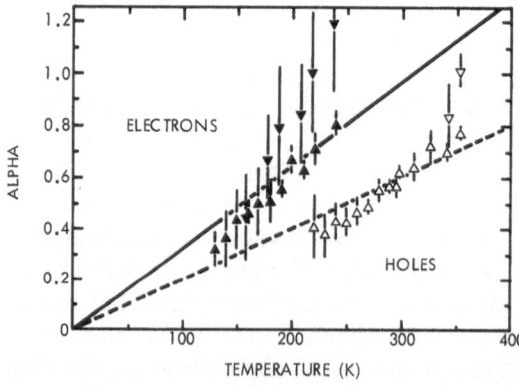

Fig. 6.16. Temperature dependence of the dispersion parameter α for electrons and holes determined from the exponent in the power law current decay [6.88]. Data obtained from the first branch of the current decay ($t < t_\mathrm{T}$) is plotted with upward pointing symbols and the second branch ($t > t_\mathrm{T}$) as downward pointing symbols

[6.33] and an assumed density of states at the mobility edge of 10^{21} cm^{-3} eV^{-1}, it is easy to show that the characteristic energy of the exponential that connects the two is about 300 K.

Further evidence for exponential band tails in a-Si : H is provided by spin-density measurements of *Dersch* et al. [6.98 b] in doped material, who also reported ≈ 300 K for the conduction-band tail width.

The analysis of the multiple-trapping model presented in the preceding section breaks down when α is close to unity. In this regime, a substantial fraction of the photo-injected electrons remain in free states above the mobility edge so that the density of free electrons cannot be neglected relative to the density of trapped electrons. A straightforward extension of the argument in Sect. 6.3.1 that takes the free-carrier density into account [as in (6.3)], shows that the time dependence of the drift mobility has the form

$$\mu_D(t) = \mu_0 \frac{\alpha(1 - \alpha)}{(\omega_0 t)^{1 - \alpha} - \alpha^2} . \tag{6.30}$$

This expression is the source of the model curve in Fig. 6.13 for α close to unity. It reduces to the previous results [(6.29) and (6.5)] in the high and low temperature limits.

The temperature dependence of the hole drift mobility can also be interpreted in terms of the multiple-trapping model [6.98 c]. We note that if the larger activation energy for the hole drift mobility (0.35–0.40 eV) [6.71, 86, 88] relative to the electrons (0.15–0.19 eV) [6.14, 45, 70, 88] is attributed entirely to the difference in the width of the band tails, then the valence band tail should be twice as wide as the conduction band tail. This estimate (≈ 0.05 eV) is in good agreement with the $T_c = 500$ K inferred from the temperature dependence of the slope of the photo current decays in the time-of-flight experiments [6.86, 88, 98 c].

The temperature dependence of the hole drift mobility can be fit by $\omega_0 = 1.6 \times 10^{12}$ s^{-1}, $\mu = 0.7$ cm^2/V s and $T_c = 500$ K [6.86, 88]. We note that the larger value of the attempt rate ω_0 and lower free-carrier mobility μ, relative to the corresponding values for the electrons, is an indication that the holes are more strongly coupled to the lattice than the electrons. Finally, we note that the width of the valence band localized state distribution, inferred from the hole mobility data, is remarkably close to the 0.05 eV characteristic width of the optical absorption tail [6.93 a, 93 b, 94].

6.4.2 Computer Simulations

Except for certain limiting cases such as $\alpha = \frac{1}{2}$, the multiple trapping model of dispersive transport is difficult to solve analytically. Other models are no easier to solve. In order to understand the physical processes controlling the

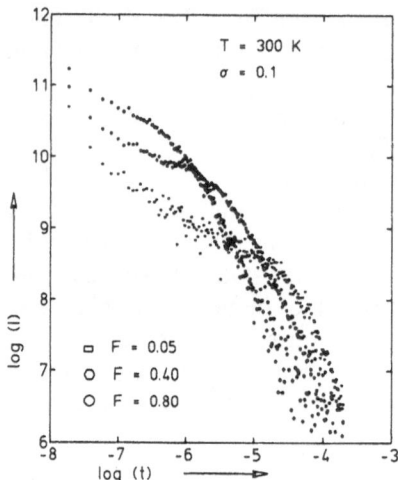

Fig. 6.17. Monte Carlo simulations of a time-of-flight experiment for a Gaussian distribution of trap states of the form $N(\varepsilon) \sim \exp\left(-\varepsilon^2/2\sigma^2\right)$ for $\sigma = 0.1$ eV and three different values of the applied field $E = 10^6 F$ [V/cm] [6.100]

charge transport, one needs to find out how model assumptions affect quantities that can be measured experimentally. One solution is the simulation of the charge transport by computer experiments on model systems – the Monte Carlo technique. This technique has been used extensively by *Silver* and his collaborators [6.47, 67, 99, 100].

Schonherr et al. [6.100] applied it to the multiple-trapping problem where the trap distribution is Gaussian rather than exponential. In the former case, the current transients are hard to distinguish from those for an exponential distribution of traps, as illustrated in Fig. 6.17. The difference shows up at long times, where the dispersion disappears in the Gaussian case. The dispersion disappears when $\varepsilon^*(t_T)$ gets far enough down into the trap distribution that the logarithmic slope of the density of states is greater than $(kT)^{-1}$. In this case the "instantaneous" α is greater than unity and the transport is nondispersive.

The Monte Carlo method has also been applied to the situation where the trap distribution is exponential but α is close to unity [6.67]. According to theory, the dispersion must disappear for $\alpha > 1$ at infinite time. *Silver* et al. [6.67] have shown that for typical experimental time scales, significant dispersion remains even at $\alpha = 1.2$. For example, the current in the initial branch of the time-of-flight experiment ($t < t_T$) is not constant at $\alpha = 1.2$, as one would expect for diffusive transport. Instead it decays roughly logarithmically with time.

6.4.3 Photoconductivity Decay

As discussed in Sect. 6.2.2, the amplitude of the photoconductivity produced by a flash of light of known intensity provides information about the transport process even when the electrodes are ohmic or too far apart for a transit time to be observable.

a) Picosecond Experiments

Fast time-resolved photoconductivity measurements show that the drift mobility is thermally activated even at a time interval as short as 25 ps after the carriers are injected into the band states [6.44, 45]. *Johnson* et al. [6.44] showed that the activation energy of the photocurrent is 58 meV after 25 ps. Their data is shown in Fig. 6.18. It is not unreasonable to attribute the photocurrent signal entirely to electrons, since as noted in the previous section, the electron drift mobility is significantly larger than the hole mobility.

The photocurrent activation energy can be compared with what one would expect for the electrons from the conduction band tail parameters determined by time of flight. According to theory (6.5), the drift mobility at a fixed delay time t_D should have a temperature dependence of the form $(\omega_0 t_D)^{T/T_c}$. Over a small temperature range this temperature dependence can be regarded as a thermal activation energy $\Delta = akT \ln \omega_0 t$. For a measurement time $t = 25$ ps, $\omega_0 = 4 \times 10^{11} \text{s}^{-1}$, $T_c = 312$ K and $T = 225$ K, the calculated activation energy is 32 meV, somewhat smaller than the experimental value of 58 meV. If the attempt frequency were $3 \times 10^{12} \text{s}^{-1}$, the calculated value would agree with experiment. Note that the model predicts a progressive reduction in the activation energy at low temperatures, in qualitative agreement with the experimental data in Fig. 6.18.

The magnitude of the initial mobility (0.8–1.4 cm^2/V s) was found to be approximately the same for both glow discharge and unhydrogenated evaporated material [6.44]. This in itself may not be surprising, since it is possible that the states near the mobility edge are not strongly affected by dangling bond type defects. However, the magnitude of the mobility is surprising,

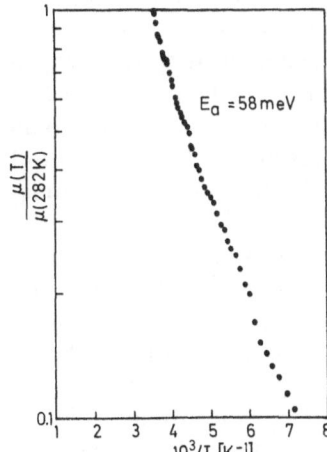

Fig. 6.18. Picosecond photoconductivity in an undoped glow discharge sample. Temperature dependence of the initial (25 ps) photocurrent [6.44]

since it is nearly the same as the time-of-flight value measured at 100 ns. It is instructive to estimate the reduction in drift mobility that one would expect from 25 ps to 100 ns. The ratio of the drift mobilities at two times t_1 and t_2 is given by

$$\frac{\mu_D(t_1)}{\mu_D(t_2)} = \frac{1 + \alpha + \alpha \ln(\omega_0 t_2)}{1 + \alpha + \alpha \ln(\omega_0 t_1)} \qquad (6.31)$$

when $\alpha \simeq 1$. This expression is derived from the arguments presented in Sect. 6.3.2 and the assumption that the density of extended states is a constant close to the mobility edge and joins smoothly with the density of localized states. For $t_1 = 25$ ps and $t_2 = 100$ ns, the ratio of the drift mobilities is about three. Strictly speaking, the experimental data is not inconsistent with this ratio, since the fast photoconductivity data is claimed to be accurate only to a factor of two, and the highest quoted value is 1.4 cm²/V s [6.44]. Other relevant factors are sample differences, surface effects (the photoconductivity is generated in the top 1000 A of the film) and anisotropy in the transport properties of the film. In a film with a columnar morphology [6.110], for example, one might expect the charge transport properties in the plane of the film to be degraded relative to the charge transport through the thickness of the film.

b) Photoconductivity Decay ($t > 10$ ns)

The ratio of the picosecond drift mobility to the drift mobility at 10 ns is close to what one would expect if one compares the picosecond results with the photoconductivity decay measurements of *Hvam* and *Brodsky* [6.41, 42]. These authors infer an initial drift mobility at 20 ns of 0.2 cm²/V s. In this comparison, note that both measurements are in the plane of the film rather than perpendicular as in time-of-flight measurements.

Although the initial part of the photocurrent decay is determined by the time dependence of the drift mobility, at long times the decay is influenced by deep trapping at defects and recombination centers [6.102]. In undoped material the electron lifetime is normally too short for the recombination-free initial part of the power law current decay to be resolved at room temperature with a 20 ns time resolution. However in phosphorous (boron) doped material, the electron (hole) lifetime can be sufficiently long for time-dependent transport effects to be observable.

Hvam and *Brodsky* [6.41, 42] have studied the photoconductivity decay in phosphorous doped material as a function of dopant concentration and excitation intensity for films deposited at 230° and 300 °C. The decay was monitored from 30 ns to 10 ms following excitation by a 10 ns flash of light. Some of this data is shown in Fig. 6.19. In the film deposited at 230 °C with gas phase dopant concentration $[PH_3]/[SiH_4] = 10^{-4}$, the photocurrent exhibited a power law decay from 100 ns to 100 μs, as shown in Fig. 6.19. The tempera-

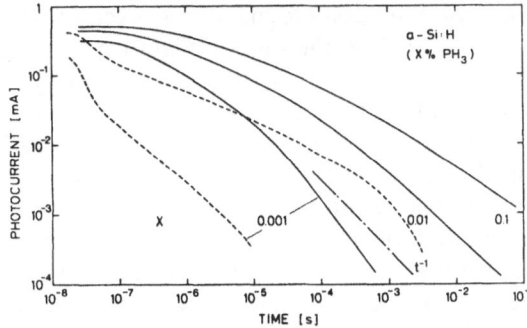

Fig. 6.19. Photoconductivity decay in material doped with phosphorous as indicated, following a 10 ns excitation flash. The data represented by the solid lines and dashed lines corresponds to films deposited at 300 °C [6.42] and 250 °C [6.41], respectively

ture dependence of the exponent $(\alpha - 1)$ in the power law was found to obey the $\alpha = T/T_c$ relation expected for the multiple-trapping model [6.41]. Measurements at room temperature on material deposited at 300 °C [6.42], also shown in Fig. 6.19, show less well-defined power law behavior and a slower decay rate at short times.

The significance of this difference is not clear. Also unclear is the strong dependence of the photocurrent decay on dopant concentration. Since the trend is for a more rapid decay at lower dopant concentration, it is possible that the electron lifetime in the lightly doped samples is sufficiently short that recombination plays a role even in the initial part of the photocurrent decay in Fig. 6.19. However, the experimental concerns mentioned in Sect. 6.3.1 associated with space charge effects and the ohmicity of the contacts for both carriers, should be resolved before a detailed interpretation is developed.

If these complications can be ignored, at long times the density of recombination centers and recombination will dominate the decay process. In this limit, the photocurrent decay is limited by the rate of thermal emission from deep traps. The long time photoconductivity decay is analogous to the second branch of the photocurrent transient in a time-of-flight experiment, and has the same $t^{-\alpha-1}$ time dependence. A difference, pointed out by *Scher* [6.102],

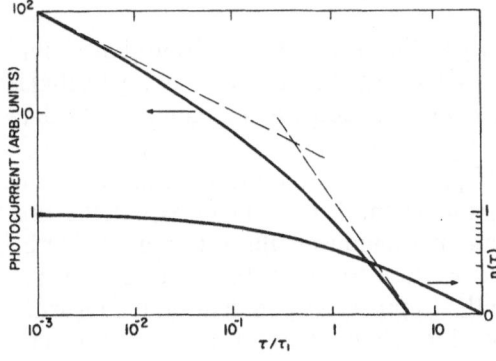

Fig. 6.20 Theoretical calculation of the photocurrent and the number of surviving carriers as a function of time for carriers undergoing dispersive transport $(\alpha = \frac{1}{2})$ and recombination [6.102]. The analogous time-of-flight photocurrent decay is shown in Fig. 6.2

is that the change in slope at the recombination lifetime occurs over a wider time interval than the corresponding break in the time-of-flight experiment (Fig. 6.20). The reason for this difference can be understood qualitatively as follows. In the time-of-flight experiment, all the electrons must drift the same distance to get to the back contact. In this sense the transit time is well defined. On the other hand, in the recombination controlled decay, some electrons start off close to a recombination center and some far away, so that there is a broad range of recombination times.

It is worth noting that the recombination scheme we have outlined here does not apply at low temperatures, where recombination proceeds by direct tunneling between localized states [6.89].

6.4.4 Photoinduced Absorption

The photoinduced absorption decay is complementary to the photoconductivity decay in the sense that the photoinduced absorption is sensitive to all the photogenerated carriers, whereas the photoconductivity is sensitive only to the free carriers. The relationship between the two in the multiple-trapping model has been discussed by *Orenstein* et al. [6.103] in the context of a-As$_2$Se$_3$. In the time regime prior to the onset of recombination, for which the photoconductivity decay is dominated by transport effects, the photoinduced absorption is constant provided the trapping states all have the same optical cross section. After recombination has taken over, that is, when capture by recombination centers is more probable than by deep traps at energies $\varepsilon > \varepsilon^*(t)$, the photoinduced absorption decays according to $t^{-\alpha}$. This time dependence is due to the decay of the charge located in deep traps:

$$n_t(\varepsilon > \varepsilon^*(t)) \simeq fN_{t0} \int\limits_{\varepsilon^*}^{\infty} e^{-\varepsilon/kT_c}\, d\varepsilon$$

$$\simeq fkTN_{t0}(\omega_0 t)^{-\alpha} . \tag{6.32}$$

Recall that the photoconductivity decay has the form $t^{-1-\alpha}$ in the same time regime.

Although these arguments suggest that the induced absorption should decay more slowly than the photoconductivity, the measured decay [6.104] is actually very similar to the photoconductivity decay [6.42] (compare the data for the 300 °C film in Fig. 6.21 with Fig. 6.19). The measurements are both on phosphorous doped material ([PH$_3$]/[SiH$_4$] = 10^{-3}) and both at about the same excitation intensity $\sim 10^{17}$ photons cm^{-3}. One might expect the first branch of the photoconductivity and photoinduced absorption decays to be similar in the special case when α is close to unity. In this case, the free electron density is proportional to the trapped electron density, if a logarithmic correction is ignored. The behavior at long times for $\alpha \simeq 1$ has not been

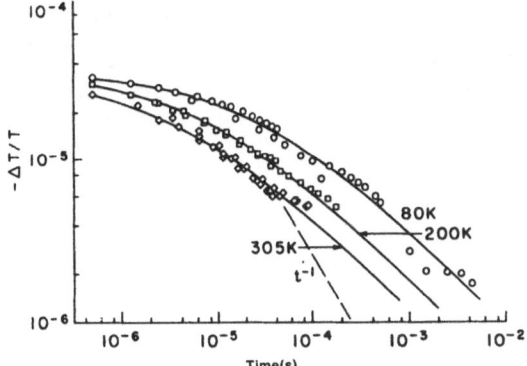

Fig. 6.21. Photoinduced absorption following a 10 ns excitation flash for glow discharge material doped with 0.1% PH$_3$ [6.105]

calculated theoretically. Clearly more experimental work needs to be done in this area, particularly on the relationship between the photoinduced absorption and the photoconductivity decay.

Photoinduced absorption can also be used to probe the location of the distribution of trapped holes in photo-excited a-Si:H. According to the preceding discussions of the time evolution of the trapped carrier distribution, it should be possible to photoexcite electrons from the valence band into the relatively narrow distribution of holes trapped near $\varepsilon^*(t) = kT \ln \omega_0 t$ in the valence band tail. This induced absorption process will exhibit threshold behavior of the form [6.105, 106]

$$\Delta\alpha(\varepsilon) \sim [\varepsilon - \varepsilon^*(t)]^{1/2}/\varepsilon \tag{6.33}$$

if the trapped hole distribution is narrow. In this expression, ε is the photon energy. *Vardeny* et al. [6.105] have shown that the induced absorption spectrum is in good agreement with (6.33) over the 0.7–1.3 eV range of their measurements. The induced absorption spectrum is shown in Fig. 6.22. The absorption threshold was found to depend linearly on temperature and logarithmically on time as expected from the time and temperature dependence of $\varepsilon^*(t)$, as illustrated in Fig. 6.23.

An advantage of the photoinduced absorption technique is that there are no electronic limitations to the time resolution of the measurement, in contrast to the situation with photoconductivity. Fast induced absorption measurements can give information about how fast the hot carriers thermalize into the localized states near the mobility edge. Picosecond induced-absorption measurements have been pioneered by *Vardeny* and *Tauc* [6.107–109]. They used a pump-probe technique based on a passively mode-locked laser with a photon energy of 2 eV, optical pulse duration of 0.6–0.8 ps and pulse energy of 1–2 nJ. The magnitude of the induced absorption (8–60 cm^{-1}) was comparable with values observed in steady-state experiments [6.106].

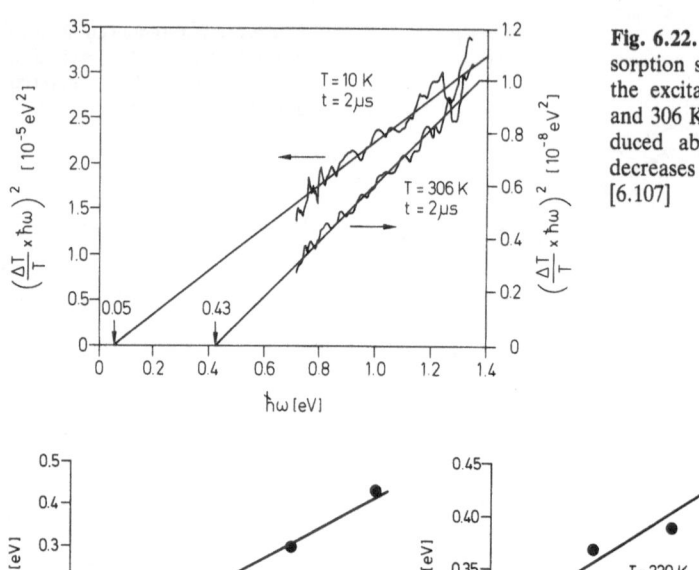

Fig. 6.22. Photoinduced absorption spectrum 2 µs after the excitation flash at 10 K and 306 K. Note that the induced absorption threshold decreases with temperature [6.107]

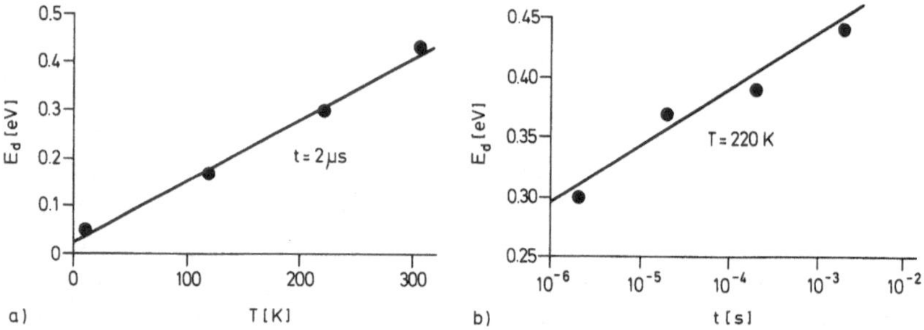

Fig. 6.23. Temperature (a) and time (b) dependence of the induced absorption threshold, following a 10 ns excitation flash [6.107]

The photoinduced absorption shows an initial peak value which decays in 0.7–1.2 ps to a background that is constant for the 20 ps time limit of the experiment. The fast excess absorption at the beginning is attributed to the larger optical cross section of hot carriers relative to thermalized carriers. Presumably the hot carriers thermalize directly from the band states into the localized band edge states in the initial fast process. Capture rates of this order ($\sim 10^{12}\,\text{s}^{-1}$) are consistent with the rates inferred indirectly from the picosecond photoconductivity experiments [6.45] discussed above.

6.4.5 Other Techniques

a) Traveling Wave

An interesting method for measuring the drift mobility in disordered semiconductors has been developed by *Adler* et al. [6.110a] and *Fritzsche* and *Chen* [6.110b]. In this technique the electron drift mobility is inferred from the dc current induced by a traveling electric field, generated by a surface acoustic wave on a LiNbO$_3$ crystal placed close to the a-Si:H film, as illus-

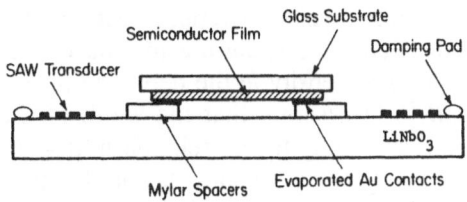

Fig. 6.24. Sample configuration for mobility measurements by the traveling surface-acoustic wave technique [6.110 a]

trated in Fig. 6.24. The traveling wave imparts a net dc drift velocity to the electrons given by $(\mu_D E)^2/V_s$, where μ_D is the drift mobility, E is the electric field produced by the acoustic wave and V_s is the surface acoustic wave velocity. The drift mobility is inferred from the electric field $(\mu_D E^2/V_s)$ required to null out the dc current induced by the traveling wave and the known values of E and V_s.

In this technique, the frequency ω of the acoustic wave defines the time scale for the drift mobility measurement, just as the transit time t_T defines the time scale in the time-of-flight experiment. For a frequency $\omega^{-1} = 8$ ns (20 MHz), *Janes* et al. [6.111] found drift mobilities of 0.5 and 0.08 cm²/Vs for two different samples, values that are consistent with time-of-flight results.

b) Space-Charge Layer Relaxation

In principle, carrier transport properties can be inferred from the relaxation time of a Schottky barrier in response to a voltage pulse, although in crystalline materials the carrier mobilities are normally too high and the response time is too short for the technique to be useful in the study of transport properties. However, this disadvantage disappears in amorphous semiconductors where the response times are much longer. *Crandall* [6.112] has applied this technique to Schottky diodes on 0.2% P doped a-Si : H.

His interpretation requires that the following two conditions be met. First, in order that the observed response be representative of the behavior of the space charge layer, the conductivity of the bulk of the material away from the depletion layer must be high enough that its dielectric relaxation

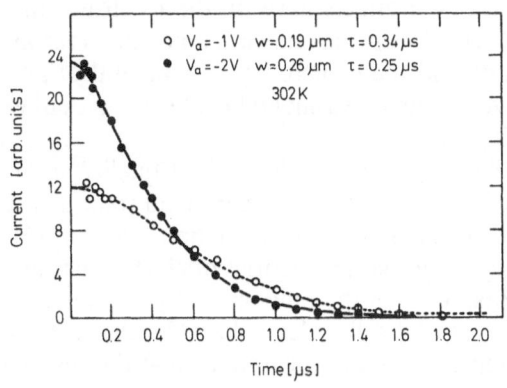

Fig. 6.25. Current transients associated with the space charge layer expansion, in response to a reverse bias voltage pulse. The theoretical fits are based on an electron drift mobility of 6.5×10^{-3} cm²/Vs and the assumptions outlined in the text [6.112]

time is short compared with the junction relaxation time. Secondly, the density of free and shallow trapped electrons responsible for the charge transport must be equal to the space charge density in the depletion layer.

With these assumptions, *Crandall* [6.112] shows that the width of the depletion region w expands with time according to the relation $w(t) = w_0$ tanh (t/τ) when an initially flat-band junction is reverse biased. The characteristic time $\tau = Lw_0/\mu_D V_0$, where w_0 is the steady-state depletion width, L is the sample thickness and V_0 is the built-in potential. The transient current associated with the expansion of the depletion region is illustrated in Fig. 6.25 for two different bias voltages. The calculated curves through the data are based on a tanh (t/τ) dependence of the depletion width on time and a drift mobility of 6×10^{-3} cm^2/V s.

This value is significantly smaller than time-of-flight values, albeit the latter are for undoped material. One conclusion is that heavy phosphorous doping greatly reduces the electron drift mobility. On the other hand, a possibility that is hard to rule out without the absolute magnitude of the current in Fig. 6.5 is that the mobile and shallow trapped charge density in the bulk material is only a small fraction of the total charge density in the depletion layer.

6.5 Unanswered Questions

The fundamental difference between amorphous and crystalline semiconductors as far as the transport properties are concerned is the existence of a high density of localized states near the band edges in amorphous semiconductors. Although some information about these states is available for a-Si : H from the experiments discussed in this chapter, for example, much more is not known. For example, what is the nature of the disorder that determines the width of the localized state distributions? Can this disorder be controlled? Why is the conduction band edge sharper than the valence band edge in a-Si : H? How does the capture cross section of the localized states depend on energy? What role do lattice relaxation effects have on the localized state binding energies? Although theoretical work has suggested possible reasons for the relative widths of the tails and their microscopic structural origins [6.113–115], at present there is very little experimental evidence that either confirms or disproves the theories.

Hydrogenated amorphous silicon is probably the highest quality amorphous semiconductor known from the point of view of density of states in the gap, dopability, stability, absence of geminate recombination, etc. [6.116]. As such it is an attractive prototype material for theories of electronic transport in amorphous semiconductors. Most of the fundamental problems concern the mobility edge and the nature of the disorder-induced localized states near the mobility edge. For example, there are suggestions that the shallow localized states actually arise from chemically distinct defects analogous to the valence alteration pairs in chalcogenides [6.117], rather than from disorder-induced band tails.

Another unresolved issue is the extent to which a-Si:H will be used in practical applications. To date, most of the commercial interest in the material has been stimulated by its potential as a low cost photovoltaic material [6.118]. In this application the solar conversion efficiency is likely to be the biggest hurdle. The solar cell efficiency is fundamentally limited by the same localized states that control the charge transport process [6.119, 120]. Other promising applications of a-Si:H are in thin film transistors [6.121], vidicon targets [6.122] and electrophotography. Transport properties play an important role in all of these applications.

6.6 Conclusion

A wide range of time-dependent charge transport data in a-Si:H can be explained remarkably well by a simple multiple-trapping model. In this model the time dependence of the charge transport phenomena arises from the progressive thermalization of electrons and holes in exponential distributions of localized states near the band edges which in the simplest interpretation are band-tail states. The ubiquitous power law decays observed in time of flight, photoconductivity and photoinduced absorption arise from a trade-off between the tendency of carriers to thermalize to lower energy and the progressively smaller density of states at low energies.

Acknowledgment. I am grateful to my colleagues B. Abeles, G. D. Cody, J. Klafter, T. D. Moustakas, P. D. Persans, A. Rose, C. Roxlo, H. Witzke and C. R. Wronski for many helpful discussions.

References

6.1   H. Fritzsche: Sol. Energy Mater. **3**, 447 (1980)
6.2   T. D. Moustakas: J. Electron. Mater. **8**, 391 (1979)
6.3   W. Paul, D. A. Anderson: Sol. Energy Mater. **5**, 229 (1981)
6.4   P. Nagels: "Electronic Transport in Amorphous Semiconductors" in *Amorphous Semiconductors*, ed. by M. H. Brodsky, Topics Appl. Phys., Vol. 36 (Springer, Berlin, Heidelberg, New York 1979), Chap. 5
6.5   P. G. LeComber, W. E. Spear: "Doped Amorphous Semiconductors," in [Ref. 6.4, Chap. 9]
6.6   J. Mort, D. M. Pai (eds.): *Photoconductivity and Related Phenomena*, (Elsevier, Amsterdam 1976)
6.7   H. Fritzsche, M. Tanielian: AIP Conf. Proc. **73**, 318 (1981)

6.8   H. Overhof, W. Beyer: Phil. Mag. B **43**, 433 (1981)
6.9   W. E. Spear, D. Allan, P. LeComber, A. Ghaith: Phil. Mag. **41**, 419 (1980)
6.10 G. H. Dohler: Phys. Rev. B **19**, 2083 (1979)
6.11 W. Beyer, H. Mell, H. Overhof: J. Phys. (Paris) C 4, Suppl. 10, **42**, 103 (1981)
6.12 J. Tauc: In *Optical Properties of Solids,* ed. by F. Abeles (North-Holland, Amsterdam, 1970)
6.13 G. D. Cody, T. Tiedje, B. Abeles, B. Brooks, Y. Goldstein: Phys. Rev. Lett. **47**, 1480 (1981)
6.14 P. G. LeComber, W. E. Spear: Phys. Rev. Lett. **25**, 509 (1970)
6.15 W. E. Spear, Haifa Al-Ani, P. G. LeComber: Phil. Mag. B **43**, 781 (1981)
6.16 P. Viktorovich, G. Moddel, W. Paul: AIP Conf. Proc. **73**, 186 (1981)
6.17 R. S. Crandall: Phys. Rev. Lett. **44**, 749 (1980)
6.18 P. G. LeComber, D. I. Jones, W. E. Spear: Phil. Mag. **35**, 1173 (1977)
6.19 J. Dresner: Appl. Phys. Lett. **37**, 742 (1980)
6.20 A. R. Moore: Appl. Phys. Lett. **37**, 327 (1980)
6.21 H. Mell, B. Movaghar, L. Schweitzer: Phys. Stat. Sol. (b) **88**, 531 (1978)
6.22 R. Haberkorn, W. Dietz: Solid State Commun. **35**, 505 (1980)
6.23 E. A. Schiff: AIP Conf. Proc. **73**, 233 (1981)
6.24 D. Weller, H. Mell, L. Schweitzer, J. Stuke: J. Phys. (Paris) C 4, Suppl. 10, **42**, 143 (1981)
6.25 W. Spear: J. Non-Cryst. Solids **1**, 197 (1969)
6.26 G. Pfister, H. Scher: Adv. Phys. **27**, 747 (1978)
6.27 H. Scher, E. W. Montroll: Phys. Rev. B **12**, 2455 (1975)
6.28 C. R. Wronski, B. Abeles, G. D. Cody, T. Tiedje: Appl. Phys. Lett. **37**, 96 (1980)
6.29 T. Tiedje, C. R. Wronski, B. Abeles, J. M. Cebulka: Sol. Cells **2**, 301 (1980)
6.30 P. Viktorovitch, G. Moddel: J. Appl. Phys. **51**, 4847 (1980)
6.31 T. Tiedje, T. D. Moustakas, J. M. Cebulka: Phys. Rev. B **23**, 5634 (1981)
6.32 J. D. Cohen, D. V. Lang: Phys. Rev. B **25**, 5321 (1982)
6.33 D. V. Lang, J. D. Cohen, J. P. Harbison: Phys. Rev. B **25**, 5285 (1982)
6.34 R. S. Crandall: J. Phys. (Paris) C 4, Suppl. 10, **42**, 43 (1981)
6.35 M. A. Lampert, P. Mark: *Current Injection in Solids* (Academic, New York 1970)
6.36 T. Tiedje, B. Abeles, D. L. Morel, T. D. Moustakas, C. R. Wronski: Appl. Phys. Lett. **36**, 695 (1980)
6.37 S. Ashok, A. Lester, S. J. Fonash: IEEE Trans. Lett. EDL-**1**, 200 (1981)
6.38 E. A. Schiff: Phys. Rev. B **24**, 6189 (1981)
6.39 D. H. Auston, A. M. Johnson, P. R. Smith, J. C. Bean: Appl. Phys. Lett. **37**, 371 (1980)
6.40 See, for example, S. L. Shapiro (ed.): *Ultra-short Light Pulses,* Topics Appl. Phys., Vol. 18 (Springer, Berlin, Heidelberg, New York 1977)
6.41 J. M. Hvam, M. H. Brodsky: J. Phys. (Paris) C 4, Suppl. 10, **42**, 551 (1981)
6.42 J. M. Hvam, M. H. Brodsky: Phys. Rev. Lett **46**, 371 (1981)
6.43 W. Fuhs, M. Milleville, J. Stuke: Phys. Stat. Sol. (b) **89**, 495 (1978)
6.44 A. M. Johnson, D. H. Auston, P. R. Smith, J. C. Bean, J. P. Harbison, A. C. Adams: Phys. Rev. B **23**, 6816 (1981)
6.45 A. M. Johnson, D. H. Auston, P. R. Smith, J. C. Bean, J. P. Harbison, A. C. Adams: AIP Conf. Proc. **73**, 248 (1981)
6.46 H. Scher: "Theory of Time-Dependent Photoconductivity in Disordered Systems" in *Photoconductivity and Related Phenomena,* ed. by J. Mort, D. M. Pai (Elsevier, New York 1976)
6.47 M. Silver, L. Cohen: Phys. Rev. B **15**, 3276 (1977)
6.48 J. Noolandi: Phys. Rev. B **16**, 4466, 4474 (1977)
6.49 F. W. Schmidlin: Phys. Rev. B **16**, 2362 (1977)
6.50 A. I. Rudenko, V. I. Arkhipov: J. Non-Cryst. Solids **30**, 163 (1978)
6.51 M. Pollak: Phil. Mag. **36**, 1157 (1977)
6.52 J. M. Marshall: Phil. Mag. B **43**, 401 (1981)
6.53 H. Imgrund, H. Overhof: J. Phys. (Paris) C 4, Suppl. 10, **42**, 83 (1981)

6.54 T. Tiedje, A. Rose: Solid State Commun. **37**, 49 (1980)
6.55 J. Orenstein, M. Kastner: Phys. Rev. Lett. **46**, 1421 (1981)
6.56 A. Rose: unpublished
6.57 A. Rose: *Concepts in Photoconductivity and Allied Problems* (R. E. Krieger, Pub., Huntington, New York 1978)
6.58 T. R. Waite: Phys. Rev. **107**, 463 (1957)
6.59 C. H. Henry, D. V. Lang: Phys. Rev. B **15**, 989 (1977)
6.60 V. I. Arkhipov, A. E. Rudenko: Fiz. Tekh. Poluprovodn. **13**, 1352 (1979) [English transl. Sov. Phys. Semicond. **13**, 792 (1979)]
6.61 A. J. Snell, K. D. Mackenzie, P. G. LeComber, W. E. Spear: Phil. Mag. **40**, 1 (1979)
6.62 J. Beichler, W. Fuhs, H. Mell, H. M. Welsch: J. Non-Cryst. Solids **35/36**, 587 (1980)
6.63 R. S. Crandall: J. Elect. Mat. **9**, 713 (1980)
6.64 G. F. Leal Ferreira: Phys. Rev. B **16**, 4719 (1977)
6.65 See, for example, M. Abramowitz, I. A. Stegun (eds.): *Handbook of Mathematical Functions* (Dover, New York 1972)
6.66 P. N. Butcher, J. D. Clark: Phil. Mag. **42**, 191 (1980)
6.67 M. Silver, L. Cohen, D. Adler: Appl. Phys. Lett. **40**, 261 (1982)
6.68 G. D. Cody, T. Tiedje, B. Abeles, T. D. Moustakas, B. Brooks, Y. Goldstein: J. Phys. (Paris) C 4, Suppl. 10, **42**, 301 (1981)
6.69 A. Ruppert, B. Abeles, J. P. deNeufville, R. Schriesheim: Bull. Am. Phys. Soc. **26**, 387 (1981)
6.70 P. G. LeComber, A. Madan, W. E. Spear: J. Non-Cryst. Solids, **11**, 219 (1972)
6.71 A. R. Moore: Appl. Phys. Lett. **31**, 672 (1977)
6.72 M. P. Rosenblum: Bull. Am. Phys. Soc. **26**, 313 (1981)
6.73 P. Nielsen, V. L. Dalal: Bull. Am. Phys. Soc. **26**, 313 (1981)
6.74 P. E. Vanier: Bull. Am. Phys. Soc. **27**, 269 (1982)
6.75 M. Silver, K. S. Dy, I. L. Huang: Phys. Rev. Lett. **27**, 21 (1971)
6.76 J. F. Peyre, J. Baixeras, D. Mencaraglia, P. Andro, C. Longeaud: J. Phys. (Paris) C 4, Suppl. 10, **42**, 163 (1981)
6.77 C. R. Wronski, R. E. Daniel: Phys. Rev. B **23**, 794 (1981)
6.78 T. Datta, M. Silver: Solid State Commun. **38**, 1067 (1981)
6.79 T. Tiedje, T. D. Moustakas, J. M. Cebulka: J. Phys. (Paris) C 4, Suppl. 10, **42**, 155 (1981)
6.80 P. B. Kirby, W. Paul, P. Jacques, J. L. Brebner: Proc. Conf. on Tetrahedrally Bonded Amorphous Semiconductors, AIP Conf. Proc. **73**, 207 (1981)
6.81 P. B. Kirby, W. Paul, S. Ray, J. Tauc: Solid State Commun. **42**, 533 (1982)
6.82 T. D. Moustakas, D. A. Anderson, W. Paul: Solid State Commun. **23**, 155 (1977)
6.83 W. E. Pickett, D. A. Papaconstantopoulos, E. N. Economou: J. Phys. (Paris) C 4, Suppl. 10, **42**, 769 (1981)
6.84 J. Dresner, D. Szostak, B. Goldstein: Appl. Phys. Lett. **38**, 998 (1981)
6.85 P. B. Kirby: private communication
6.86 D. Allan: Phil. Mag. B **38**, 381 (1978)
6.87 J. Mort, S. Grammatica, J. C. Knights, R. Lujan: Sol. Cells **2**, 451 (1980)
6.88 T. Tiedje, J. M. Cebulka, D. L. Morel, B. Abeles: Phys. Rev. Lett. **46**, 1425 (1981)
6.89 R. A. Street: Adv. Phys. **30**, 593 (1981)
6.90 D. J. Dunstan, F. Boulitrop: J. Phys. (Paris) C 4, Suppl. 10, **42**, 331 (1981)
6.91 Wei-Chung Chen, B. J. Feldman, J. Bajaj, F.-M. Tong, G. K. Wong: Solid State Commun. **38**, 357 (1981)
6.92 R. W. Collins, M. A. Paesler, G. Moddel, W. Paul: J. Non-Cryst. Solids **35/36**, 681 (1980)
6.93a B. Abeles, C. R. Wronski, T. Tiedje, G. C. Cody: Solid State Commun. **36**, 537 (1980)
6.93b C. B. Roxlo, B. Abeles, C. R. Wronski, G. D. Cody, T. Tiedje: Solid State Commun. **47**, 985 (1983)
6.94 W. B. Jackson, N. M. Amer: Phys. Rev. B **25**, 5559 (1982)
6.95 N. F. Mott: Rev. Mod. Phys. **50**, 203 (1978)

6.96   N. F. Mott, E. A. Davis: *Electronic Processes in Non-Crystalline Materials,* 2nd ed. (Clarendon Press, Oxford 1979)

6.97a   T. Tiedje (published here for the first time)

6.97b   V. I. Arkhipov, M. S. Iovu, A. I. Rudenko, S. D. Shutov: Phys. Stat. Sol. (a) **54,** 67 (1979)

6.98a   D. V. Lang: "Space-Charge Spectroscopy in Semiconductors" in *Thermally Stimulated Relaxation in Solids,* ed. by P. Braunloch, Topics Appl. Phys., Vol. 37 (Springer, Berlin, Heidelberg, New York 1979), Chap. 3
        J. Bourgoin, M. Lannoo: *Point Detects in Semiconductors* II, Springer Ser. Solid-State Sci., Vol. 35 (Springer, Berlin, Heidelberg, New York 1983)

6.98b   H. Dersch, J. Stuke, J. Beichler: Phys. Stat. Sol. (b) **105,** 265 (1981)

6.98c   T. Tiedje, B. Abeles, J. M. Cebulka: Solid State Commun. **47,** 493 (1983)

6.99   M. Silver, L. Cohen, D. Adler: Solid State Commun. **40,** 535 (1981)

6.100   G. Schonherr, H. Bassler, M. Silver: Phil. Mag. B **44,** 47 (1981)

6.101   J. C. Knights, R. A. Lujan: Appl. Phys. Lett. **35,** 244 (1979)

6.102   H. Scher: J. Phys. (Paris) C 4, Suppl. 10, **42,** 547 (1981)

6.103   J. Orenstein, M. A. Kastner, V. Vaninov: Phil. Mag. B **46,** 23 (1982)

6.104   D. E. Ackley, J. Tauc, W. Paul: Phys. Rev. Lett. **43,** 715 (1979)

6.105   Z. Vardeny, P. O'Connor, S. Ray, J. Tauc: Phys. Rev. Lett. **44,** 1267 (1980)

6.106   P. O'Connor, J. Tauc: Solid State Commun. **36,** 947 (1980)

6.107   S. Ray, Z. Vardeny, J. Tauc: J. Phys. (Paris) C 4, Suppl. 10, **42,** 555 (1981)

6.108   Z. Vardeny, J. Tauc: Phys. Rev. Lett. **46,** 1223 (1981); **47,** 700 E (1981)

6.109   Z. Vardeny, J. Strait, D. Pfost, J. Tauc, B. Abeles: Phys. Rev. Lett. **48,** 1132 (1982)

6.110a R. Adler, D. Janes, B. J. Hunsinger, S. Datta: Appl. Phys. Lett. **38,** 102 (1982)

6.110b H. Fritzsche, K.-J. Chen: Phys. Rev. B **28,** 4900 (1983)

6.111   D. Janes, S. Datta, R. Adler, B. J. Hunsinger: AIP Conf. Proc. **73,** 222 (1981)

6.112   R. S. Crandall: J. Appl. Phys. **52,** 1387 (1981)

6.113   J. Singh: Phys. Rev. B **23,** 4156 (1981)

6.114   F. Yonezawa, M. H. Cohen: *Fundamental Physics of Amorphous Semiconductors,* ed. by F. Yonezawa, Springer Ser. Solid-State Sci. **25** (Springer, Berlin, Heidelberg, New York 1981), p. 119

6.115   S. Abe, Y. Toyozawa: J. Phys. Soc. Jpn. **50,** 2185 (1981)

6.116   R. A. Street: this volume, Chap. 5

6.117   D. Adler: Sol. Cells **2,** 199 (1980); J. Phys. (Paris) C 4, Suppl. 10, **42,** 3 (1981)

6.118   D. E. Carlson: in *The Physics of Hydrogenated Amorphous Silicon* I, ed. by J. D. Joannopoulos, G. Lucovsky, Topics Appl. Phys. Vol. 55 (Springer, Berlin, Heidelberg, New York 1983), Chap. 6

6.119   T. Tiedje: Appl. Phys. Lett. **40,** 627 (1982)

6.120   B. Abeles, G. D. Cody, Y. Goldstein, T. Tiedje, C. R. Wronski: Thin Solid Films **90,** 441 (1982)

6.121   K. D. Mackenzie, A. J. Snell, I. French, P. G. LeComber, W. E. Spear: Appl. Phys. A **31,** 87 (1983)
        S. Kishida, Y. Narake, Y. Uchida, M. Matsumura: J. Appl. Phys. Jpn. **22,** 511 (1983)

6.122   See, for example, S. Oda, K. Saito, H. Tomita, I. Shimizu, E. Inoue: J. Appl. Phys. **52,** 7275 (1981)

7. Vibrational Properties of Amorphous Alloys

G. Lucovsky and W. B. Pollard

With 31 Figures

7.1 Introduction

There is considerable interest in the vibrational properties of amorphous silicon (a-Si), and of a-Si alloys containing hydrogen (a-Si:H), fluorine (a-Si:F), chlorine (a-Si:Cl), oxygen (a-Si:O), nitrogen (a-Si:N), carbon (a-Si:C) and germanium (a-Si:Ge), as well as ternary alloys containing various pairs of alloy atoms, as in a-Si:H:F, a-Si:Ge:H and a-Si:H:O. Recent studies of pure a-Si (and a-Ge) have emphasized small, but significant changes in the Raman spectrum of samples that have been prepared in different ways and/or subjected to different thermal annealing cycles. The changes have been associated with aspects of the local atomic structure, both the short-range order, e.g., the spread in the distribution of the bond angles, and the intermediate-range order, as reflected in the distribution of dihedral angles, and in the various aspects of the ring statistics. Other studies, and those emphasized in this chapter, have sought to delineate the detailed nature of the local bonding configurations at the alloy atom sites by considering the variation of the frequencies of the Raman and infrared (ir) active vibrations with the concentration of bonded alloy atoms. These studies have sought to assign reported frequencies and/or groups of frequencies to specific local atomic arrangements that are expected from the normal valence bonding requirements of the particular alloy species, as, for example, monohydride (SiH) or dihydride (SiH$_2$) groups in a-Si:H alloys. There has been a general consensus regarding the majority of these assignments and some controversy regarding others. In many instances model calculations and/or considerations of the local symmetry at the alloy atom sites have served to verify the qualitative and/or quantitative aspects of a particular assignment. In other instances, there are sufficient ambiguities in the interpretations so that several specific and important assignments have yet to be completely resolved. This is particularly so in a-Si:H and a-Si:F alloys with high H and F atom concentrations, > 20–50 at.%. In some of these cases, theoretical calculations have either resolved the controversy or at least identified definitive ways to test the validity of competing assignments. Of particular interest is the very recent use of deuterium substitutions in a-Si:H alloys.

The qualitative behavior of the vibrational modes involving alloy atoms in a-Si alloys parallels very closely the behavior of alloys atoms in a crystalline solid. The incorporation of alloy atoms in a host a-Si network produces certain types of vibrations in which the atomic displacements are localized on the alloy atom and its immediate neighbors [7.1, 2]. The difference in bonding between an alloy atom in an a-Si host and an alloy atom in a crystalline silicon (c-Si) host is in the way the atom bonds to its immediate neighbors, and not in the qualitative aspects of these local mode vibrations. In the c-Si host the alloy atoms are either constrained to substitute for Si-atoms of the host and hence be contained in a site that has a tetrahedral bonding character, or to reside at interstitial sites which also have a limited number of specific local symmetries. On the other hand, the incorporation of alloy atoms in an a-Si host is visualized as a very different process where the alloy atom resides at a bonding site dictated by its own valence bonding requirements. This in turn determines the local symmetry at the alloy atom site, including the number and spatial arrangement of the nearest neighbor Si-atoms. With regard to the local modes, these depend on the symmetry at the alloy atom site, whether it is a terminal or singly-coordinated species such as H, F or Cl, a bridging or twofold-coordinated species such as O, or has a higher coordination as for N where the coordination is threefold, or C and Ge, where it is fourfold, The other important factors in determining the local mode behavior are the same as in the crystalline phase, the mass of the alloy atom relative to the mass of the host Si-atoms and the magnitude of the short-range interatomic forces at the alloy atom sites relative to similar forces involving the Si-atoms in the remainder of the host network.

The vibrational modes involving significant alloy atom motion will either occur at frequencies which are in excess of the highest vibrational frequency of the host solid, or at frequencies which are within the range of the vibrational frequencies of the host. The first type of vibration is termed a local mode, where the word local specifically relates to its relative frequency of vibration. There are two types of modes in the second class. If the host has gaps or pseudo-gaps in its vibrational spectrum, then alloy atom vibrations can occur within the frequency regime of these gaps; in this case the vibrations are designated as gap modes. If the host has no gap or pseudo-gap, then certain alloy atom vibrations will have frequencies of vibration that are degenerate with vibrational modes of their host. These types of vibrations are designated as resonance modes. Since a-Si displays a continuum of vibrational modes, extending from very low frequencies to about 500 cm^{-1} and hence no gap region, this means that virtually all alloy atoms will have one or more vibrations that are resonance modes. In this chapter we will emphasize the local modes. They are easiest to determine experimentally, mostly by ir absorption, and there are theoretical approaches for describing their properties that are relatively easy to use and that give very accurate qualitative and quantitative results. The most important aspects of the alloy atom with regard to its vibrational properties in an a-Si host are: (i) the local coordination and symmetry at the alloy atom bonding site, (ii) the mass of the alloy atom relative to the mass of the host Si-atoms, and (iii) the differences between the short-range bonding forces at the alloy atom site and those of the host network. All of these contribute to the character of the alloy atom vibrations and in particular serve to determine the number of vibrations that retain a local mode character and the number that are resonance modes.

In this chapter we will emphasize the experimental and theoretical aspects of alloy atom motions in a a-Si host. We will consider both binary and ternary alloy systems and note the unique character of certain ternary systems such as a-Si:H:O and a-Si:H:N, where both alloy atoms can be bonded to a common Si-site. We will further emphasize the quantitative and qualitative understanding of the local mode properties and point out what is known about resonance modes in those instances where they have been reported. In order to deal with the alloy atom problem, it is necessary to first consider the vibrational properties of the host network in which the alloy atoms are contained. This is accomplished in Sect. 7.2 where we treat "pure" a-Si, describing the results of Raman and ir spectroscopy, and the interpretations of these spectra in terms of several different and complementary models. We next consider the theory of alloy atom vibrations in Sect. 7.3, describing two complementary approaches which, when applied to the same chemical system, yield a complete description of the vibrational properties. One of these techniques is based on the exact solution of a finite cluster that includes the alloy atom and a sufficent number of Si neighbors so as to isolate the localized vibrations at the alloy atom site and its immediate Si neighbors from dynamical effects that are related to the motions of atoms at the boundary of the cluster. A cluster that satisfies this condition is designated as a cluster of intermediate size. The second method is based on an infinite aperiodic system consisting of the alloy atom and its immediate Si neighbors in a finite cluster, but with the terminal atoms attached to infinite Si Bethe lattices.

Section 7.4 deals with the general aspects of alloy atom incorporation. It begins with a description of the types of alloy atom bonding sites as they are defined by the coordination-dependent local symmetry of the alloy atom species. This general description is then extended to a detailed discussion of the alloy atom vibrations that are associated with specific atomic species: H, F, O, and C, etc. Section 7.5 considers the more extensively studied binary systems and Sect. 7.6 treats ternary alloys. The procedure that we shall follow in the these sections is to present the experimental results and then identify those assignments that are firmly established, including the basis for the particular assignment. Finally we consider those instances where there is controversy. We describe the nature of the controversy as well as the arguments that have led to the differing interpretations. This particular approach is designed to give the reader a framework in which to review past studies and a critical prespective for judging new results and interpretations. We have not made an attempt to include every reference that deals with the vibrational properties of a-Si alloys. We have given a sufficient number of references to identify the main points of a particular discussion, including representative experimental data and theoretical and model calculations. There are some topics we have chosen to omit completely. One of these deals with the use of ir spectroscopy as a tool for quantitative analysis, i.e., the use of the integrated ir absorption as a means of determinging the concentration of a particular alloy species. Recently, *Solar Cells* has published a two-volume set which contains all of the references to a-Si up to 1980. The reader is referred to these two volumes to obtain additional references on those subjects we have included in this review, and all of the references on those we have omitted. The volumes are indexed by both author and subject matter [Sol. Cells, *4*, (3, 4) (1982)].

7.2 Properties of Pure Amorphous Silicon

In recent years, there has been a renewal of interest in the local atomic structure of pure a-Si. The designation "pure" refers to samples free of large numbers of impurity or alloy atoms, and generally prepared by evaporating or sputtering Si in a relatively clean deposition system. In a practical sense, when considering the vibrational properties of the system, the term "pure" refers to a-Si samples with impurity atom concentrations below about 0.1 at.%, i.e., below the limit of detection by either ir absorption spectroscopy or Raman scattering. This is a very different definition of a model system than can be employed when discussing the electronic properties. The attention being paid to the vibrational properties of this material is a consequence of the experimental and theoretical interest in the properties of alloys which can be tailored to yield desirable electronic and/or photoelectronic properties. Specifically, the nature of the modifications to the host a-Si network by the incorporated alloy atoms cannot be understood without a precise picture of the pure a-Si host system as a reference system for specifying changes in the short and intermediate-range order. Moreover, as pointed out in the introduction, a knowledge of the vibrational states of the pure host a-Si system is an essential prerequisite for understanding the nature of the additional vibrational states induced by the alloy atoms. This is true in a general way for both local and resonance modes; however, the extent to which we can describe the character of the resonance modes is more strongly dependent on the detailed nature of the host network, in particular the topology of the network that is defined by its connectivity and ring statistics.

Previous studies and review articles have been concerned with the structure of the a-Si network and the way in which it is revealed in the vibrational spectra. These studies have emphasized interpretations of the ir and Raman spectra in terms of the vibrational density of states (VDOS) of a-Si, which is assumed to be intimately related to the VDOS of c-Si. In the context of this assumption, they have used a nomenclature in which the features found in the vibrational spectra of a-Si are described in terms of their "parentage" in the crystalline phase. Other structural models have also been used to generate VDOS's with the aim of relating specific features in the VDOS to particular aspects of the bonding, including any departures from idealized short-range order to the inclusion of specific types of intermediate-range order that are manifested in the ring statistics. In this context there is general agreement that a-Si can be visualized as a continuous random network based on the interconnection of fourfold-coordinated Si-atoms. This network is assumed to have three types of structural disorder that in turn are responsible for its amorphous nature: (i) bond angle variations, (ii) dihedral angle variations, and (iii) topological disorder that is described by variations in the ring statistics. In reality, these three types of disorder are not independent so that the modelling of a network cannot be carried out in a way that simply reflects the degree of disorder assumed for any of the three structural "variables". Moreover, the a-Si network is known to contain a relatively large number of native defects in the form of dangling bonds. In the real material these are associated with internal microvoids of varying size. The number of defects can be as high as 10^{21} cm^{-3} in an unannealed evaporated sample, but can be reduced by about two to three orders of magnitude through careful control of deposition and annealing procedures. The role that these defects play in the nature of the random network structure is not completely understood.

Raman scattering and ir absorption measurements have served as an indirect experimental measure of the VDOS of the pure a-Si host. In this context, it is important to note that since long-range order and periodicity are absent in the amorphous material, all of the vibrational modes can contribute to both the ir and Raman response. The particular degree of Raman or ir activity is then determined by the type of vibration and the way that it is influenced by certain aspects of the short and intermediate-range order. Hence, the measured ir and Raman spectra are expected to display features associated with structure in the VDOS, but modulated by frequency-dependent matrix elements. This is born out by comparisons between the ir and Raman spectra and a model VDOS generated by introducing a reasonable degree of broadening into a calculation of the VDOS of c-Si. In Fig. 7.1 we show the Raman spectrum of pure a-Si as measured by *Smith* and co-workers [7.3], and the broadened VDOS of the crystalline material. The Raman spectrum is characterized by three distinct bands: (i) a relatively sharp and dominant peak centered at about 480 cm^{-1}, (ii) a broad shoulder center about 300 cm^{-1} with a hint of structure near 250 cm^{-1}, and finally, (iii) a low frequency peak centered at about 200 cm^{-1}. The ir absorption spectrum of a-Si as measured by *Shen* et al. [7.4] is displayed in Fig. 7.2. The spectrum is qualitatively similar to the Raman spectrum in the sense that each spectrum is a continuum extending up to about 500 cm^{-1} and shows features in three similar spectral regimes. The ir spectrum displays more structure than the Raman spectrum with six features being identified: (i) a high frequency peak at about 465 cm^{-1}, (ii) a shoulder near 390 cm^{-1}, (iii) a well-defined peak centered about 300 cm^{-1}, (iv) a dip, similar to that seen in the Raman spectrum, near 210 cm^{-1}, (v) a low frequency peak at 180 cm^{-1}, and finally (vi) a shoulder at about 130 cm^{-1}. Even though the number of features appears to

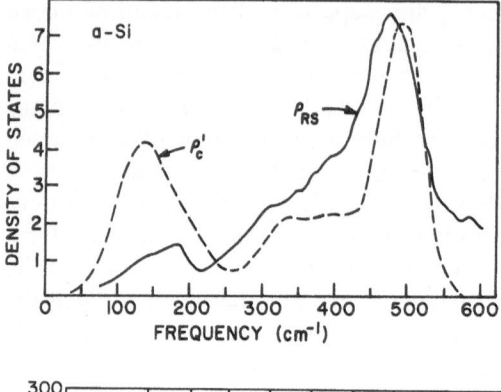

Fig. 7.1. Comparison between the Raman scattering spectrum of a-Si (—) and a broadened one-phonon DOS for c-Si (---) (from [7.3])

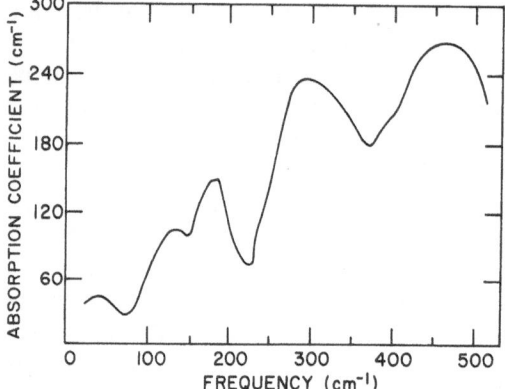

Fig. 7.2. IR absorption spectrum of a-Si (from [7.4])

be larger than in the Raman spectrum, the features in the ir spectrum fall into the same three major bands as the Raman spectrum: the so-called TO band in the region of 400–500 cm^{-1}, the LO-LA band between 210 and 400 cm^{-1}, and the TA band centered at about 200 cm^{-1}. Recently, it has been reported by several groups [7.5, 6] that the relative strength of these three bands in the Raman spectrum can be changed by either changing the deposition and/or annealing conditions. These changes occur primarily in the width and exact position of the high frequency TO-like band, which can occur in a frequency range from about 460 to 480 cm^{-1}, and in the position and relative strength of the low frequency TA-like band. Subtle changes in intensity also occur within the LO-LA band.

A number of different theoretical approaches have been used to generate a VDOS states for pure a-Si and to determine the relationship between this VDOS and the vibrational spectra. As indicated above, these descriptions of the vibrational spectra have also presumed an intimate relationship between the VDOS of a-Si and the VDOS of the crystalline phase. It is therefore deemed useful to display the crystalline VDOS [7.7], as we do in Fig. 7.3, and identify the origins of the major features in terms of the types of vibrational motions. The dominant features in the crystalline VDOS are found to

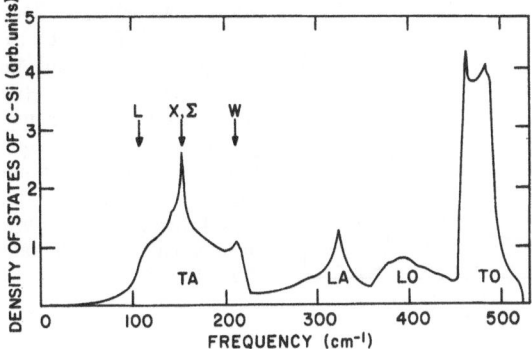

Fig. 7.3. Vibrational DOS calculated for c-Si (from [7.7]). The major bands are designated as TA, LA, LO and TO, and the structure within the TA band is identified with specific critical points

occur in four bands that are centered about frequencies of 480 cm^{-1}, 390 cm^{-1}, 325 cm^{-1} and 160 cm^{-1}, respectively, and are labelled as transverse optic (TO), longitudinal optic (LO), longitudinal acoustic (LA) and transverse acoustic (TA), respectively. The similarities between the ir and Raman spectra of a-Si and the VDOS of the crystalline phase have been taken to imply that the major features of the c-Si VDOS are then also retained in a-Si. Model calculations have attempted to define the particular elements of short and intermediate-range order that are retained in the amorphous phase, and hence are the origin for the similarities reported. It has been demonstrated that the amorphous and crystalline forms of Si exhibit the same short-range order in the context of the Si-Si bond length and the local tetrahedral coordination. It has also been determined that there are bond angle variations within the amorphous phase and that these may be as large as ten degrees. The absense of certain types of features in the radial distribution function has been taken as evidence for the lack of order beyond bond angle, or three-body correlations. It should be noted, however, that all models for the topology of the amorphous phase include small rings consisting of either five, six or seven atoms, or some combination of all of these types. One of the most useful theoretical techniques for studying the correlation between local atomic structure and both vibrational and electronic DOS's has been the Bethe Lattice technique [7.8]. We now review the aspects of these correlations that are related to a description of the vibrational properties.

This relationship can be established by first comparing the crystalline VDOS with the VDOS that is calculated for an a-Si Bethe lattice. The Bethe lattice is an infinite aperiodic network of Si atoms with the same short-range order as the crystal (bond length and tetrahedral coordination), but which contains no closed rings of bonded atoms. In this context the Bethe lattice removes any local atomic structure that is correlated specifically with phased sequences of particular dihedral angles. For a more detailed discussion of the Bethe lattice, see Sect. [7.3]. We include in Fig. 7.4 the VDOS of the a-Si Bethe lattice and compare it to the VDOS of the crystalline phase. The

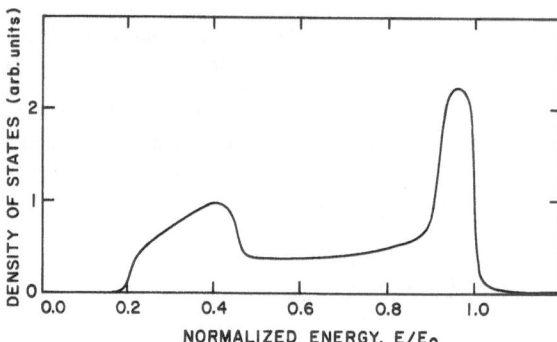

Fig. 7.4. Vibrational DOS for the a-Si Bethe lattice. This is calculated using the Born Hamiltonian (see [7.8]). E_0 is the maximum phonon energy and is equal to $(8\alpha/m)^{1/2}$. α is the bond-stretching force, β is the bond-bending force and m is the Si atomic mass. The ratio $\beta/\alpha = 0.6$

comparison indicates that two of the four bands of the crystalline VDOS are reproduced via the Bethe lattice calculation; these are the high frequency TO band and the low frequency TA band. The LA and LO bands of the crystalline VDOS states are not present in the Bethe lattice VDOS as distinct features. Instead, the VDOS of the Bethe lattice is essentially constant in the regime between the TA and TO-like peaks. This lack of LA and LO-like features is not the consequence of a particular choice of force constants or the choice of a particular way to construct the Bethe lattice. Since the Bethe lattice in effect only includes the local coordination at the Si sites and no aspects whatsoever of intermediate-range order, this allows us to conclude that the presence of the TA and TO-like peaks in the Bethe lattice VDOS, as well as the presence of TA and TO-like peaks in the vibrational spectra of a-Si, arise solely from the fourfold-coordination and tetrahedral bonding environments at the Si atom sites. This means that the dominant features of the Raman spectrum and two of the major features of the ir spectrum are determined entirely by the short-range order.

A subtle aspect of this comparison is in the postion of these peaks in the Bethe lattice, the experimental spectra and the crystalline VDOS. The center of the TO-like peak in the three spectra is virtually the same, provided we consider the experimental data for films that have relatively low densities of dangling bonds, i.e., films containing some bonded H or films that are annealed at temperatures in excess of about 350 °C [7.6]. The position of the TA peak is a different story. It is about the same in the Bethe lattice VDOS and the amorphous spectra, but about 40 cm^{-1} lower in the crystalline phase. The exact position of the TA phonon feature in c-Si has received considerable theoretical study and it is well known that this type of feature is in part determined by long-range periodic interactions and that these serve to depress the frequency at which the feature occurs. Secondly, the failure of the Bethe lattice to yield any distinct features in the LA and LO-like regime suggests that these features derive from elements of intermediate range that are absent in the Bethe lattice, e.g., closed rings of atoms that are known to be present in the crystalline phase and assumed to be present in the amorphous phase as well. Therefore, in order to give an adequate explanation of the features in the ir and Raman spectra and the VDOS of the amorphous phase, it will be necessary to consider the effects of both short and intermediate-range order on the calculated VDOS and further, the effects of quantitative disorder of the specific structural parameters such as bond lengths, bond angles, dihedral angles and the number and specific character of the rings of atoms that must be present.

There have been many theoretical studies which have addressed the problem of calculating the VDOS and the ir and Raman response for tetrahedrally coordinated amorphous semiconductors [7.9–14]. Most of these have employed numerical techniques to determine the VDOS for a particular continuous random network (CRN) that was constructed with some set of constraints imposed, e.g., only even-membered rings, etc. Such studies are important because they provide a direct comparison with experimental data and hence can give considerable insight to the matrix elements responsible for the ir and Raman activity and, in particular, for the different degrees of relative activity for particular bands within each of the two spectra. The comparisons also provide a test for the validity of the assumptions that had been used in the construction of the network itself, e.g., as mentioned above, any assumptions that were made about the nature of the types, and distributions of rings of atoms.

One of the more comprehensive of this class of studies is the one performed by *Alben* and co-workers [7.9]. Using a Keating representation of the near-neighbor interatomic forces and "simple" semi-empirical models for the ir and Raman activities, they calculated the VDOS, ir absorption and Raman scattering spectra for several different CRN and microcrystalline models. Each of the structures was characterized by different degrees of short and intermediate-range order involving bond angles, dihedral angles and topological variations. The results of their study confirm that the TO-like modes centered near 480 cm^{-1} and the TA-like mode centered near 180 cm^{-1} occurring in both the ir and Raman spectra are derived from the fourfold-coordination of the Si atoms and are retained independent of the particular degree of disorder. In contrast, as we have noted above, any structure in the LA-LO region of the VDOS and ir and Raman response, is strongly influenced by the topology assumed in a particular model, including the nature of the ring statistics. Similar calculations by *Meeks* [7.12] and *Beeman* and *Alben* [7.11] of the VDOS of a-Ge support this last point. The studies in [7.11, 12] paid particular attention to the effects of disorder on both the longitudinal and transverse-like phonons, including those in the LA-LO frequency regime. The effects of disorder are illustrated in Fig. 7.5 where we display the VDOS, as calculated in [7.11] for three different CRN structures (b–d). The deviations from tetrahedral symmetry, bond angle distortions and increasing ratios of odd to even-membered rings, increase in going from the top to the bottom of the figure. There are four significant changes to note in going from the VDOS of the crystalline phase (Fig. 7.5a) to the CRN with the largest degree of distortion (Fig. 7.5d). These include: (i) a broadening of

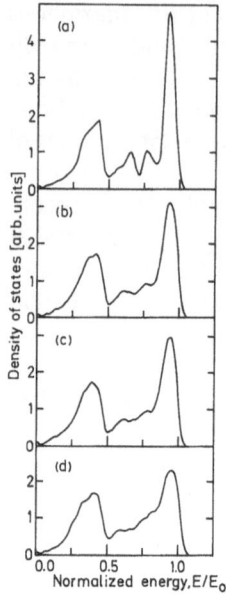

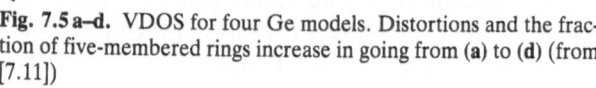

◀ **Fig. 7.5 a–d.** VDOS for four Ge models. Distortions and the fraction of five-membered rings increase in going from (**a**) to (**d**) (from [7.11])

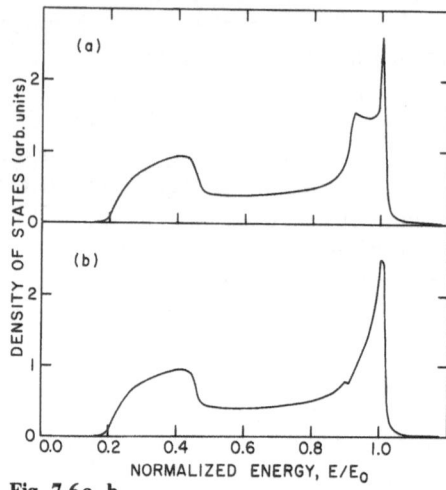

Fig. 7.6 a, b.

Fig. 7.6 a, b. LDOS for an atom in the center of a cluster consisting of a distorted tetrahedron. Si-Bethe lattices are attached to the outside atoms (see [7.9, 10]). (**a**) The central atom is Nr. 9 of the Connell-Temkin CRN. (**b**) The central atom is Nr. 1 of the same model

the high frequency TO band, (ii) an increase of states between otherwise distinct LA and LO bands with the eventual loss of all structure between the TO and TA peaks, (iii) the filling in of the dip in the VDOS at the top of the TA band, and (iv) a shift in the peak at the center of the TA band to lower frequency. The origin of these changes can be correlated and understood by examining the specific effects of bond angle, dihedral angle and topological disorder separately.

Studies on amorphous semiconductors suggest that dihedral angle variations in three-dimensional CRN's by themselves produce insignificant changes in the VDOS. These changes are not accompanied by changes in the ring statistics, i.e., the degree of topological disorder. In contrast, bond angle distortions can produce significant effects. In Fig. 7.6 the local DOS for two different atoms, taken from the Connell-Temkin model [7.9, 10] for a-Si and embedded in the Si Bethe lattice, are displayed. The most important effect of bond angle distortions is an increase in the width of the TO-like band which is accompanied by a shift of the center of the band to higher frequency. There is a corresponding slight shift in the position of the TA-lke band to lower frequency. The most important thing to note is that bond angle fluctuations do not produce features in the LA-LO region of the spectrum. The origin of the features in this spectral regime is, however, revealed in the work of *Thorpe* [7.14] and *Yndurian* and *Sen* [7.13].

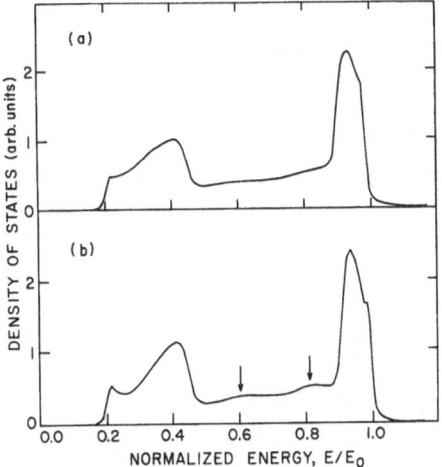

Fig. 7.7. (a) LDOS for an atom with a single sixfold ring of bonds passing through it. **(b)** LDOS for an atom with six sixfold rings of bonds passing through it. The ring configurations are derived from the diamond lattice and have Si-Bethe lattices attached to the exterior atoms

The studies reported in [7.13, 14] deal with the specific effects of topological disorder. They demonstrate conclusively that the structure in the LA-LO frequency regime is due to the presence of closed rings of bonds in the network structure of the amorphous solid. In particular, the two peaks in the VDOS states spectrum (Figs. 7.5a–c) can be linked to the presence of six-membered rings of atoms, while the absence of such peaks in Fig. 7.5d is due to an increase in the percentage of five-membered rings. This is illustrated in Fig. 7.7a, where we show the local DOS (LDOS) of an atom in a cluster in which there is only one six-membered ring of atoms. Although there are some changes in the VDOS near the TO and TA-like peaks, there are no features evident in the LO-LA frequency regime. This is not the case, however, for the LDOS spectrum shown in Fig. 7.7b for an atom with six different six-membered rings passing through it. The calculated spectrum yields two broad resonances centered near 300 and 400 cm^{-1}, respectively. These states are clearly modes of vibration associated with the large number of six-membered rings passing through a common atom. The spectrum in Fig. 7.7a is similar to the results obtained by *Thorpe* [7.14] for a single six-membered ring embedded in an "effective medium", while the spectra in Fig. 7.7b is suggestive of the results obtained by *Yndurian* and *Sen* [7.13] who examined the LDOS of atoms in larger clusters that were taken from the diamond structure in which twelve six-membered rings pass through the central atom. Their LDOS spectra exhibit two clearly defined peaks in the LA-LO region. Their results also suggest that the strength of the resonances in this frequency regime is proportional to the number of rings that pass through each atom. This in turn may provide some measure of the extent of delocalization of these particular resonance vibrations. In this context, the major difference between five and six-membered rings is in the particular types of features they generate; six-membered rings yield two features in this regime, whereas five-member rings yield only one.

So far we have shown that the one-phonon VDOS, the ir absorption and the Raman scattering of pure a-Si all display features in three well-defined frequency regimes which bear a direct relationship to the well understood VDOS of crystalline Si. The model calculations have clearly demonstrated that the dominant spectral features in the TO and TA bands are determined primarily by the local coordination of the Si atoms, i.e., the tetrahedral symmetry of the local bonding environment. The exact frequencies of the peaks in the Raman and ir spectra and the VDOS are sensitive to both bond angle distortions and the the network topology. In contrast, the structure in the LA-LO frequency regime (from 200–400 cm^{-1}) is derived entirely from closed rings of bonded atoms. Six-membered rings give rise to two features, whereas five-membered rings give rise to only one. Furthermore, the vibrations that yield these spectral resonances are delocalized over many adjacent rings of atoms and their strength depends on the number of rings passing through a given atom. Variations in both the ring topology and the bond angles at the Si sites contribute to broadening and shifting of the TA and TO-like peaks.

7.3 Amorphous Silicon Alloys: General Theory of Vibrational Properties

In the last section we presented an overview of the vibrational properties of pure a-Si including the experimental ir and Raman spectra and the theoretical basis that has been established to interpret the spectra. The theoretical calculations for the most part consisted of involved numerical calculations based on CNR's with different specific bond angle and topological considerations. Although the calculations were detailed, there were a number of significant conclusions that could be made that are independent of particular force constant models, or particular models of the ir and Raman activities. These were enumerated specifically at the end of the last section. It turns out that the methods that were used to establish the correlations between spectral features and particular aspects of intermediate and short-range order suffer from serious disadvantages when applied to the question of the vibrational modes of alloy atoms. First of all, many approaches used for pure a-Si require construction of large CRN's, and hence to find the vibrational properties, it is necessary to either diagonalize large matrices or to solve a large system of linearly coupled algebraic equations. The size and complexity of the network can also limit the types of force fields and ir and Raman matrix elements that can be employed. Secondly, and more importantly, such methods do not provide direct information about the correlations between the local bonding environments at the alloy atom sites and the associated local and resonance vibrations involving significant motion of the alloy atom and its immediate Si neighbors. Moreover, at least for the local modes, it appears that the network topology, as reflected in the ring statistics, is not a particularly important parameter. Let us begin our discussion of the vibration properties of alloy atoms by considering a number of relevant questions, and then developing a conceptual framework which leads us to the desired answers.

What are the frequencies, symmetry-character and associated atomic displacements of the vibrational states produced by the introduction of significant numbers ($>$ 1 at.%) of alloy atoms? How localized are these vibrations in the vicinity of the alloy atom sites? From what is known about local modes of impurity atoms in crystalline systems, it appears reasonable to assume that similar types of vibrations involving alloy atoms in an amorphous system will also be localized spatially in the vicinity of the alloy atom. This would imply that the properties of the alloy atom vibrations will be determined more by the local bonding environment and symmetry at the alloy atom site itself, rather than by the global properties of the entire amorphous network. As noted earlier, these modes are generally associated with the incorporation of alloy atoms with an atomic mass which is smaller than that of the host system. The effects can be amplified if, in addition, the local force constants are greater than the corresponding types of forces between atoms of the network. Both of these conditions are met for virtually all of the alloy systems of interest, where the alloy atoms are H, D, F, O, N and C. The only systems in which the mass of the alloy atoms is larger, involve Ge and Cl; these then require special consideration. As noted earlier, there is generally more than one type of vibration associated with each alloy atom that is added to the network. Generally, a given alloy atom will exhibit one or more local modes with the remaining vibrations occuring within the frequency spectrum of the host, i.e., as resonance modes. The only exception to this is for the case of H in the so-called monohydride bonding geometry where both the bond-stretching and bond-bending vibrations occur in the local mode regime, i.e., at frequencies in excess of 500 cm^{-1}. For the other alloy systems, the resonance modes are expected to be delocalized to a greater extent than the local modes and hence to be somewhat more sensitive to the details of the network topology. The approach we will follow is based on an assumption that the alloy atom local modes are sufficiently well-localized to allow us to employ theoretical models which are specifically tailored to give information about local vibrations.

The approach that has been most productive to date consists of isolating the local bonding arrangement at the alloy atom site and embedding it in an environment that facilitates ease of computation and interpretation. This type of calculation is then expected to yield a valid representation of what happens in the real system. The keys are the way in which the local bonding geometry is determined and the way in which the alloy atom and its immediate Si neighbors are treated in the dynamical calculations. Two theoretical models have emerged which are complementary in nature and which have proved their value in describing the vibrational properties in two of the most important alloy systems: a-Si:H and a-Si:F [7.15, 16]. They have also been applied to a number of other systems with equal effectiveness. Two of the systems most recently studied involve ternary alloys, in which the model calculations have identified bonding arrangements in which both alloy atoms are bonded to the same Si site, a-Si:H:O and a-Si:H:N [7.17, 18]. The two complementary methods are the cluster-Bethe-lattice method (CBLM) and an approach based on the use of clusters of intermediate size (CIS), where intermediate size is defined operationally in terms of the dynamical properties of the particular system.

In the CIS method, clusters of approximately ten to thirty atoms, including the alloy atom of interest and its immediate Si neighbors, are extracted from the network and treated as a large molecule. The bonding at the alloy atom site is assumed to be in a geometric configuration that is determined by the normal valence bonding requirements of the particular species. For example, H and F are assumed to terminate the dangling-bond position on a Si atom whose other three neighbors conform to an idealized tetrahedral geometry. The bond length is determined from empirical considerations; the known bond lengths in molecules containing the particular alloy group, modified if necessary to include the nature of the Si next-nearest neighbors. The force constants for a particular calculation are determined in a similar manner using a well-established empirical scaling procedure that has been shown to hold for amorphous solids as well as discrete molecules [7.19–21]. Complementary considerations hold for the environments assumed to exist at twofold-coordinated bridging O alloy atoms sites, etc. The best way to illustrate the way the "intermediate size" is defined is by the choice of a particular example.

A typical cluster that has been used in calculating the vibrational modes of the monohydride group, SiH, as in a-Si:H alloys with relatively low hydrogen concentrations is shown in Fig. 7.8a [7.22]. The cluster consists of an SiH unit that is attached to twelve additional Si atoms. The cluster contains no rings of bonded atoms so that nine of the atoms are, in effect, surface atoms which can be characterized as being "free" of the usual constraints that Si atoms are subject to in the real solid. It is well established that the SiH group has two optic mode vibrations; one of these is a bond-stretching mode where the H atom moves in the direction of the SiH bond, and the second is a doubly degenerate bond-bending mode where the H atom moves in a direction that is at right angles to the SiH bond. Since the mass of the H atom is very small compared to that the Si atom, the displacement of the Si atom is significantly less than that of the H atom. The dynamical solution gives the displacements of all the atoms of the cluster for the two local modes in question. We can designate the nature of the Si atoms in the cluster by assigning them a shell number that indicates the extent to which they are removed from the H atom. The immediate neighbor Si atom is in the first

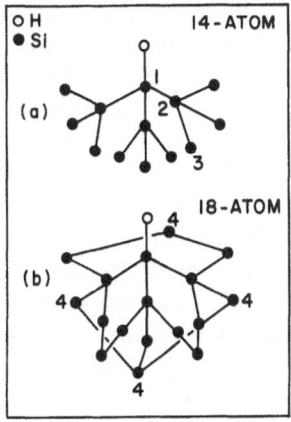

Fig. 7.8. (a) 14- and (b) 18-atom clusters used in the CIS method to calculate the vibrational properties of SiH (and SiF) groups in an a-Si host (from [7.22]). The numbers 1, 2, 3 and 4 represent the shells of Si atoms defined with respect to the H atom of the SiH group

shell which simply consists of that atom. The second-neighbor Si atoms form the second shell, containing three atoms, and the third-neighbors form the third shell, containing the remaining nine atoms of the cluster. The dynamical calculations give the atomic displacements of all of the atoms of the cluster for the 36 molecular modes of vibration of the cluster ($36 = 3N - 6$, where N is the number of atoms in the cluster).

The solution of the dynamical problem requires the diagonalization of a $3N \times 3N$ matrix. An examination of the atomic displacements associated with both of the SiH local modes indicates that the only significant atomic displacements involve those of the H atom and its immediate Si neighbor. There is some motion of the Si atoms in the second shell, but the eigenvectors are typically an order of magnitude smaller than the displacement of the Si atom in the first shell. The atoms in the third shell display even smaller motions. The concept of intermediate size is defined by these relative displacements and, in effect, determined by a test of how sensitive the local mode frequencies and eigenvectors are to the nature of the freee atoms on the boundary of the cluster. For this particular example, replacing the nine third-shell Si neighbors by atoms with a significantly larger mass, which in effect changes the boundary condition from one having freely moving atoms to one in which the boundary atoms are constrained to have at most very small displacements, does not change the properties of the local modes. This means that the 14-atom cluster shown in Fig. 7.8a is adequate for determining the properties of an isolated SiH group in an a-Si network, and thereby qualifies as a cluster of intermediate size in the context of the models we are describing. Isolated, as used above, means that there are no H atoms on any of the near-neighbor Si sites, for example, the additional twelve atoms of the cluster. Other aspects of boundary conditions and the effects of network topology can be studied by modifying the cluster shown in Fig. 7.8a to include rings of atoms as well. This change in cluster geometry is shown in Fig. 7.8b.

The cluster in Fig. 7.8b is constructed from the cluster in Fig. 7.8a by adding four Si atoms that generate a molecule in which every Si atom is a member of a closed ring of bonded atoms. This process introduces a fourth shell of Si atoms as well. The additional four atoms can also be given masses much greater than the Si atom mass and thereby change the boundary conditions on the surface atoms of the cluster from "free" to "fixed". The molecular dynamics calculations for the local SiH vibrations of the cluster in Fig. 7.8b gives identical results to those found for the cluster in Fig. 7.8a. This means that either cluster satisfies the criteria for being of appropriate intermediate size so as to yield a correct description of the local mode vibrations of the impurity atom being studied. This result is not at all surprising for an SiH group in which the alloy atom has a much smaller mass than the Si atoms of the network. However, as pointed out in [7.22], the same cluster is adequate for the bond-stretching vibration of the SiF group, even if the mass of the F atoms is now much larger relative to that of Si. For the SiF group, the bond-bending mode occurs at a frequency of less than 500 cm^{-1} so that it is a resonance mode rather than a local mode. For this case the properties of the bending mode are not independent of the detailed nature of the cluster and the motion extends to further shells of Si atoms. This in effect is the limitation of the CIS method. It is adequate for local modes in which the atomic displacements are generally constrained to one or at most two shells of Si atoms. For these cases there are no dynamical effects due to the presence of topological factors such as closed rings of bonds. For resonance modes the decay in the magnitude of the atomic displacements is more gradual and the nature of the network topology becomes a factor in the calculation. This, in effect, makes it more difficult to define a representative cluster in which the calculated properties of the resonance modes are completely independent of the boundary conditions, i.e., the motion of the atoms on the perimeter of the cluster and their inclusion or lack of inclusion in closed ring configurations of bonds. Nevertheless, the resonance mode frequences that are obtained from a cluster satisfying the local mode criteria discussed above are generally close to features that are found in the spectra and associated with the incorporation of the alloy atoms. A test for this is the calculation of the bond-bending mode of the SiF group. Using the same force constants and retaining the Si mass for all atoms of the cluster, the frequencies of the bond-bending modes calculated using the clusters in Figs. 7.8a, b differ by less than 10%. Similar differences exist for the calculated values of resonance mode frequencies in other systems of interest.

The CBLM method [7.23–25] represents an improvement over the CIS method in the sense that it removes the necessity to identify boundary value conditions on the terminal atoms of the cluster, i.e., those atoms in the last shell of Si neighbors. This is accomplished by attaching to each of these atoms a Si Bethe lattice. The Bethe lattice is an infinite aperiodic network of atoms in which each atom has the same coordination as that of the real system it replaces. Unlike the real system, it has, however, no closed rings of

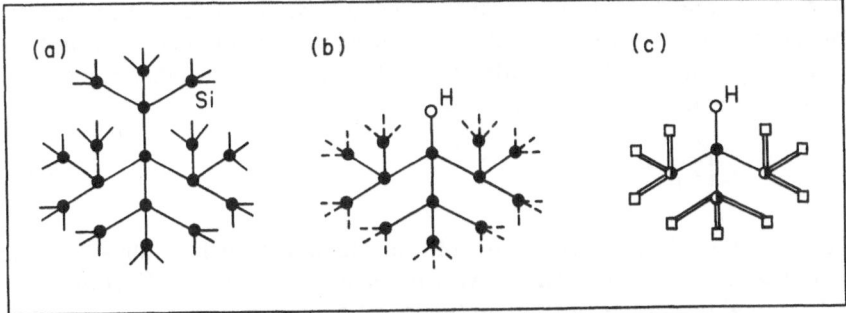

Fig. 7.9 a–c. Schematic illustration of the CBLM. (a) A portion of the Si-Bethe lattice. Each atom is fourfold-coordinated. The solid lines represent the nearest-neighbor chemical bonds. (b) The SiH group of Fig. 7.8a embedded in the Si-Bethe lattice. The dashed lines indicate the infinite extent of the Bethe lattice structure. (c) The attachment of the Bethe lattice to the surface bonds of a cluster containing an SiH group. This is equivalent to replacing the self-energies of the surface atoms of the cluster and their nearest-neighbors by "effective" self-energies (squares for the nine atoms of the third shell and half-filled circles for the three atoms of the second shell). The squares and half-filled circles signify different values of the "effective" self-energies. The nearest-neighbor interactions are then replaced by "effective" interactions and represented by the double lines

bonds. In this sense, the Si Bethe lattice (Fig. 7.9a), apart from topological differences, is identical to the crystalline and amorphous phases of silicon out to third neighbors. When attached to the surface dangling bonds of a finite molecular cluster such as those shown in Figs. 7.8a, b, the Bethe lattice serves to simulate the general effects of the a-Si host from which the cluster was removed. The cluster and the Bethe lattice then form a system in which the local bonding of the alloy atoms is preserved and in which all of the neighboring Si atoms have the same bonding environment as in the real system. Moreover, the CBLM allows for an exact calculation of the LDOS for each atom of the cluster. It is also possible to calculate the ir and Raman response through the use of models for the ir effective charge, and the Raman polarizability. The CBLM does not readily yield the eigenvectors associated with particular vibrational frequencies; as such it complements the CIS technique which yields both eigenvectors and eigenfrequencies. To demonstrate the complementary nature of the both methods in describing similar types of vibrations, we now outline the procedure that is used to calculate the DOS functions.

To accomplish this, we first give the general procedure used for the solution of the a-Si Bethe lattice. The equation of motion for the local vibrational Green's function G_{00} for the site O can be written as

$$(\omega^2 - D)G_{00} = 1 + \sum_j \left(D_j G_{j0} + \sum_{k \neq j} T_{jk} G_{k0} \right), \tag{7.1a}$$

where D_j is the dynamical matrix connecting the atom O and its nearest neighbors along bond j. T_{jk} is the dynamical matrix connecting O with its second neighbors along the bonds j and k. D is the dynamical matrix connecting atom O with itself and is given by

$$D = -\sum_j \left(D_j + \sum_{k \neq j} T_{jk} \right) . \tag{7.1b}$$

The solution for G_{00} is obtained by introducing molecular-like fields Φ_j along the bonds j connecting atom O with its nearest neighbors; Φ_k connecting nearest-neighbor atoms other than O along bonds k; Φ_{jk} connecting second-neighbor atoms connected by nearest-neighbor bonds j and k. After some algebraic manipulations, one can show that

$$G_{00}' = \left\{ \omega^2 - D_{00} - \sum_j \left[D_j \Phi_j + \sum_{k \neq j} T_{jk} (\phi_{kj} + \phi_k \Phi_j) \right] \right\}^{-1} , \tag{7.2}$$

where

$$\Phi_j = \frac{1}{\omega_{jj}} \left[D_j + \sum_{j' \neq j} T_{jj'} \Phi_{j'} + \sum_{k \neq j} \left(D_k + \sum_{l \neq k} T_{kl} \phi_l \right) \phi_{kj} \right] \tag{7.3a}$$

$$\phi_k = \frac{1}{\omega_{kk}} \left[D_k + \sum_{k' \neq k} T_{kk'} \phi_{k'} + \sum_{l \neq k} \left(D_l + \sum_{m \neq l} T_{lm} \phi_m \right) \phi_{lk} \right] \tag{7.36}$$

$$\phi_{kj} = \frac{1}{\omega_{kk}} \left(T_{kj} + \sum_{k' \neq k} T_{kk'} \phi_{k'j} \right) , \tag{7.3c}$$

and

$$\omega_{kk} = \omega^2 - D_{kk} - \sum_{l \neq k} \left[D_l \phi_l + \sum_{m \neq l} T_{lm} (\phi_{ml} + \phi_m \phi_l) \right] . \tag{7.3d}$$

The set of equations (7.3a–d) can be solved by numerical iteration and then the solution for G_{00} is obtained. The Bethe lattice DOS, which is the LDOS of atom O, is then given by

$$\varrho(\omega) = -\frac{2m\,\omega}{\pi} \text{Im} \{ \text{Tr} \, G_{00} \} \tag{7.4}$$

The DOS spectrum of the Si Bethe lattice is shown in Fig. 7.4. As noted above, this spectrum reproduces those features in the VDOS that depend only on the fourfold-coordination of the network atoms, in particular the two peaks that have been labelled TA and TO. These are the dominant features in the VDOS spectrum. To generate the LA and LO-like features, it is necessary to include rings of bonded atoms as well as the average network coordination. This has not as yet been accomplished in a satisfactory way.

Figure 7.9b shows one of the clusters (Fig. 7.8a), which we discussed in connection with the CIS method, embedded in a Si Bethe lattice. The LDOS of each atom of the cluster can be obtained in a manner that is similar to that used to calculate the Bethe lattice DOS as outlined above. Here, we attach appropriate first and second-neighbor fields to each of the cluster's surface atoms. This is equivalent to replacing the actual self-energies of each surface atom and its nearest neighbors in the cluster by "effective" self-energies and their respective nearest-neighbor interactions by "effective" interactions which are functions of the molecular fields. This is shown schematically in Fig. 7.9c. This process allows the LDOS of each atom in the cluster to be calculated analytically.

To illustrate the complementary nature of the CIS and CBLM, we show the calculated LDOS spectra for the SiH group in Fig. 7.10. The figure shows (i) the LDOS on the H atom, (ii) the LDOS of the Si atom in the first shell, i.e., the Si atom of the SiH group, (iii) the LDOS for one of the three second-neighbor Si atoms of the second shell, and finally, (iv) the LDOS and one of the nine Si atoms of the third shell. Also shown in the figure are schematic representations of the atomic displacements of the H and Si atoms for the bond-stretching vibration at 2000 cm^{-1} and the bond-bending vibration at 630 cm^{-1}. These two modes are both local modes, occuring at frequencies in excess of the highest vibrational frequencies in the DOS of the Si host network. Note that the contributions to the LDOS obtained via the CBLM are in agreement with the eigenvectors that were determined in solving for the vibration properties of the finite cluster, CIS method. The bond-stretching vibration (2000 cm^{-1}) involves motion of the H and Si atoms along the direction of the SiH bond. The relative displacements reflect the fact that the mass of the H atoms is small with respect to that of the Si atom. This is also reflected in the contributions to the LDOS for the H atom and the Si atom in

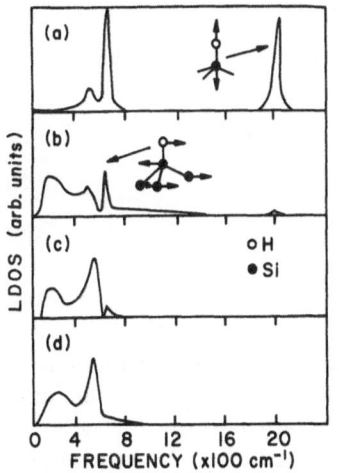

Fig. 7.10 a–d. LDOS of (a) the H atoms and (b–d) for three Si atoms in the first, second and third shells of Si neighbors, respectively, of the SiH group (from [7.15]). Also shown are the atomic displacements for the two local modes as calculated using the CIS method. The LDOS spectra are from a calculation using the CBLM (from [7.15])

the first shell. The CIS calculation gives essentially no Si atom motion in either the second or third shell. This is also reflected in the complete absence of any contributions to the LDOS near 2000 cm^{-1} for these Si shells. The results for the bond-bending mode (630 cm^{-1}) are different, but only in a quantitative way. The molecular calculation indicates relatively larger displacements of the Si atoms in both the first and second shells. This is indicated in the schematic diagram of the atomic displacements of the bond-bending mode (Fig. 7.10b). The contributions the LDOS, as obtained from the CBLM calculation, clearly reflect these differences. The comparisons illustrated in Fig. 7.10 were based on calculations in which the same force constants were used in both calculations. This is reflected not only in the "equivalence" between the eigenvector and LDOS contributions, but in the fact that the eigenfrequencies obtained from the CIS molecular calculation are at the positions of the peaks in the LDOS spectra as well. Hence, by applying these two techniques to the same systems, it is possible to obtain a complete specification of the vibrational properties of the local modes that are associated with the incorporation of alloy atoms. The calculations yield (i) the frequencies of the vibrational modes, (ii) the atomic displacements, which in turn yield the symmetry character of the vibration and hence relate to the relative activity in the ir absorption and Raman scattering spectra, (iii) the degree of localization of the modes on the alloy atom and its Si nearest and next-nearest neighbors, and (iv) the LDOS spectra for each atom in the cluster. As indicated above, the same force fields can be used in each of these methods. The force field that has been used in [7.15, 16], where these complementary methods were first discussed, is the valence force field or VFF representation which includes both two-body and three-body near-neighbor forces [7.26, 27]. The majority of the Bethe lattice calculations that were applied to pure a-Si employed only two-body or nearest-neighbor forces as in the Born model. The inclusion of three-body forces represents a distinct improvement over the earlier work.

In the VFF representation the elastic energy per atom, considering up to three-body forces, is given by

$$U = \frac{1}{2}\sum_j K_r(\Delta r_j)^2 + \frac{1}{2}\sum_{j>k} K_\theta(r_0\Delta\theta_{jik})^2 + \sum_{j>k} K_{rr}(\Delta r_j)(\Delta r_k)$$
$$+ \sum_{j>k} K_{r\theta}(\Delta r_j)(r_0\Delta\theta_{jik}) ,$$

(7.5)

where r_j is the scaler change in the length of nearest-neighbor bond j about atom i, θ_{jik} is the change in the angle formed by bonds j and k about i, and r_0 is the equilibrium bond length. The first term in [7.5] corresponds to the two-body bond-stretching force (K_r) between pairs of atoms. The three remaining terms correspond to the three-body forces between triads of atoms: bond-bending (K_θ), simultaneous stretching of neighboring bonds ($K_{rr'}$) and simul-

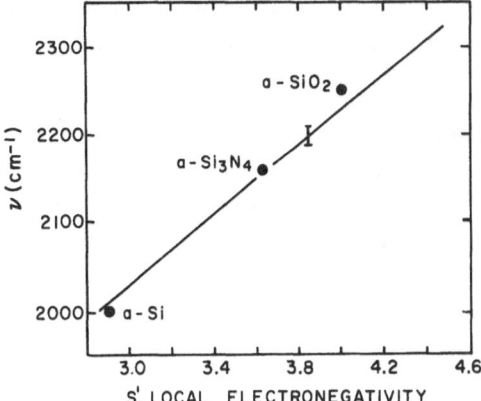

Fig. 7.11. Frequencies of SiH bond-stretching vibrations as a function of the local electronegativity S'. The solid line is a linear regression analysis of molecular data; the points indicate the frequencies for the SiH vibration in three different local bonding environments (from [7.30])

taneous stretching and bending $(K_{r\theta})$. Reliable force constants can be obtained from an empirical chemical bonding model which is based on an analysis of molecular data [7.19, 22]. The model takes into account changes in the two and three-body forces that are induced by the changes in the local bonding chemistry in the neighborhood of the alloy atom sites. The scaling variable is one that reflects the degree of local charge transfer and hence changes in the bond lengths. The key to the application of this approach is to use the same scaling variable to analyze the molecular data as to describe the bonding environment of the alloy atom in the a-Si host. Scaling variables that have been used include (i) electronegativity sums of second neighbors [7.19], (ii) partial charges [7.28, 29], and (iii) average "effective electronegativities" [7.30]. Each variable gives similar results when applied to the a-Si alloy systems. To illustrate the way this technique is used, we consider the bond-stretching frequency of the SiH group. This has a value of 2000 cm^{-1} which is significantly lower than the SiH stretching frequency in any substituted silane molecule. Figure 7.11 indicates the origin of the difference and the earliest success of this model. The figure indicates an analysis of molecular data where the scaling variable is chosen to the average electronegativity of the Si atom in the particular molecule. The procedure used to obtain this is outlined in [7.30]. The straight line is a linear regression analysis of the molecular data and the error bars indicate the degree of uncertainty in this analysis. The straight line is given by

$$\nu(\text{Si–H}) = 1433 + 200\,S' \pm 12\,\text{cm}^{-1}, \tag{7.6}$$

where S' is the effective electronegativity of the Si atom. *Lucovsky* [7.30] outlines a procedure for calculating the value of S' for the Si atom to which the alloy atom is attached.

Also included in the figure are the frequencies of the SiH vibrations in (i) a-Si, (ii) a-Si$_3$N$_4$ and (iii) SiO$_2$. The low frequency for the SiH vibration in a-Si simply reflects the fact that the Si next-nearest neighbors, as well as the rest of the host network, provide a bonding environment for the SiH group that is very different from that in any of the substituted silane molecules [7.19]. The difference is derived from the fact that the three Si atoms of the network have electronegativities that are smaller than any of the substituting atoms or groups in the molecules. This, in effect, means that the total amount of charge withdrawn from the Si atom of the SiH group is also smaller, thereby resulting in a larger Si–H bond length and a lower stretching frequency [7.19, 22].

7.4 Symmetry Determined Vibrations at Alloy Atom Sites

This section addresses the question of the relationship between the local symmetry at an alloy atom bonding site and the vibrations that involve significant displacements of that alloy atom and its immediate Si neighbors. As we noted in the introduction, the most important difference between the way alloy atoms are incorporated into crystalline and amorphous hosts is in the local bonding. Incorporation of alloy atoms into a crystalline Si host can occur in two ways, either as substitutional impurities in which case the atoms sit in sites with tetrahedral symmetry, or as interstitial impurities wherein additional types of bonding coordination are possible. In both instances, the crystal lattice structure is the determinant factor in establishing the local environment at the alloy atom site. In amorphous materials the situation is qualitatively different in the sense that the local bonding at the alloy atom site adapts to the valence bonding requirements of that atom. This in effect makes it possible to classify the types of alloy atoms sites in a-Si by the valence bonding requirements of the alloy atoms. In this section we will restrict the discussion to isolated sites where the alloy atom is embedded in an environment that contains only Si atoms. In the next sections we will expand the discussion to consider situations where two alloy atoms can be bonded to the same Si site.

The most important alloy atoms in a-Si include species in which the local coordination of the alloy atom ranges from one to four. The examples we will explicitly consider in this section are classified by their valence bonding requirements: (i) univalent alloy atoms such as H and the halogens, (ii) divalent alloy atoms such as O, (iii) trivalent alloy atoms such as N, B and P, and finally, (iv) quatravalent alloy atoms such as C and Ge. In order to emphasize the role of the local symmetry, we will use as examples H, O, N, P and C. In latter sections of the text we will consider the bonding of F and Cl, and indicate the quantitative and qualitative differences between the halogen atom bonding and the bonding of H. Figure 7.12 includes the local symmetry

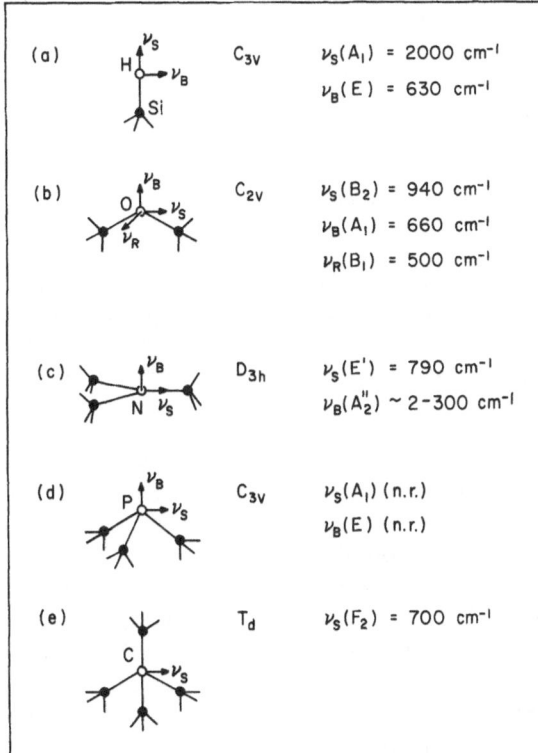

(a)	H	C_{3v}	$\nu_S(A_1) = 2000$ cm⁻¹
			$\nu_B(E) = 630$ cm⁻¹
(b)	O	C_{2v}	$\nu_S(B_2) = 940$ cm⁻¹
			$\nu_B(A_1) = 660$ cm⁻¹
			$\nu_R(B_1) = 500$ cm⁻¹
(c)	N	D_{3h}	$\nu_S(E') = 790$ cm⁻¹
			$\nu_B(A_2'') \sim 2\text{-}300$ cm⁻¹
(d)	P	C_{3v}	$\nu_S(A_1)$ (n.r.)
			$\nu_B(E)$ (n.r.)
(e)	C	T_d	$\nu_S(F_2) = 700$ cm⁻¹

Fig. 7.12 a–e. Local bonding environments for isolated alloy atoms in an a-Si host. Included are the atomic displacements of the alloy atoms, the local site symmetry, and the local mode frequencies that have been obtained from ir studies. (a) H in a-Si; (b) O in a-Si; (c) N in a-Si, planar bonding site; (d) P in a-Si, pyramidal bonding site (there are no reported values for the local modes); (e) C in a-Si

and the bonding geometry at the alloy atom sites, the displacements of the alloy atom that are determined by the symmetry of its local bonding site, and the vibrational frequencies as determined by ir absorption studies. Consider first the bonding of a single H atom, which is in turn bonded to three Si atoms. The local symmetry at the bonding site is determined by a unit that includes the Si–H bond and the three Si atoms that are attached to that group, and in effect, couple the SiH group to the network. The effective symmetry is the C_{3v}. The H atom can execute two independent motions, a bond-stretching motion in which the H atom is displaced in a direction parallel to the Si–H bond, and a bond-bending motion where the H atom displacement is perpendicular to the direction of the Si–H bond. These motions are shown in the figure with the frequencies of the local mode vibrations determined from ir absorption spectroscopy and Raman scattering [7.31, 32]. The stretching mode has A_1 symmetry and the bending mode is doubly degenerate and has E symmetry. Both vibrations are ir and Raman active. Each is observed in the ir spectrum; however, only the stretching mode is observed in the Raman spectrum. The lack of observance of the bending mode in the Raman spectrum is due to competing second-order scattering associated with vibrations of the host a-Si network. It is important to note that in addition to

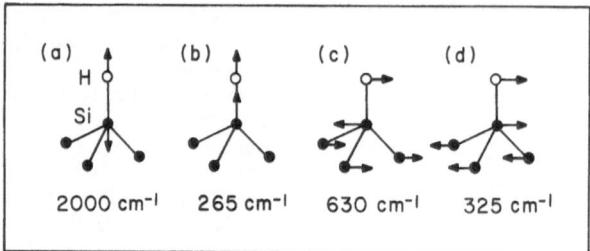

Fig. 7.13 a–d. The complete set of alloy atom vibrations for the SiH group in an a-Si host (**a**) and (**c**) are the local modes, the bond-stretching and bond-bending vibrations, respectively. (**b**) and (**d**) are the resonance modes. (**b**) has a stretching character and (**d**) a shearing character. The displacements of the Si and H atoms have opposite phases for the local modes and the same phase for the resonance modes [7.22]

the motion of the H atom, there are also displacements of the neighboring Si atoms to consider. The character of these has been indentified via calculations based on both the CIS and CBLM techniques [5.22]. The complete set of motions involving alloy and Si atom motions is shown in Fig. 7.13, and is most conveniently described in terms of in-phase and out-of-phase motions involving the displacements of the H atom as noted above and displacements of its Si neighbor that are also driven by the effective C_{3v} symmetry. The out-of-phase motions (Figs. 7.13a, c) are the local modes of the SiH group, whereas the in-phase motions (Figs. 7.13b, d) give rise to in-band resonance modes. There are two important things to note: (i) the degenerate modes involving Si and H atom motion perdendicular to the axis of the Si–H bond also involve small displacements of the second shell of Si atoms, i.e., the three Si atoms that are bonded to the Si atom of the SiH group, and (ii) that the in-phase modes will be weakly ir active due to a charge redistribution between the heteropolar Si–H bond and the three homopolar Si–Si bonds. The frequencies of the in-phase in-band resonance modes have been estimated from the CIS calculations [7.22]. The shear-type motion, (Fig. 7.13d) is calculated to have a frequency of 325 cm^{-1}, while the stretch-type motion (Fig. 7.13b) has a calculated frequency of 265 cm^{-1}. Ir absorption and Raman scattering has been reported [7.4] in a-Si:H alloys at frequencies close to those calculated; however, we point out that the calculated frequencies of the in-band resonance modes are sufficiently close to band edges of the DOS of the a-Si host to make them somewhat suspect. Clearly, future work is needed to clarify the nature of the vibrations observed. In addition to the four modes involving the in and out-of-phase displacements of the H and Si atoms, there are additional modes of vibration that involve significant displacements on the Si atom of the SiH group. One of these is a "surface" atom vibration which occurs at a frequency that is very close to the maximum frequency of vibration within the Si host network, approximately 500 cm^{-1}. All a-Si:H alloys exhibit some ir absorption in this frequency regime. It is best classified as an effect due to an alloy enhanced matrix element that changes the effective charge on the Si atom of the Si–H bond due to the heteropolar nature of that bond [7.4]. The designation surface conveys the idea that the atom that is moving is bonded to three Si atoms, as it would be for an atom on the (iii) surface of a Si crystal.

Substitution of D atoms or halogen atoms (F, Cl, etc.) does not change the symmetry at the bonding site. However, the increased mass of the alloy atom produces qualitative changes in the character of the alloy atom vibrations. For D and F the stretching modes are local modes. The bending vibration of the SiF group is an in-band resonance and the bending vibration of the Si-D group occurs just about at the top of the a-Si host network DOS [7.31] and thus introduces an interesting type of coupled mode behavior. This will be discussed in the next section. The stretching vibration of Cl also appears to be a local mode, although the experimental data is quite limited [7.33]. The bending mode, on the other hand, is clearly an in-band resonance.

Consider next the prototypical divalent alloy atom oxygen. The local bonding is assumed to be similar to that of O atoms in the crystalline and amorphous phases of SiO_2, i.e., the O atom is assumed to reside in a bridging site in which it has two Si neighbors. The bond angle assumed, for purposes of this discussion, and the calculations presented in a later part of the paper for bonding sites in a-Si : H : O alloys, is 150°. The local symmetry at the oxygen atom site yields three independent degrees of motion for the oxygen atom. The bonding geometry is shown in Fig. 7.12b as are the independent motions of the oxygen atom. These motions are best characterized with reference to the twofold symmetry axis that is the bisector of the Si–O–Si bond angle. There are two in-plane motions, one along the direction of the bisector that has been called either a symmetric stretching motion or a bending motion (we shall use the bending motion descriptive), and the second is a motion at right angles to the twofold axis, in a direction parallel to a line joining the two Si atoms. This motion has been designated as an asymmetric stretching motion, or simply a stretching motion (in those schemes in which the other in-plane motion is referred to as bending). The third motion is an out-of-plane motion in a direction that is perpendicular to the plane of the Si–O–Si bond; this is designated as a rocking motion. The bending and stretching motions of the O atom are accompanied by displacements of the two Si atoms in parallel directions, but with opposite phase. The symmetry at the bonding site of the oxygen atom is C_{2v}, and as a consequence, all three vibrations are expected to be both ir and Raman active. The frequencies of these vibrations have been determined from ir measurements performed on films of a-Si prepared by sputtering Si in an oxygen ambient [7.34] and from studies of the absorption in layers of Si rendered amorphous via O ion implantation [7.35]. The stretching vibration is found to occur at 940 cm^{-1} and the bending vibration at 660 cm^{-1}. A third vibration at 500 cm^{-1} has been assigned to the rocking vibration. Its frequency is sufficiently close to the highest frequency of the a-Si host network so that it must also involve a strong coupling to modes of the host as well. Studies on ternary alloys in the system a-Si : H : O also display a vibration at 500 cm^{-1} [7.36] whose absorption strength scales with the oxygen concentration. This establishes that the motion involves either O atom motion of the type discussed above, or that it involves significant displacements of the two Si atom bonded to the O alloy atom, or both.

So far we have considered the bonding environments of univalent and divalent alloy atoms as illustrated, respectively, by H and O. Now we turn to atoms which exhibit a threefold coordination. There are two possible types of bonding sites: (i) a planar configuration in which the alloy atom resides in the center of an equilateral triangle formed by its three Si neighbors, and (ii) a pyramidal configuration where the alloy atom is at the apex of a pyramid and its three Si neighbors define the base. We expect N, and possibly B atoms to occupy planar sites, and P and As to occupy pyramidal sites. Before we discuss anything about symmetry and atomic displacements, we note that B, As and P are active as dopants in a-Si : H alloys, in which case it has been assumed that the reside in bonding sites with a tetrahedral coordination. In this type of environment, the mechanism by which B acts as an acceptor and P and As act as donors, completely parallels that of the same atoms in a c-Si host. It has been observed by NMR [5.37, 38] that B and P are not 100% efficient as dopants and that the electronically inactive atoms reside in sites with a threefold coordination. For the case of As, an electronic phase transition has been observed in alloys containing about 1 at. % of As. Samples with less than 1 at. % As display a relatively high conductivity indicative of doping by the As atoms, whereas samples with higher As concentrations become highly insulating [7.39]. Raman and EXAFS studies indicate a threefold coordination for As in the regime of higher As alloys [7.39, 40]. The absence of a 100% efficient doping by As atoms in the lower concentration regime is then taken to imply threefold-coordinated, electrically inactive As atoms as well as fourfold-coordinated As atoms which are active as donors.

We first discuss the planar bonding geometry with particular reference to a-Si : N alloys and then go on to indicate the changes in the vibrational-properties that are derived from a change to a pyramidal geometry. Referring to Fig. 7.12c, the N atom can execute two independent types of motion. These are a doubly degenerate in-plane mode and a nondegenerate out-of-plane mode. The in-plane motion gives rise to an ir active vibration that has been observed to occur at a frequency of about 790 cm^{-1} in material rendered amorphous by N ion implantation [7.41]. The other out-of-plane vibration is expected to occur at a lower frequency that is within the spectrum of the host a-Si, and hence is a resonance mode. It has not been observed to date. A complete specification of the vibrational properties at the N site requires a consideration of the atomic displacements on the neighboring Si atoms as well.

Figure 7.14 displays the four characteristic vibrations associated with a N atom and its four Si neighbors. This group of atoms has been designated as

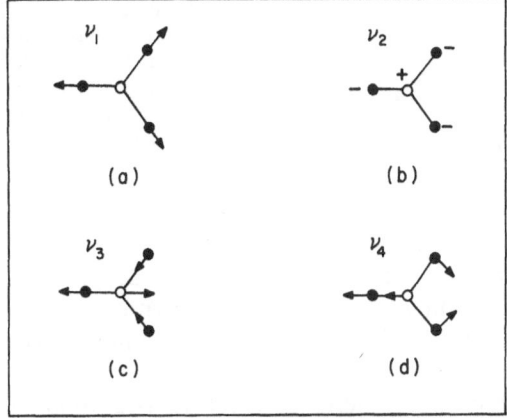

Fig. 7.14 a–d. Skeletal vibrations of the Si_3N group. The solid circles are the three Si atoms and the open circle is the N atom. (a) The breathing mode; (b) the out-of-plane mode; (c) the asymmetric stretching mode; (d) the in-plane bending mode [7.18]

the skeletal Si_3N group and its vibrational properties have been discussed in relation to the molecule $N_3(SiH_3)_3$ [7.42]. The symmetry of this molecular cluster is D_{3h}. The vibrations split into nondegenerate modes with A_2' and A_2'' symmetry character and doubly degenerate modes with E' character. The four vibrations are shown in the figure. Two of these involve displacements of the N atom that are the same as those shown in Fig. 7.12c. The E' mode is an asymmetric stretching vibration and involves net displacements of the N atom and its three Si neighbors which have an opposite phase. The mode is ir active and also Raman active and depolarized. The mode with an out-of-plane N displacement has A_2'' symmetry, is ir active and has not been observed in either the amorphous phase or in molecules containing the skeletal Si_3 group. The E' in-plane stretching mode, as noted above, has a frequency of 790 cm^{-1}. The other two vibrations involve only displacements on the Si atoms. One of these is an in-plane breathing mode with A_2' symmetry. This mode is Raman active and polarized and has a frequency of about 500 cm^{-1} as reported [7.42] for the $N(SiH_3)_3$ molecule. The remaining mode is an in-plane bending mode with E' symmetry. This mode is expected to be ir active, but it has not as yet been observed. Estimates of its frequency are 200 cm^{-1}. There are also rigid translations and rotations of the Si_3N group that are possible because of the bonding of the three Si atoms to the host network. These are mentioned for completeness, but they have not been observed or discussed elsewhere.

Figure 7.12d displays the vibrations of the pyramidal site for a threefold-coordinated atom such as P or As. The atomic motions are qualitatively similar to the vibrations of the planar configuration. The local symmetry is C_{3v} and all four vibrations involve displacements of the alloy atom and its three Si neighbors. Two of the modes, those shown in Figs. 7.15a, b, are bond-stretching vibrations. The threefold-coordinated atom motion is along the symmetry axis of the pyramid for the A_1 mode and perpendicular to this axis for the E mode. Both modes are ir and Raman active. The Raman

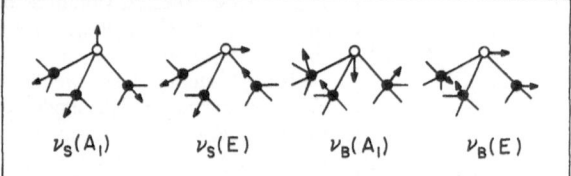

Fig. 7.15 a–d. The normal modes of a pyramidal bonding group. (**a**) The symmetric stretching vibration; (**b**) the asymmetric stretching vibration; (**c**) and (**d**) are bending vibrations

response of the A_1 mode is polarized, whereas the E mode is depolarized. The other two modes are bond-bending vibrations, with activities in the ir and Raman spectra that are the same as those of the stretching modes with the same symmetry.

Finally, Fig. 7.12e gives the local geometry at a fourfold-coordinated alloy atom site; in this example, C. If we consider the five-atom cluster consisting of the C atom and its four Si neighbors, then the local symmetry at the alloy atom site is tetrahedral, T_d. There is only one vibration, a triply-degenerate mode with F_2 symmetry, that involves motion of the C atom. This has been observed in a-Si : C alloys produced by ion implantation of C atoms into c-Si substrates [7.43]. The frequency of this vibration is 700 cm^{-1}. There are three other vibrations involving displacements of the four Si atoms in the cluster. One of these is a breathing mode of the Si atoms. It has A_1 symmetry

Table 7.1 Properties of Local Bonding Environments in an a-Si Host

Coordination	Symmetry	Alloy atom	Local mode vibrations Symmetry	Frequency [cm^{-1}]
Single				
(a) Terminal	C_{3v}	H	$v_s(A_1)$	2000
			$v_b(E)$	630
		D	$v_s(A_1)$	1460
			$v_b(E)$	510
		F	$v_s(A_1)$	830
Twofold				
(a) Bridging	C_{2v}	O	$v_s(B_2)$	940
			$v_b(A_1)$	660
			$v_r(B_2)$	500
Threefold				
(a) Planar	D_{3h}	N	$v_s(E')$	790
			$v_b(A_2'')$	2–300
(b) Pyramidal	C_{3v}	P	$v_s(A_1)$	nr^a
			$v_b(E)$	nr
Fourfold				
(a) Tetrahedral	T_d	C	$v_s(F_2)$	700

^{a} nr = not reported

and is Raman active and polarized. It has not been reported. The other two modes are bending vibrations, a mode with E symmetry that is Raman active (depolarized), and one with F_2 symmetry that is both ir and Raman (depolarized) active. These modes have not been reported, but are expected to have frequencies that are below 300 cm^{-1} and therefore make them resonance modes. The A_1 breathing mode is expected to have a frequency close to 500 cm^{-1} making it somewhat difficult to observe in the Raman spectrum because of the very stong scattering from the modes of the host network in the same frequency regime.

Table 7.1 summarizes the results we have presented in this section. It includes the symmetry-character of the local bonding arrangement involving the alloy atom and its immediate Si neighbors, the types of vibrations that give rise to local modes and the frequencies of those vibrations as reported for: (i) the univalent alloy atoms H, D and F, (ii) the divalent alloy atom D, (iii) the trivalent alloy atom N in a planar bonding site, (iv) the trivalent atom P in a pyramidal bonding site, and (v) the quatravalent alloy atom C in a tetrahedral site. The classification of alloy atom modes in terms of the symmetry at their bonding sites will be used as a starting point for the discussion of more general bonding of alloy atoms which follows in the next two sections.

7.5 Vibrations in Binary Amorphous Silicon Alloys

There have been many studies of the vibrational properties of a-Si binary alloys. The alloy systems separate into two groups: those studied over a wide range of composition where it becomes likely by statistics alone that more than one alloy atom will attach to the same Si site, and those systems that have been studied in a lower alloy composition range in order to identify the local atomic coordination at the bonding sites of isolated atomic species in the a-Si host network. We will not consider research on stoichiometric compounds involving Si as one of the atomic species, such as, for example, in SiO_2 or Si_3N_4. We will only briefly discuss studies that have considered an entire alloy composition regime, where the focus was not on a property of alloyed a-Si. Two such systems include a-Si : C and a-Si : Ge. The binary alloy systems that have received the most attention are those systems which are of interest in some aspect of a-Si solar cell or transistor device technology. As such, the two most important and extensively studied systems are a-Si : H and a-Si : F. For each of these systems we will first present the experimental data for the range of alloys studied, including all of the relevant ir and Raman signatures. We shall then discuss the structural interpretations, first noting those assignments where there is a general acceptance, and then those assignments where important questions remain to be answered. This approach is designed to give the reader a framework in which to review past studies and a critical prespective for judging new experimental results and interpretations. In presenting the experimental data we will emphasize the studies of local mode vibrations where there is a sufficiently sound theoretical basis for quantitative as well as qualitative interpretation. We will note measurements of resonance modes and indicate those cases where the qualitative nature of the modes are understood.

7.5.1 a-Si : H Alloys

The most extensively studied system to date has been a-Si : H. This films have been prepared in a number of different ways including glow discharge decomposition of SiH_4, reactive sputtering of Si in a H-containing ambient and CVD using SiH_4 as well as higher silanes. These films can have concentrations of bonded H ranging from 2 at.% to an excess of 50 at.%, where the amount of incorporated H is a function of the deposition parameters. There is not a one to one correlation between the amount of bonded H in a film and the features in the ir absorption spectrum which have been correlated with any one of the possible local bonding arrangements involving Si sites with one or more H atoms attached. In this sense the fraction of H that is bonded in a particular way is not simply H concentration dependent, but rather depends on a large number of deposition parameters which include the temperature of the substrate on which the film is condensed, the amount of available H and the amount of input power that is used to drive a particular plasma, CVD or sputtering technique. In spite of these factors, there are a number of very important trends. It has proven possible to track features in the spectra in a more or less systematic way and use this information to provide a guideline for interpretation in terms of the local bonding. Since we are dealing with heteropolar bonds, the primary experimental tool has turned out to be ir absorption spectroscopy rather than Raman scattering. While some of the high frequency local modes are readily observable in the Raman spectrum, second-order scattering from the a-Si host makes it virtually impossible to detect alloy atom vibrations at frequencies below about 900 cm^{-1}. The spectral range between 500 and 900 cm^{-1} is very rich in alloy atom vibrations for systems containing H as well as the majority of the other alloy atoms we will focus on; hence the limited use of Raman scattering spectroscopy.

Before discussing the experimental data and proposed assignments, it is worthwhile to consider two more general aspects of alloy atom incorporation and vibrational behavior. Figure 7.16 includes a schematic representation of the way univalent alloy atoms such as H, D, F, Cl, etc., can be incorporated into an amorphous Si host, and Fig. 7.17 is a schematic representation of the symmetry-determined atomic displacements of the univalent atom for three of the bonding geometries shown in Fig. 7.16. The configurations shown in

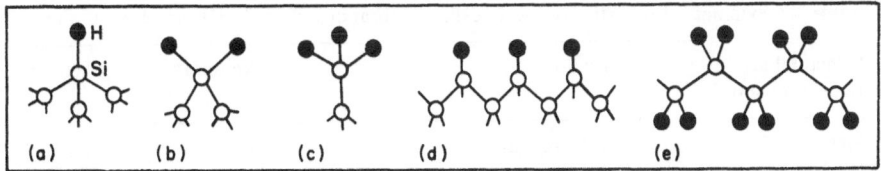

Fig. 7.16 a–e. Local bonding geometries for univalent atoms. (a) Isolated bonding sites; (b) bond sites on internal surfaces; (c) bonding sites in polymer configurations (from [7.22])

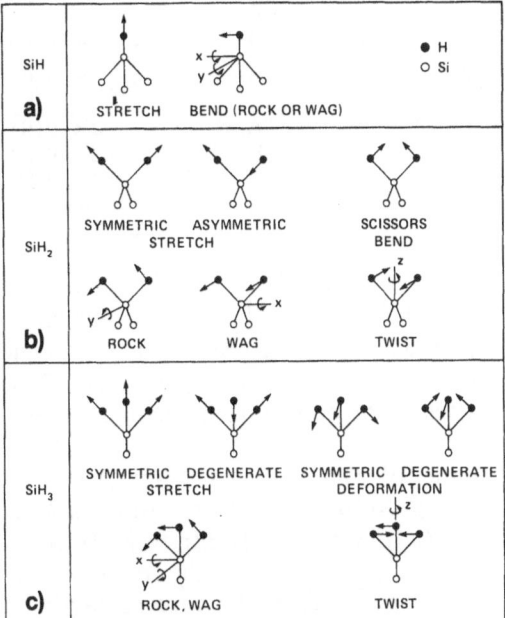

Fig. 7.17 a–c. Atomic displacements of H atoms in local and resonance mode vibrations in (a) SiH groups (b) SiH$_2$ groups and (c) SiH$_3$ groups (from [7.32])

Fig. 7.16a are for isolated bonding sites which can accommodate either one, two or three univalent atoms. If the univalent atom is H, then these sites are designated as monohydride, dihydride and trihydride, respectively. Figure 7.16b indicates two of the many possible surface bonding configurations in which the univalent species can occur as near neighbors, terminating the available dangling bonds that are associated with a particular surface geometry. Finally, Fig. 7.16c indicates two different conformations that are based on the polymerization of SiX$_2$ groups. The first is the trans conformation and the second represents a random arrangement of the SiX$_2$ groups. We will designate these polymerized forms as (SiH$_2$)$_n$ for the polysilane material and use corresponding formulas and descriptives for the halogen compounds. Figure 7.17 gives the displacement vectors of the symmetry determined vibrations at SiH, SiH$_2$ and SiH$_3$ sites. The activities of these vibrations are also indicated in Table 7.2. The spectral features reported for a-Si:H (and

Table 7.2 Symmetry Properties of the Localized Vibrations of SiH, SiH$_2$ and SiH$_3$ Groups

Structural group (local symmetry)	Vibration	Symmetry	Activity
SiH (C$_{3v}$)	Bond-stretching	A$_1$	IR, R(P)
	Bond-Bending	E	IR, R(D)
SiH$_2$ (C$_{2v}$)	Symmetric Stretch	A$_1$	IR, R(P)
	Asymmetric Stretch	B$_1$	IR, R(D)
	Scissors Bend	A$_1$	IR, R(P)
	Wagging	B$_2$	IR, R(D)
	Twisting	A$_2$	R(D)
	Rocking	B$_1$	IR, R(D)
SiH$_3$ (C$_{3v}$)	Symmetric Stretch	A$_1$	IR, R(P)
	Asymmetric Stretch	E	IR, R(D)
	Symmetric Deformation	A$_1$	IR, R(R)
	Degenerate Deformation	E	IR, R(D)
	Rocking (or Wagging)	E	IR, R(D)
	Twist	A$_2$	Inactive

D) and a-Si : F have been associated with the types of bonding arrangements designated in Fig. 7.16, and specific frequencies have been assigned to some of the vibrations whose displacement vectors are indicated in Fig. 7.17. Additional features have been assigned to the polymerized bonding arrangements shown in Fig. 7.16c.

The pioneering experimental studies of ir absorption and Raman scattering in a-Si : H were done by the IBM and Xerox research groups: *Brodsky* et al. [7.31] and *Knights* et al. [7.44] and *Lucovsky,* et al. [7.32], respectively. Ir absorption spectra from [7.32] are shown in Fig. 7.18 and the frequencies of the local mode vibrations are shown in a schematic representation in Fig. 7.19. It is compiled from the data displayed in Fig. 7.18 and in the references cited above. By studying the changes in the character of the ir absorption spectra that are correlated with different amounts of bonded hydrogen, and by comparing the features in the ir spectra of a-Si : H alloys with the ir bands in molecules that contain SiH, SiH$_2$ and SiH$_3$ groups, it is possible to identify in an unambiguous way those features associated with sites containing one H atom, the SiH or monohydride group, from those sites containing either two or three H atoms, as in SiH$_2$, (SiH$_2$)$_n$ and SiH$_3$ [7.31]. The features assigned to the SiH group are those at 2000 cm^{-1} and 630 cm^{-1}. As discussed in the last section, the 2000 cm^{-1} absorption is due a bond-stretching vibration (Figs. 7.13a, 17) and the absorption at 630 cm^{-1} is due to the bond-bending vibration, also shown in these figures. The 2000 cm^{-1} feature is also observed in the Raman spectra, while the feature at 630 cm^{-1} is masked by second-order scattering from the a-Si host. Changes in the spectra that correlate with additional bonded H atoms, as in the dihydride, trihydride and polysilane arrangements, are a shift of the bond-stretching vibrations to

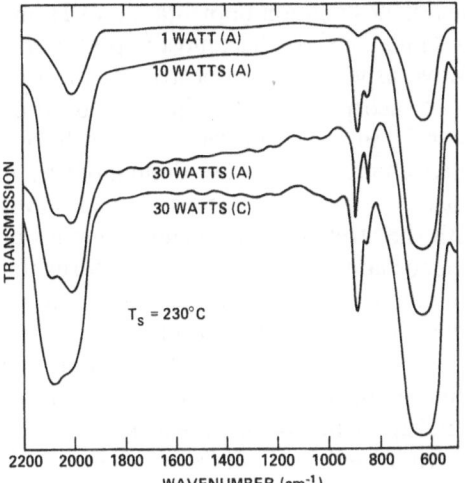

Fig. 7.18. Ir transmission for a-Si films produced by the glow discharge decomposition of SiH_4. (*A*) and (*C*) indicate samples produced, respectively, on anode and cathode substrates. The absorptions near 2000–2100 cm^{-1} are bond-stretching in character. Absorptions near 600 cm^{-1} are due to bending, rocking and wagging motions. Absorptions between 800 and 900 cm^{-1} are also due to bending and wagging motions, but only for Si sites with more than one H atom attached, as in SiH_2 and $(SiH_2)n$ (from [7.32])

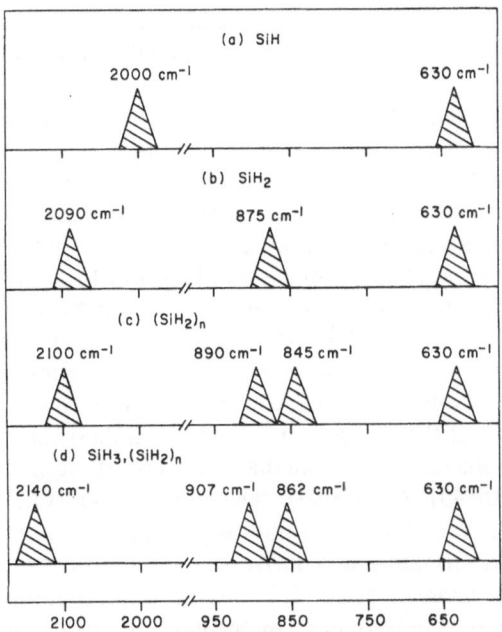

Fig. 7.19 a–d. Schematic representation of ir data for a-Si : H alloys. The absorptions are organized into groups that have been assigned to different local bonding configurations. The most recent interpretations favor (**d**) being attributed to polysilane, rather than SiH_3 (from [7.15, 22])

frequencies between 2090 cm^{-1} and 2140 cm^{-1} and the appearance of new bands between 830 and 920 cm^{-1}. We first discuss the monohydride bands and then discuss the controversy that has surrounded the assignments of the particular features in the 830–920 cm^{-1} regime.

Films deposited in a variety of ways, including the glow discharge decomposition of silane, diode and magnetron sputtering of Si in a H-containing ambient; CVD using SiH_4 and higher silanes and ion implantation of H into a-Si, all yield films with qualitatively similar spectra. The most important parameters in determining the qualitative aspects of the spectra are the substrate temperature and the amount of incorporated H. Films produced on substrates held at temperatures generally in excess of 250–300° and having bonded H concentrations of less than about 15 at.% are dominated by ir absorption bands at 2000 cm^{-1} and 630 cm^{-1}, i.e., the modes of absorptions that are associated, respectively, with the stretching and bending modes of the SiH group. The only point that needs additional discussion concerns the relatively low frequency of the stretching mode. In substitued silane molecules, the stretching frequency of the SiH group varies from 2315 cm^{-1} in $SiHF_3$ to 2120 cm^{-1} in $SiH(CH_3)_3$ [7.19]. The major factor in determining this frequency is the relative electronegativity of the three atoms or groups that are back-bonded to the SiH group [7.19, 22]. It has been shown that the SiH bond-stretching frequencies vary linearly with the sum of the electronegativities of these atoms or molecular groups [7.19–21] and further, that these scaling relationships hold in amorphous solids as well [7.19]. The low value of the bond-stretching frequency of the SiH group in an a-Si host network is then derived from the fact that the three Si atoms back-bonded to the SiH group have a lower electronegativity sum than do any of the substituting atoms or groups in the silane molecules that have been studied. The scaling relationship can also be cast in other variables defining the nature of the local bonding, the partial charge on the Si atom of the SiH group [7.28], or the effective electronegativity of the same atom [7.30]. Moreover, the scaling laws carry over to SiH_2, SiH_3, SiF and SiF_2 groups [7.22, 30]. The bond-bending vibrations also scale with the local chemical bonding, but the relationship have not been quantified to the same extent as for the bond-stretching modes. The first reference to scaling of the bond-stretching frequencies is in [7.31]. The scaling relations, in effect, provide a means of obtaining values for the near-neighbor force constants which can then be employed in model calculations. The vibrations of the SiH group have been studied using small clusters of 5 atoms [5.44], clusters of intermediate size (14–18 atoms) [4.15, 22], and by the CBLM [7.15, 45]. All of these calculations support the assignment given above and give essentially the same displacement vectors for the H and Si atoms. Further confirmation was derived from the studies on deuterated samples performed by *Brodsky* et al. [7.31]. They found that the bond-stretching feature was depressed in frequency by the expected factor of about 1.4 when D was substituted for H. The bond-bending vibration shows a comparable shift to lower frequencies, but also shows a qualitative change that has recently been explained in terms of a coupling between the SiD bending vibration and modes of the host network that include significant displacements of the Si atom of the SiD group as well [7.46].

The features in the vicinity of 830–920 cm^{-1} and the shifts in the bond-stretching frequency to higher wave number occur together and appear in films with higher bonded H concentrations and in films deposited on substrates held at lower temperatures [7.31, 32, 44]. There is no controversy concerning the assignment of the features in the absorption due to the vibrations of the isolated SiH$_2$ group. The SiH$_2$ group has six modes of vibration and they have been assigned accordingly: (i) the absorption at 2090 cm^{-1} has been assigned to the symmetric and asymmetric stretching vibrations; (ii) the absorption at 875 cm^{-1} has been assigned to the scissors bending mode; (iii) the band at 630 cm^{-1} has been assigned to the bond wagging mode. The twisting mode is not ir active and the rocking mode occurs as an in-band resonance mode [7.16]. These assignments have evolved in time, particularly the assignment of the bond rocking mode, and were finally clarified by calculations based on the CIS and CBLM methods [7.16, 22]; also, see Fig. 7.20. The modes that have received the most attention and discussion have been a doublet that develops in the 830–920 cm^{-1} regime. It was originally believed to reflect the presence of both SiH$_2$ and SiH$_3$ groups [7.31]. Recently it has been demonstrated via direct synthesis of polysilane molecules [7.47, 48] and by model calculations using empirically scaled and realistic force constants that both components of the doublet are due the vibrations of the SiH$_2$ group in polysilane chains [7.16, 22].

The features that are under discussion occur in two groups as shown Fig. 7.19. A doublet at 845–890 cm^{-1} is correlated with a bond-stretching feature at 2100 cm^{-1}, and a second doublet with components at 862–907 cm^{-1} is correlated with a bond-stretching vibration occurring between 2120 and 2140 cm^{-1}. The second group of features is also accompanied by a decrease in the relative amplitude of the absorption band at 630 cm^{-1} [7.44]. We will not discuss the history of the various assignments made for the components of the doublet, but will note that both of the original assignments involved SiH$_3$, as well as SiH$_2$ and (SiH$_2$)n groups [7.31, 32]. These

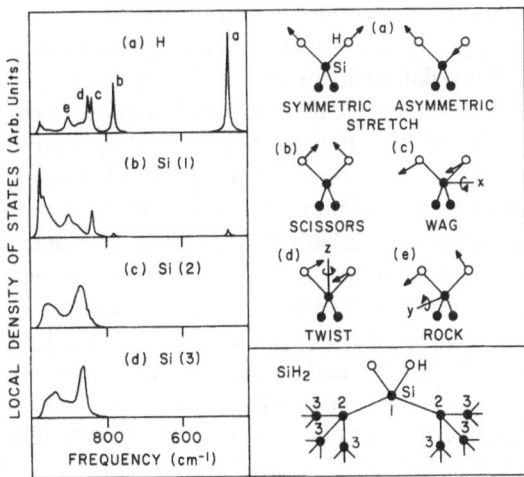

Fig. 7.20. LDOS spectra of the SiH$_2$ configuration. The numbers give the shell designations of the Si atoms and the letters in (a) indicate the particular types of vibrations as designated on the right side of the figure (from [7.15])

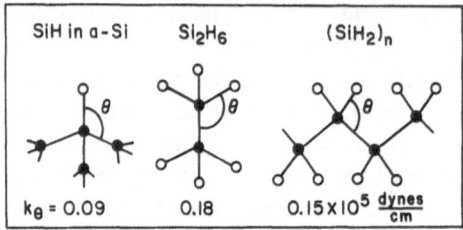

Fig. 7.21. Local atomic arrangements for the Si-Si-H bonding triad for the SiH bond in a-Si : H alloys, the disilane molecule and the interior of a polysilane chain (from [7.22])

assignments were supported by qualitative and quantitative comparisons between the vibrations of SiH_2 and SiH_3 groups in various silane molecules, and the vibrations occurring in the a-Si:H alloys. The key factors that have finally led to the accepted interpretation of the origin of these spectral features involve (i) calculations based on the scaling of bond-bending as well as bond-stretching force constants [7.15, 22], (ii) synthesis of polysilane compounds and the study of their ir absorption spectra [7.47], and finally, (iii) studies of the deuterated samples, where different degrees of splitting between the two components of the doublet in H and D containing samples are explained by dynamical calculations based on chemically scaled, realistic force constants [7.46]. The most important force constant turns out to be the three-body bond-bending force constant k_θ (Si–Si–H) [7.22]. The local atomic structure indicated in Fig. 7.21 shows this group in three different situations: (i) the isolated SiH group in a-Si : H alloys (ii) the disilane molecule Si_2H_6, and (iii) the interior of a polysilane chain. The force constant for (i) is obtained from a fit to the bond-bending frequency of 630 cm^{-1} for the SiH group in a-Si : H alloys, where a value of 0.09×10^5 dyne/cm is found. This value is close to the bond-bending force for the tried Si–Si–Si. Fits to the vibrational frequencies of the disilane molecule yield 0.18×10^5 dynes/cm [7.49]. The differences between the two values quoted above have been associated with the number of H atoms bonded to the second Si atoms of the Si–Si–H triad [7.22, 30]. This atom has no H neighbors for the first case and three for the second. For an SiH_2 group in the center of a polysilane chain, there are two H's and linear interpolation leads to a value of 0.15×10^5 dynes/cm. This value has been employed in calculations based on the CIS and CBLM methods [7.15, 16, 22] and yields the doublet in the frequency regime under discussion; this is evident in Fig. 7.22 from [7.15]. The low frequency component of the doublet is due to a wagging motion, as originally proposed on the basis of qualitative comparisons with polymerized polyethylene chains [7.32, 44], and the higher frequency component is due to a scissors type motion. The wagging mode is shifted from 630 cm^{-1} for the isolated SiH_2 group to 845 cm^{-1} in polysilane chains, and the scissors mode is similarly shifted from 875 to 890 cm^{-1}. Finally, the in-band rocking mode is shifted up to a frequency of about 630 cm^{-1} and its lower ir activity then accounts for the observed decrease in the relative ir absorption in this band [7.44]. The second triad of features at 862, 907 and 2120–2140 cm^{-1} has been attributed

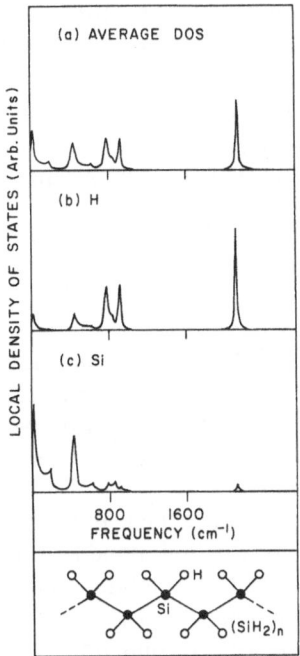

Fig. 7.22 a–c. DOS spectra for interior H and Si atoms of a polysilane chain. The figure shows the average density of states and the LDOS for a H and a Si atom (from [7.15])

to a different conformation of the polysilane chain [7.22]. These higher frequency modes dominate in samples containing the most H and produced at the lowest substrate temperatures. The higher frequencies and sharper lines have been attributed to increased stero-regularity, i.e., partial crystallinity that is manifested in parallel allignment of the polysilane chains [7.16]. Finally, there is essentially no evidence in any of the ir spectra for a significant concentration of SiH_3 bonds in any of the a-Si:H alloys [7.22].

There have been a number of studies of the morphology of a-Si:H films [7.50, 51]. These have clearly demonstrated that films displaying the characteristic polysilane absorptions (usually identified as the doublet structure at 845, 890 cm^{-1}) have a columnar structure and are diphasic in character. Raman scattering experiments, coupled with an etching technique that removes material between the columns but does not etch the material in the columns as rapidly, have established that the columnar material is predominantly an a-Si:H alloy with about 5–10 at.% H and that the material in the columns is polysilane [7.52]. This phase separation is not surprising when one takes into account the very different local bonding in the two different phases discussed above.

There have also been studies of the changes in the ir spectra produced by thermal annealing and ion implantation [7.53, 54]. We will not emphasize these studies in detail. Basically the same types of SiH and SiH_2 absorption bands are noted. There are indications from these studies that there may be yet another local bonding configuration that can produce an ir absorption band at a frequency of 2090 cm^{-1} [7.54, 55]. For the cases of films produced by diode sputtering or subject to post-deposition ion bombardment, this 2090 cm^{-1} absorption is not accompanied by absorption at 875 cm^{-1} or the polysilane doublet and therefore appears to be associated with some type of monohydride bonding arrangement. It is then likely that this SiH group is apart of a radiation damage defect complex [7.54].

7.5.2 a-Si:F Alloys

We next consider the vibrational properties of a-Si:F alloys as they are revealed through studies of the ir absorption. Before discussing the spectra, we first identify the changes with respect to a-Si:H alloys that simply derive from the differences in the atomic masses of H and F and from the increased strength of the Si–F bond relative to the Si–H bond [7.22].

The mass ratio of H to Si is 0.03 and that of F to Si is 0.68; the bond energies are 3.4 eV for the Si–H bond and 6.0 eV for the Si–F bond [7.22]. The larger F mass means that (i) there will be more relative Si motion in the SiF vibrations than in the corresponding SiH vibrations, and (ii) that the corresponding vibrations will be lower in frequency in a-Si:F compared to a-Si:H alloys. The increased bond strength in part compensates for the increase in the F mass, but the net effect is still a significant decrease in the relative Si–F frequencies. The combined effects of these two factors is that the SiF bond-stretching vibrations are local modes at frequencies between 825 and 1025 cm^{-1}, while all of the other types of vibrations associated with SiF and SiF$_2$ configurations are resonance modes with frequencies less than 400 cm^{-1} [7.22].

A second manifestation of the increased mass of F relative to H is in the frequency splitting between the asymmetric and symmetric stretching modes of SiF$_2$ and SiF$_3$ groups. If M is the mass ratio $m(F)/m(Si)$ and k is the ratio of the three-body to two-body bond-stretching frequencies $k(Si-F)_r/k(Si-F)_{rr'}$, then the ratio of frequencies $\nu_a/\nu_s =$ for the SiF$_2$ group is given by

$$\left(\frac{\nu_a}{\nu_s}\right)^2 = \left(\frac{1 + 2M\sin^2\theta}{1 + 2M\cos^2\theta}\right)\left(\frac{1-k}{1+k}\right) \tag{7.7}$$

2θ is the Si–F–Si or Si–H–Si bond angle assumed to be equal to the tetrahedral bond angle of 109.47°. For a mass ratio of 0.68 and $k = 0.05$ for Si–F–Si bonds, the calculated frequency ratio is about 1.10. A similar calculation for the splitting of the stretching vibrations of the SiH$_2$ group yields a value of 1.003. The calculation for the SiH$_2$ groups supports the fact that no splitting of the stretching modes is evident in the experimental data. The calculated fractional splitting corresponds to 6 cm^{-1}, whereas the line width appears to be at least 20 cm^{-1}. For the case of the same type of vibrations in the a-Si:F alloys, the splitting of the modes is observable in accordance with the calculated ratio of 1.10 [7.22].

There have been a number of studies of SiF vibrations in a-Si:F and a-Si:H:F alloys [7.56–59]. We will restrict the discussion here to the binary alloys in the a-Si:F system. There is a consensus regarding the frequency of the SiF group bond-stretching mode; this occurs at a frequency of 830 cm^{-1}. The bond-bending mode of the same group is a resonance mode and its frequency has been estimated to be approximately 300 cm^{-1} [7.57]. The 830 cm^{-1} vibration is observed in all of the a-Si:F alloy samples, those with relatively low F concentrations, approximately 5–10 at.%, and those with substantially higher concentrations as well. The force constant for this vibration has been determined by scaling the frequencies of SiF bond-stretching vibrations in silane molecules with the local bonding chemistry and then applying a similar analysis to the amorphous solid state [7.22, 30]. The vibrational properties of the SiF or monofluoride group have also been calculated using the CBLM and CIS techniques [7.15, 22, 45]. These calculations support the comments made earlier relative to the increase in the size of the displacements of the Si atoms and the bond-bending vibration being an in-band resonance mode.

At least four additional vibrations have been observed in the bond-stretching frequency regime in alloy samples having higher F concentrations (> 20 at.%). Typical spectra are shown in Fig. 7.23. Most of these features have been assigned to SiF$_2$ groups; however, there has been much controversy with respect to the assignment of the mode at 1015 cm^{-1} which dominates in samples with the highest F concentrations. *Shimada* et al. [7.59] have assigned this vibration to the asmmetric bond-stretching vibration of the SiF$_3$ group with the symmetric vibration being at 838 cm^{-1}. This assignment has not held up for two reasons, one theoretical and the second experimental: (i) the ratio the frequencies of the asymmetric and symmetric components is larger than is expected from an analysis of the vibrational properties of SiX$_3$ groups [7.22]; (ii) the group at the Max Planck Institute in Stuttgart

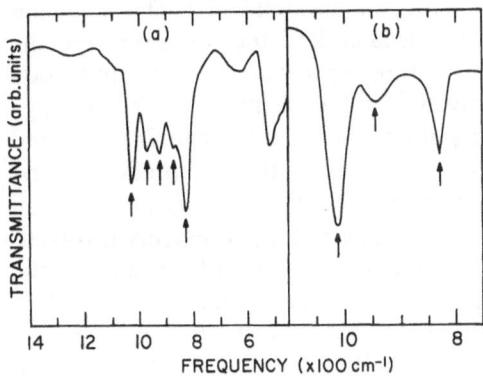

Fig. 7.23 a, b. Ir absorption spectra for a-Si:F alloys. **(a)** for a film displaying "sharp" features in the bond-stretching frequency regime. The arrows indicate the five discrete features discussed in the text (from [7.59]). **(b)** For a film exhibiting columnar growth and a diphasic structure (from [7.58, 61])

(MPI) has shown that the integrated absorption strength in these two modes behaves differently as a function of annealing time, implying that the two absorptions cited above are not derived from vibrations of the same local bonding group [7.57]. The MPI group, in a study of a-Si : H : F alloys (to be discussed in greater detail later in this chapter), has assigned the vibration to the asymmetric bond-stretching vibration of the SiF_4 molecule, a species that they believe is trapped in internal voids in the amorphous phase. This assignment would require a polarized Raman mode at 800 cm^{-1}, the symmetric stretching mode of the SiF_4 molecule. This feature is not observed in the Raman spectrum of samples in which the 1015 cm^{-1} ir absorption dominates; hence, the assignment must be regarded as questionable [7.60]. A Japanese group has made a similar assignment, based on the fact that upon thermal annealing at relatively high temperatures, SiF_4 molecules are evolved [7.61]. The association between the gaseous species that are evolved in a thermal decomposition and the local bonding groups or molecules present in the solid state is also highly questionable. *Lucovsky* [7.22] has proposed an alternative explanation which is based on the same mechanism that applies in a-Si : H alloys and explains in a quantitative way the frequency differences between the wagging modes in isolated SiH_2 groups and SiH_2 groups in polysilane chains.

Figure 7.24 indicates three bonding environments for SiF_2 groups in an a-Si : F host [7.22]: (i) as an isolated SiF_2 group, (ii) as an SiF_2 group in the interior region of a fluorine substituted polysilane chain $(SiF_2)_n$, and (iii) at the boundary between an $(SiF_2)_n$ chain and an a-Si host network. The local bonding environments in the vicinity of the SiF_2 groups differ with respect to the nature of the atoms that are back-bonded to the Si atom; only Si atoms in the first case, other SiF_2 groups in the second and a combination of Si and SiF_2 groups in the third. Each of these environments produces a different degree of charge transfer between the Si and F atoms of the SiF_2 group under consideration and hence will produce different pairs of symmetric and asymmetric bond-stretching frequencies. Figure 7.25 is an analysis of molecular data which gives the variation of the two bond-stretching frequencies of the SiF_2 group as a function of the electronegativity sum of the atoms or groups of atoms that are back-bonded to the group in question [7.22]. The solid lines are a linear regression analysis of the molecular data; the points shown in the diagram are the frequencies of the features reported in the ir absorption spectrum of a-Si : F alloys with relatively high F concentrations, in the range of 10 to 50 at.% [7.58, 59]. Two pairs of vibrations, at 827 and 870 cm^{-1} and 920 and 965 cm^{-1}, were assigned to SiF_2 groups in the interpretations first proposed with the experimental data [7.58, 59]. The analysis presented in [7.22] served to identify the specific and different environments involved, namely, that the lower frequency pair was due to isolated SiF_2 groups and the higher frequency pair to SiF_2 groups that are at the boundary of an $(SiF_2)_n$ chain and an a-Si host. The absorption band that has caused the most discussion is the one at 1015 cm^{-1}, as was already noted above. The empirical

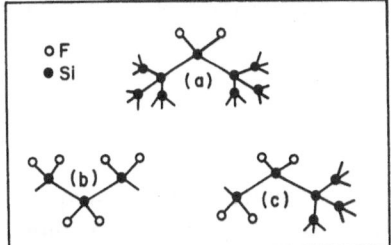

Fig. 7.24 a–c. Local bonding arrangements for SiF_2 groups at (**a**) an isolated bonding site. (**b**) a bonding site in the center of an $(SiF_2)_n$ chain, and (**c**) a bonding site at the termination of an $(SiF_2)_n$ polymer chain. The respective electronegativity sums of the atoms back-bonded to the Si atom of the group are 5.24, 8.86 and 7.01 (from [7.22])

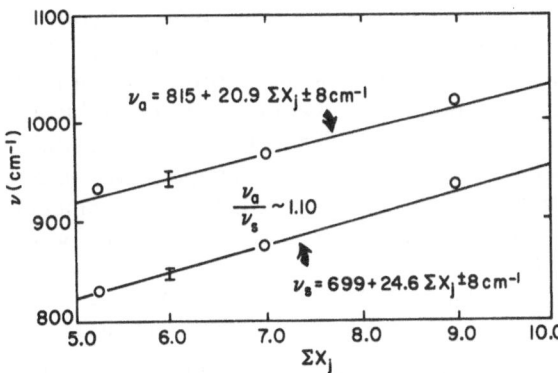

Fig. 7.25. Induction relationships for the asymmetric (*a*) and symmetric (*s*) vibrations of the SiF_2 group. The solid lines are from an analysis of molecular data and the points represent the data in Fig. 7.23a (from [7.22])

relationships in Fig. 7.25 resolve this question and demonstrate that the 1015 cm^{-1} absorption is associated with SiF_2 units in the interior of an $(SiF_2)_n$ chain.

There are two qualitatively different kinds of ir absorption spectra for a-Si:F films with relatively high F concentrations (> 25 at.% F); one which exhibits all of the absorptions discussed above and one which is dominated by absorption bands at 830 cm^{-1} and 1015 cm^{-1} with a broad and weaker feature at about 930 cm^{-1}. These are shown in Figs. 7.23a, b. The film in which the second type of absorption spectrum occurs (Fig. 7.23b) has been shown to be diphasic [7.61] with a characteristic columnar structure similar to that found in a-Si:H films with similarly high H content [7.50, 51]. The interpretation given in [7.61] is that the 830 cm^{-1} band is associated with the columns which are assumed to contain lightly alloyed (5–10 at.% F) a-Si:F, and that the connective tissue between the columns contains SiF_4 molecules. The assignment of the 1015 cm^{-1} vibration to the F substituted silane molecule has been shown to be questionable and the explanation that is preferred is derived from the model calculations presented in [7.15, 22], *i.e.* that the 1015 and 930 cm^{-1} features are the asymmetric and symmetric bond-stretching modes and SiF_2 groups in the interior of an $(SiF_2)_n$ chain. This assignment then establishes a completely parallel behavior in a-Si:H and a-Si:F films. The factor limiting the amount of either atom in a single phase system is the competition of a second polymeric phase; polysilane in one case and fluorine substituted polysilane in the second. This in turn yields the diphasic behavior that shows up in the columnar structure.

7.5.3 Other Binary Alloy Systems

A number of binary alloy systems have been investigated by the technique of ion implantation into either crystalline or amorphous Si hosts [7.35, 41, 43, 54]. If the implanted species did not render the Si amorphous, then a pre-implant of Si was used for this purpose. We shall not discuss the experiments done with H, other than to point out that these experiments reveal the monohydride bonding group and an additional absorption at 2090 cm^{-1} that is not related to SiH$_2$ bonding or the incorporation of near-neighbor strongly electronegative atoms such as O or F, but rather is due to a defect configuration that is metastable [7.54]. The metastability is identified by the fact that this absorption disappears after a low temperature thermal anneal which does not remove H from the sample [7.55]. This environment is assumed to be the same one that has been discussed by the Harvard group in their sputtered material [7.55].

The ion implantation studies done on C, N and O yield the ir active vibrations expected from the symmetry considerations discussed in Sect. 7.4. The C implantation yields a single absorption band centered at 700 cm^{-1} [7.43] that is due to the triply-degenerate bond-stretching vibration of the C atom (Fig. 7.12e). The N implantation also yields one absorption band, a broad feature centered at about 790 cm^{-1} [7.41]. This feature is assigned to the asymmetric in-plane bond-stretching vibration (Fig. 7.12c). The other ir active modes of this bonding site are assumed to be resonance modes and therefore are not readily observed due to the competing absorption of the a-Si host. The O implant yields three features, at 940, 660 and 500 cm^{-1} [7.35]. These are assigned to the three vibrations involving O atom displacements: the asymmetric stretching vibration, the symmetric stretching (or bending) vibration and the out-of-plane rocking vibration (Fig. 7.12b). The progressive increase of the stretching frequency from 700 to 790 to 940 cm^{-1} reflects a strengthening of the bond between the Si and the implanted atom that is a function of the their electronegativity difference. Another factor that adds to this ionic contribution is a back donation of electrons from occupied non-bonding states of the N and O atoms to empty orbitals with d-like symmetry on the Si atom [7.22]. This particular combination of effects is also the factor that promotes the very large value of the bond-stretching force constants for the various Si-F groups [7.22].

Other binary systems studied included a-Si:C [7.62, 63] and a-Si:Ge [7.64]. The focal point of the a-Si:C studies has been the local bonding in the compound composition SiC. Is the amorphous phase chemically ordered or does it contain a statistical distribution of bonds, Si–C, C–C and Si–Si? The spectra indicate the presence of all three bond types and the experiments indicate that the particular distribution of the bond types is related to the way the sample is prepared [7.62]. A further complication in this alloy system is derived from the observation that the bonding coordination of the C atoms changes from a fourfold coordination that is characteristic of the diamond structure to a threefold coordination that is characteristic of the graphite structure and that the point of conversion depends on the method of preparation. It is different for glow discharge deposited films and for sputtered films [7.63]. The two C environments are distinguished through their different signatures in the electron energy loss spectrum [7.63].

Alloys in the a-Si : Ge system have also been studied by Raman scattering spectroscopy [7.64]. They display spectra which show the three types of vibrations that are anticipated for an alloy system where the distribution of bond types is simply determined by the composition and the coordinations of the two atomic species. The spectra show features indicative of Si–Ge, Ge–Ge and Si–Ge bonds. These vary in intensity according to the composition of the alloy, with no effects indicative of clustering of like atoms, or phase separation.

7.6 Amorphous Silicon Ternary Alloys

This section of the chapter deals with ternary alloys with one of the components being a-Si. Each of the systems studied to data has also included H as a second component. These alloy systems display two types of behavior for the alloy regime close to a-Si, *i.e.,* for relatively small additions of each of the other two components. One of these is exemplified by the a-Si : H : F alloys studied by the MPI group [7.57]; these exhibit features due to both SiH and SiF local bonding groups, with no features indicative of both of the alloy atoms H and F being bonded to a common Si site. In contrast, other alloy systems, notably a-Si : H : O, exhibit features that can only be interpreted by local arrangements involving O and H atoms bonded to the same Si site [7.17, 65, 66]. We will discuss both types of alloy systems, but emphasize the changes in the vibrational modes that occur when both alloy atoms bond to a common Si site.

7.6.1 a-Si : H : F Alloys and a-Si : H : Cl Alloys

a-Si : H : F alloys were studied initially by the ECD group [7.56] and later by the group at MPI [7.57]. The first reported values for the frequencies of SiF bond-stretching vibrations were derived from the studies at ECD. Figure 7.26, taken from [7.57], is a schematic representation of the MPI absorption data. Indicated in the diagram are the assignments proposed by the MPI group. The assignment of the local mode at 1015 cm^{-1} to the asym-

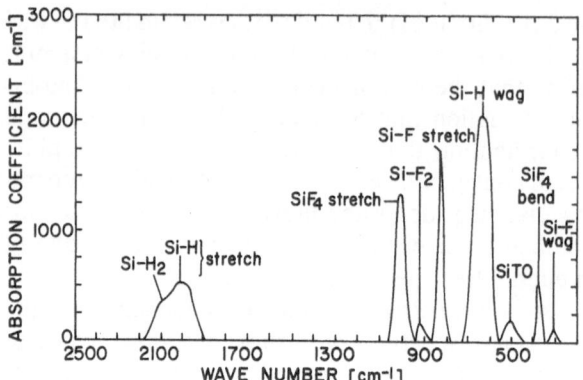

Fig. 7.26. Ir absorption for an a-Si:H:F alloy. The assignments are those proposed in [7.57]

metric stretching mode of the SiF_4 molecule, and the in-band resonance mode at 380 cm^{-1} to a bending mode of the same molecule have been discussed above where it was pointed out that a more likely explanation for the origin of the absorption at 1015 cm^{-1} was that it was due to the asymmetric stretching mode of an SiF_2 group in the interior of an $(SiF_2)_n$ chain. Additional evidence against a model that is based on SiF_4 molecules is derived from recent studies of a-Si:F alloys generated by ion implantation of F and SiF_n ($n = 1$–3) ions into crystalline Si substrates. These indicate a conversion of the material to the amorphous form and the presence of correlated absorptions at 1015, 930 and 380 cm^{-1} [7.67]. It is not likely that implantation by the molecular ions discussed above could yield SiF_4 molecules in sufficient number to account for the relatively strong ir absorption that is reported. One of the most interesting aspects of the MPI study of the a-Si:H:F alloys was that it showed no evidence for H and F atoms being attached to the same Si site. There have been some studies done of the a-Si:H:Cl system [7.33] which indicate several absorption bands between the Si–Cl stretching vibrations near 540 cm^{-1} and the Si–H features at 840 and 885 cm^{-1} that could be due to simultaneous attachment of H and Cl atoms to the same Si site. These are the vibrations at 740 and 795 cm^{-1}. More work must, however, be done before assignments of this sort can be made definitive.

7.6.2 a-Si:H:O Alloys

Significant amounts of unintentionally added oxygen are frequently found in a-Si:H alloys prepared by either glow discharge decomposition of silane or by reactive sputtering. The source of the oxygen is believed to be from water molecules that are adsorbed on the walls and internal components of the sample chamber, or carried into the system with the processing gas, or in extreme cases from leaks in the system. In any event, the incorporation of oxygen in amounts of the order of 0.5 at.% is easily detected via a broad absorption band centered at about 980 cm^{-1} [7.34, 65]. There is no evidence for any OH bonding groups in these materials even though the source of the oxygen is primarily from water molecules. In order to gain a better understanding of the bonding of oxygen atoms in a-Si : H alloys and to determine its effect on the electronic and photoelectronic properties of the films, a number of studies were undertaken in which oxygen was intentionally added to the processing gases [7.34, 36, 65, 66]. These studies included alloy compositions which could be characterized as a-SiO$_2$: H, which we will not discuss in this review.

As noted in the introduction to this section, the a-Si : H : O system is one in which there is evidence for H and O atoms bonded to the same Si site. This is illustrated in Fig. 7.27 taken from [7.68]. The sample in question contains approximately 10 at.% H and 5 at.% O. The figure contains pointers that indicate the positions of the absorption bands in binary alloys of a-Si : H and a-Si : O with comparable concentrations of the respective alloy atoms. It is clear that the ir absorption spectrum in Fig. 7.27 is not simply the sum of the ir absorption generated by isolated Si–O and Si–H bonding groups. This is evident in (i) the shift of the SiH bond stretching frequency from 2000 cm^{-1} in a-Si : H alloys with monohydride bonding to 2090 cm^{-1} in the ternary system, (ii) a similar shift of the asymmetric stretching mode of O in a-Si : O alloys from 940–980 cm^{-1}, and (iii) in the appearance of another new vibration at a frequency of 780 cm^{-1}. A recent study tracked the absorption strength in each of these bands as a function of the oxygen concentration and has established a linear relationship between the absorption and the concen-

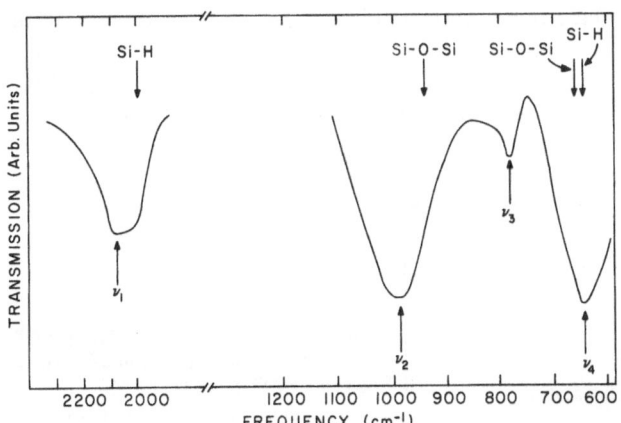

Fig. 7.27. Ir absorption for an a-Si : H : O alloy. The markers at the top of the figure are the frequencies of vibrations reported in films of a-Si : H and a-Si : O with respective concentrations of H and O that are about the same as those in the ternary alloy: 10–15 at.% H, and 5 at.% oxygen. The features denoted by ν_1, ν_3 and ν_3 only occur in the alloy films and in the text are shown to be associated with alloy environments in which both H and O atoms are bonded to the same Si atoms (from [7.68])

tration [7.36]. This study also established a similar scaling for the band centered at 495 cm^{-1}. We will restrict our emphasis to the three features cited initially, and simply indicate that the 495 cm^{-1} band has been associated with the out-of-plane rocking motion of the O atom [7.35].

The first indication of bonding of O and H atoms to the same Si site came from the systematic shifts in the frequency of the SiH bond-stretching vibration that were due to either one, two or three oxygen atoms back-bonded to the Si atom of the SiH bond [7.19, 65, 66]. The chemical bonding induced frequency shifts were predicted on the basis of the same model that was used to explain the unusually flow frequency of the SiH bond stretching vibration in a-Si:H alloys [7.19], and were found to agree very accurately with the frequencies observed in the a-Si:H:O alloys [7.65, 66]. These observations then establish a basis for more detailed studies of the local bonding environments at Si sites with both O and H neighbors [7.17, 68]. The theoretical models are derived from the details of the geometry of the bonding site.

Figures 7.28a, b indicate the local bonding geometry at isolated H and O atom bonding sites. Figures 7.28c, d indicate two of the possible bonding environments in which a Si atom has both H and O nearest neighbors. The structural factor that discriminates between the various possible geometric arrangements of the H and O atoms is the orientation of the Si–H bond relative to the plane defined by the O atom and its two Si neighbors. An argument, given in [7.17], is that the energetically favored arrangements are those in which the Si–H bond is coplanar with the triad of atoms defining the Si–O–Si bond. This argument is made on the basis of a minimization of the repulsion between the nonbonding electron pairs of the oxygen atom and the electrons that comprise the Si–H bond. Two orientations are then possible, the so-called cis-bonding geometry shown in Fig. 7.28c and the trans-bonding geometry shown in Fig. 7.28d. Using force constants obtained from the chemical bonding models discussed earlier in the chapter [7.19, 22] and the CIS and CBLM techniques, the local modes have been calculated for these

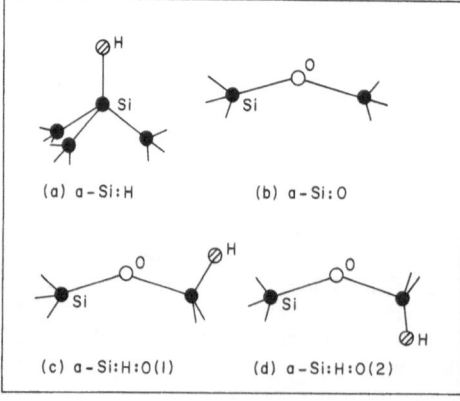

(a) a–Si:H   (b) a–Si:O

(c) a–Si:H:O(I)   (d) a–Si:H:O(2)

Fig. 7.28 a–d. Local bonding geometries: (a) isolated SiH site in a-Si:H; (b) isolated Si–O–Si group in a-Si:O; (c, d) two bonding arrangements in which H and O atoms are bonded to a common site as in a-Si:H:O. The SiH bond lies in the plane of the Si–O–Si triad in each arrangement; in (c) the O and H atoms are in CIS conformation and in (d) they are in a TRANS conformation (from [7.17, 36, 68])

two configurations and for a third configuration in which the Si–H bond is in a plane that is perpendicular to the plane of the Si–O–Si triad [7.68]. These calculations have also been applied to the cases wherein D, F and Cl atoms have been substituted for the H atom. Table 7.3 gives the results of these calculations for the H and D atom configurations, and Fig. 7.29 gives the atomic displacements for the modes involving H.

Three of the local mode vibrations are insensitive to the orientation of the Si–H bond. These are the SiH bond-stretching vibration which with the chemically adjusted force constants, is predicted to occur at 2090 cm^{-1} the asymmetric stretching mode of the O atom which has a predicted frequency of 980 cm^{-1} and the out-of-plane SiH bond-bending vibration which has a frequency of 630 cm^{-1}. These calculations then explain three of the dominant features reported in the ir absorption of a-Si : H : O alloys in which the deposition conditions are adjusted to insure H incorporation in monohydride bonding arrangements, and in which the H concentration exceeds the oxygen concentration [7.36]. Only one of the two bonding configurations gives a vibration near 780 cm^{-1}; this is the cis conformation [7.17]. For this particular geometry, the 780 cm^{-1} vibration is a mode which couples Si–O–Si and Si–H motions that have a predominantly bond-bending character with the H motion at approximately right angles to the direction of the Si–H bond and the O motion displaced by about 10° from the direction of the bisector of the Si–O–Si bond angle. The trans conformation does not yield a coupled mode, but rather yields two bending modes, one at about 650 cm^{-1} that involves mostly bending motion of the O atom, and one at about 630 cm^{-1} which is an in-plane SiH bond-bending mode. Substitution of D for H removes the coupled mode behavior for the cis configuration and yields decoupled O and H bending modes at frequencies of 650 and 520 cm^{-1}, respectively [7.17, 68].

A comparison between the experimental data and the results of the calculations leads to the assignments proposed in [7.17, 68], i.e., that a significant fraction of the local bonding arrangements in the alloy samples are of the type shown in Fig. 7.28c. If this is the case, then we would also expect to find an almost equal number in the trans conformation as well. It has been shown

Table 7.3 Vibrational Frequencies for Local Modes in a-Si : H : O and a-Si : D : O Alloys

Vibrational character	Calculated frequencies of vibration [cm^{-1}]			
	Hydrogen Alloys		Deuterium Alloys	
	CIS	TRANS	CIS	TRANS
Uncoupled Modes				
SiH(D) Stretch	2090	2090	1505	1505
Si–O–Si Stretch	980	980	980	980
SiH(D) Out-of-Plane Bend	630	630	510	510
Coupled Modes				
SiH(D), Si–O–Si "Bend"	750	650	650	650

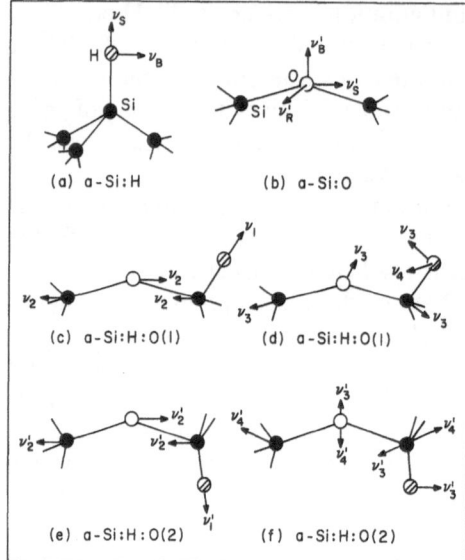

Fig. 7.29 a–f. Atomic displacements for localized vibrations: **(a)** the isolated SiH group; **(b)** the isolated Si–O–Si group; **(c–f)** the bonding groups for a-Si : H : O alloys that are shown in Fig. 7.28. The frequencies of the various vibrations are given in Table 7.3. The mode that is unique to the CIS conformation is ν_3; it is a coupled mode with significant O and H atom displacements and gives rise to the sharp absorption band at 780 cm^{-1} (see Fig. 7.27) (from [7.36, 68])

in [7.36] that the absorption strength of the feature at 780 cm^{-1} scales linearly with the amount of incorporated O over a range from about 0.5 at.% to about 15 at.%. This in turn means that the fraction of the bonding sites with the cis geometry is also constant. Very recent experiments have focused on a-Si alloys in which D has been substituted for [7.46]. These alloys do not display the 780 cm^{-1} vibration, as predicted by the CIS calculations, but instead exhibit the predicted decoupled Si–O–Si and Si–D bending vibrations near 630 cm^{-1} and 520 cm^{-1}, respectively (see Table 7.3).

There have been other studies involving the incorporation of O in alloys containing substantially larger H concentrations, films produced on low temperature (100 °C) substrates in which the H is incorporated in polysilane chains [7.36]. These films have shown to be diphasic and they do not exhibit any features indicative of O atoms being bonded to Si sites containing more than one H atom. The O is assumed to be found in the columnar material rather the connecting tissue and is in bonding environments like those discussed above. Films have also been grown in which the O concentration is substantially larger than the H concentration, essentially alloys of the form SiO$_2$: H. These alloys exhibit new vibrations which are derived from sites in which the H atom is bonded to a Si atom that has more than one O attached, as in the SiH center in a-SiO$_2$ [7.17, 36]. Studies have also been made of the post-deposition oxidation of a-Si : H alloys [7.66]. These films display SiH vibrations associated with SiH$_2$ centers in which the Si atom has at least one O neighbor.

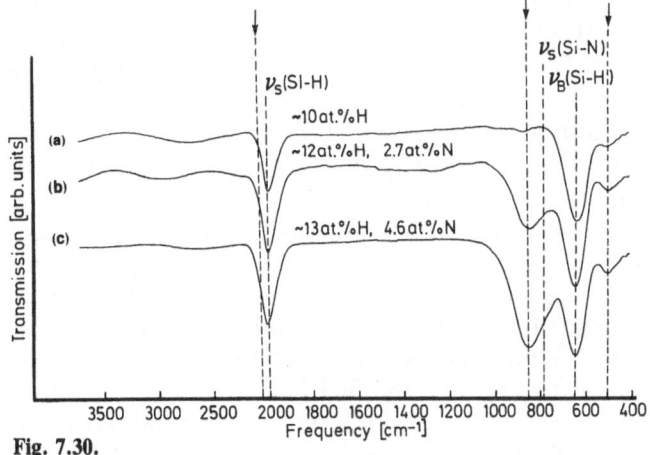

Fig. 7.30.

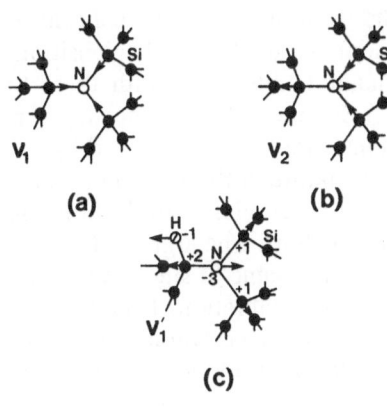

(a)

(b)

(c)

Fig. 7.30 a–c. Ir absorption spectra for a-Si : H : N alloys. The spectra labelled (a–c) are for increasing amounts of incorporated N. Markers are shown for the features occurring in a-Si : H and a-Si : N alloys. The features unique to the ternary alloys are also identified (from [7.18])

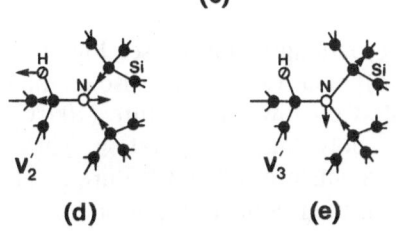

(d)

(e)

Fig. 7.31.

Fig. 7.31 a–e. Atomic displacements for (i) the local breathing (a) and stretching (b) vibrations at an isolated N bonding site in a-Si : N, and (ii) the local breathing (c) and asymmetric stretching (d–e) vibrations at sites in a-Si : H : N alloys in which H and N atoms are bonded to the same Si atom. The vibration labelled (a) is not ir active, however, the qualitatively similar vibration labelled (c) is ir active due to the presence of the H atom (from [7.18])

7.6.3 a-Si : H : N Alloys

The ir absorption spectra of a-Si : H : N alloys, studied by the ECD group [7.18], are qualitatively similar to those of a-Si : H : O in the sense that the spectra clearly indicate the bonding of H and N atoms to the same Si sites. This is shown in Fig. 7.30 taken from [7.18]. The spectra shown in this figure are for a samples deposited on substrates held at 400 °C and under conditions for which the bonded H concentration is about 15 at.% and the N concentration increases from less than 3 at.% to about 10 at.%. The evidence for bonding similar to a-Si:H:O films is derived from the character of three of the absorption bands: (i) the shift of the SiH bond-stretching vibration from 2000 to 2060 cm^{-1}; (ii) a similar shift in the frequency of the SiN asymmetric bond-stretching vibration from 790 to 840 cm^{-1}; (iii) the disorder induced Si breathing mode at 495 cm^{-1}. Figure 7.31 gives the atomic displacements of these two vibrational modes in the ternary alloys, (d, e and c), respectively. It also includes the corresponding displacement vectors for the same types of vibrations at an insolated N site, as in an a-Si:N alloy; these are (b and a), respectively. A Japanese group has studied alloys with significantly higher N concentrations in the range of 25 to 50 at.% and found qualitatively similar spectra [7.69]. The main difference between their results and those reported in [7.18] is in the shifted frequency of the SiH bond-stretching vibration. They report a frequency of approximately 2100 cm^{-1} which is indicative of a bonding environment in which there are two N atoms back-bonded to the Si atom of the SiH group. For the samples discussed in [7.18], the concentrations are such that one expects only one N atom in a similar bonding position. The N atom is assumed to reside in a planar bonding site with three Si neighbors and for the spectra shown in Fig. 7.30, one of these Si neighbors is also bonded to a H atom. This type of bonding site has been studied by the CIS method and it is found that there is no signature in the spectrum that is indicative of a particular orientation of the SiH bond relative to the bonding plane defined by the N atom and its three Si neighbors. The argument for the simultaneous attachment of H and N atoms to the same Si site was derived from the shifted frequencies of the SiH stretching vibration, from 2000 to 2060 cm^{-1}, and the SiN asymmetric in-plane stretching vibration from 790 to 840 cm^{-1}. Additional evidence is from the increase in the ir activity of the weaker absorption at 495 cm^{-1}. This absorption has been attributed in [7.18] to a disorder induced ir activity in the Si breathing vibration [see Fig. 7.31 and compare the displacements in (c and a)]. The ECD group also studied the incorporation of N into films deposited on low temperature (100°C) substrates in which the H bonding is in the polysilane configuration. These films are found to be diphasic and exhibit a single N related feature as an absorption band centered at 790 cm^{-1}. This is the frequency reported for the SiN stretching vibration in N implanted films [7.41]. The structural model suggested for the ternary alloy is one in which the columnar material contains isolated SiH and SiN bonding sites, and the connective tissue is polysilane. This is qualitatively different from the low temperature a-Si:H:O films wherein the columnar material exhibits absorption bands characteristic of the bonding of H and O atoms to a common Si site.

7.6.4 a-Si:H:C and a-Si:H:Ge Alloys

Alloy films in the a-Si:H:C system were studied by the IBM group [7.70] and films in the a-Si:H:Ge system by the Harvard group [7.71]. These studies emphasized the preferential bonding of H atoms to one of the two alloy species. In the a-Si : H : C system there is a preferential attachment of H atoms to the C sites (H concentrations of 20–30 at.%) and in the a-Si : H : Ge system the preferential attachment is to the Si sites (H concentrations of about 5–15 at.%). These results are explained by a chemical bonding model which presumes that the most probable bonding sites for the H atoms are those sites in which the bonding energy is largest [7.71]. The energies for the three bonds of interest are arranged in the following order $E(C–H) > E(Si–H) > E(Ge–H)$, so that the results obtained for the alloy films follow the predictions of the chemical bonding model [7.22, 71].

In addition to the features in the spectra due to bond-stretching vibrations involving H atoms, there are other absorption bands reflecting the types of bonds present in these two alloy systems. For example, the a-Si : H : C system shows Si–C stretching vibrations at frequencies near 700 cm^{-1}, as well as a large number of other absorptions between 600 and 1500 cm^{-1} that are due to various types of bending, wagging and rocking vibrations of CH$_2$ and CH$_3$ groups. The frequencies of these modes are in part influenced by the nature of the atoms or atom back-bonded to the group, i.e., whether they are C or Si atoms. There is no specific reference in the IBM study to C atoms being bonded in threefold-coordinated sites, as in graphite, although the results described in the section on binary alloys of a-Si : C would suggest their presence in these ternary alloys as well [7.63].

The study of the a-Si : H : Ge system [7.71] emphasized the bond-stretching modes involving H atoms. This was done in an attempt to point out the relationship of the H bonding configurations to the electrical properties of the films. The ir result was consistent with the observation of the generally poor electronic properties in the alloy films with high Ge concentrations (> 30–40 at.%). The origin of the trapping and recombination sites in these alloys are then presumed to be dangling bonds of Ge atoms that have been compensated by H attachment [7.71].

7.6.5 Other Alloy Systems

Three other alloy systems that have received some attention for their vibrational properties are the systems involving the most often used dopant atoms; a-Si : H : B [7.72–74], a-Si : H : P [7.73, 74] and a-Si : H : As [7.39]. Vibrations involving B, P, or As atoms are only evident at alloy concentrations in excess of about 5 at.%. This is in excess of the range of interest for doping. Spectra for the a-Si : H : B system exhibit the features expected in a ternary alloy in which there is no strong chemical bonding force that promotes a preferential attachment of H to either of the other two alloy atoms. This system then displays the local modes indicative of two kinds of B–H bonding groups: (i) terminal BH groups with a frequency of 2560 cm^{-1}, and (ii) H bridging groups (B–H–B) with vibrational frequencies of about 1985 cm^{-1}. The spectra also display the characteristic Si–H bond-stretching and bond-bending vibrations, as well as other vibrations associated with SiB bonds (840 cm^{-1}) and BH bending vibrations. The samples studied in [7.72] also exhibited O and C incorporation as evidenced by a large number of absorption bands between 600 and 1500 cm^{-1}. Calculations of the dynamics have not been performed to support these assignments, but it is almost certain that the B atoms are in threefold-coordinated bonding sites for the range of alloy samples that display the B–H vibrations discussed above. This follows from comparisons based on molecules containing similar bonding configurations.

The studies of ir absorption spectra for a-Si : H : P are even fewer in number, just the one undertaken by the MPI group [7.73, 74]. They reported P–H bond-stretching vibrations between 2100 and 2400 cm^{-1}. The multiplicity of bands reported would suggest several different types of bonding sites, and possibly even microcrystallinity. The studies of the a-Si : H : As system emphasized the vibrations associated with Si–Si, Si–As and As–As bonds, which were identified in Raman scattering spectra by comparisons with the end-member elements and crystalline compounds containing Si–As bonds [7.39]. These studies also identified vibrations involving H atoms, primarily Si–H bond-stretching vibrations in the Si-rich alloy regime. There was no evidence in the spectra for clustering or for crystallization of either an elemental or compound phase. As in the case of the a-Si : H : B and a-Si : H : P alloys, there are no calculations with which to compare the spectra.

7.7 Concluding Remarks

This chapter has addressed the question of the vibrational properties of a-Si alloys. We have reviewed the general theory of alloy atom vibrations using as a point of reference the theory of the localized vibrational modes of alloy atoms in crystalline hosts. We have emphasized the important differences between the incorporation of alloy atoms in crystalline and amorphous hosts by discussing the way the local atomic arrangements at alloy atom sites in an amorphous material are determined primarily by the valence bonding requirements of the alloy atom. This is to be contrasted with the alloy atom sites in a crystalline solid which are constrained by the symmetry and the geometry of the crystalline lattice. We have shown that local modes in an amorphous solid can be treated within the context of relatively simple models from which all of the vibrational properties can be calculated with a high degree of confidence. We then discussed the experimental results reported for both binary and ternary alloys in the context of this theoretical framework. For the majority of these systems, the assignment of specific types of local mode vibrations to the various absorption bands found in the ir spectrum is without question. On the other hand, we have identified situations where there was, and in some instances still is controversy regarding certain specific assignments. We have also pointed out the limitations of the same theoretical approaches in dealing with in-band resonance modes. The issue here stems from the fact that the local mode vibrations are insensitive to the details of the host network topology, whereas the resonance modes are not. Thus it appears that new approaches are needed which include the host network topology in a theoretical framework that still permits a complete and reliable determination of the local vibrational properties of all of the atoms in the immediate vicinity of the alloy atom sites. One other shortcoming that we now want to highlight is experimental in character. The spectroscopic techniques that have contributed to our knowledge of the vibrational properties and local bonding geometries of alloy atoms are only useful for alloy atom concentrations exceeding about 0.1 to 0.5 at.%, whereas there is interest in the local bonding and vibrational properties for concentrations of alloy atoms that are many orders of magnitude smaller. It is not clear whether vibrational spectroscopic techniques can ever be extended to this range of alloy atom concentrations that are of interest and importance in yielding a deeper insight into the microscopic mechanisms that determine the electronic and photoelectronic properties of a-Si alloys.

Acknowledgement. One of the authors (GL) wishes to acknowledge partial support for SERI contracts XG-L-1071-1 and XB-2-02065, and ONR contract N00014-79-C-0133.

References

7.1   D. G. Dawber, R. J. Elliott: Proc. Soc. (London) A **273,** 222 (1963)
7.2   P. Mazur, E. W. Montroll, R. B. Potts: J. Wash. Acad. Sci. **46,** 2 (1956)
7.3   J. E. Smith, M. H. Brodsky, B. L. Crowder, M. I. Nathan, A. Pinczuk: Phys. Lett. **26,** 642 (1971)
7.4   S. C. Shen, C. J. Fang, M. Cardona, L. Genzel: Phys. Rev. B **22,** 2913 (1980)
7.5   R. Tsu, M. Izu, V. Cannella, S. R. Ovshinsky, G. J. Jan, F. Pollak: J. Phys. Soc. Jpn., Suppl. A **49,** 1249 (1980)
7.6   J. S. Lannin, L. J. Pilione, S. T. Kshirrugar, R. Messier, R. C. Ross: Phys. Rev. B **26,** 3506 (1982)
7.7   W. Weber: Phys. Rev. B **15,** 4789 (1977)
7.8   J. D. Joannopoulos, W. B. Pollard: Solid State Commun. **20,** 947 (1976); J. Non-Cryst. Solids **35/36,** 1179 (1980)
7.9   R. Alben, D. Weaire, J. E. Smith, M. H. Brodsky, Phys. Rev. B **11,** 2271 (1975)
7.10 G. A. N. Connell: *Amorphous Semiconductors,* ed. by M. H. Brodsky (Springer-Verlag, Berlin, Heidelberg, New York 1979), p. 295
7.11 D. Beeman, R. Alben: Adv. Phys. **26,** 339 (1977)
7.12 P. E. Meeks: *Proceedings of the Fourth International Conference on the Physics of Non-Crystalline Solids,* ed. by G. H. Frischat (Trans. Tech. Publ., Aedermansdorf 1977), p. 586
7.13 F. Yndurian, P. Sen: Phys. Rev. B **14,** 531 (1976)
7.14 M. F. Thorpe: Phys. Rev. B **8,** 5352 (1973)
7.15 W. B. Pollard, G. Lucovsky: J. Phys. (Paris) C 4, **42,** 353 (1981)

7.16 W. B. Pollard, G. Lucovsky: Phys. Rev. B **26**, 3172 (1982)
7.17 G. Lucovsky: Sol. Energy Mater. **8**, 165 (1982)
7.18 G. Lucovsky, J. Yang, S. S. Chao, J. E. Tyler, W. Czubatyj: Phys. Rev. B **28**, 3234 (1983)
7.19 G. Lucovsky: Solid State Commun. **29**, 571 (1979)
7.20 A. L. Smith, N. C. Angelottii: Spectro. Acta **15**, 412 (1959)
7.21 H. W. Thompson: Spectro. Acta **16**, 412 (1959)
7.22 G. Lucovsky: Springer Ser. in Solid State Sci. **22**, 87 (1981)
7.23 R. B. Laughlin, J. D. Joannopoulos: Phys. Rev. B **12**, 4922 (1978)
7.24 D. Allen, J. D. Joannopoulos, W. B. Pollard: Phys. Rev. B **25**, 1065 (1982)
7.25 F. Yndurian: Phys. Rev. Lett. **37**, 1062 (1976)
7.26 M. J. Musgrave, P. Pople: Proc. Soc. (London) A **268**, 479 (1962)
7.27 R. M. Martin: Phys. Rev. B **1**, 4005 (1970)
7.28 G. Lucovsky: Sol. Cells **2**, 431 (1980)
7.29 G. Lucovsky: AIP Conf. Proc. **73**, 100 (1981)
7.30 G. Lucovsky: J. Phys. (Paris) C 4, **42**, 741 (1981)
7.31 M. H. Brodsky, M. Cardona, J. J. Cuomo: Phys. Rev. B **16**, 3556 (1977)
7.32 G. Lucovsky, R. J. Nemanich, J. C. Knights: Phys. Rev. B **19**, 2064 (1979)
7.33 S. Kalem, J. Chevallier, S. Al Dallal, J. Bourneix: J. Phys. (Paris) C 4, **42**, 361 (1981)
7.34 M. A. Paesler, D. A. Anderson, E. C. Freeman, G. Moddel, W. Paul: Phys. Rev. Lett. **41**, 1492 (1978)
7.35 G. Lucovsky, R. Tsu: Unpublished data
7.36 G. Lucovsky, S. S. Chao, J. Yang, J. E. Tyler, W. Czubatyj: Phys. Rev. B **28**, 3225 (1983)
7.37 S. Greenbaum, W. E. Carlos, P. C. Taylor: Solid State Commun. **43**, 663 (1982)
7.38 J. A. Reimer, T. M. Duncan: Phys. Rev. B (in press)
7.39 R. J. Nemanich, J. C. Knights: J. Non-Cryst. Solids **35/36**, 243 (1980)
7.40 J. C. Knights, T. M. Hayes, J. C. Mikkelsen Jr.: Phys. Rev. Lett. **39**, 712 (1977)
7.41 A. D. Yadav, M. C. Joshi: Thin Solid Films **59**, 313 (1979)
7.42 F. A. Miller, J. Perkins, G. A. Gibbon, B. A. Swisshelm: J. Raman Spectr. **2**, 93 (1974)
7.43 J. A. Borders, S. T. Picraux, W. Beezhold: Appl. Lett. **18**, 509 (1971)
7.44 J. C. Knights, G. Lucovsky, R. J. Nemanich: Phil. Mag. B **37**, 467 (1978)
7.45 W. B. Pollard, J. D. Joannopoulos: Phys. Rev. B **23**, 5263 (1981)
7.46 G. Lucovsky, J. Yang, S. S. Chao, J. E. Tyler, W. Czubatyj: Phys. Rev. B **15** (in press);
      G. Lucovsky, S. S. Chao, J. Yang, J. E. Tyler, W. Czubatyi: J. Vac. Sci. Tech. A **2** (1984)
      (in press)
7.47 P. John, I. M. Odeh, M. J. K. Thomas, J. I. B. Wilson: J. Phys. (Paris) C 4, **42**, 651 (1981)
7.48 P. John, I. M. Odeh, M. J. K. Thomas, M. J. Tricker, F. Riddoch, J. I. B. Wilson: Phil.
      Mag. B **42**, 671 (1980)
7.49 E. A. Clark, A. Weber: J. Chem. Phys. **45**, 1759 (1966)
7.50 J. C. Knights, R. A. Lujan: Appl. Phys. Lett. **35**, 244 (1979)
7.51 J. C. Knights: J. Non-Cryst. Solids **35/36**, 159 (1980)
7.52 R. J. Nemanich, D. K. Biegelsen, M. P. Rosenblum: J. Phys. Soc. Jpn. Suppl. A **49**, 1189
      (1980)
7.53 S. Oguz, D. A. Anderson, W. Paul, H. J. Stein: Phys. Rev. B **22**, 880 (1980)
7.54 H. J. Stein, P. A. Peercy: Appl. Phys. Lett. **34**, 604 (1979)
7.55 E. C. Freeman, W. Paul: Phys. Rev. B **18**, 4288 (1978)
7.56 M. A. Madan, S. R. Ovshinsky, E. Benn: Phil. Mag. B **40**, 259 (1979)
7.57 C. J. Fang, L. Ley, H. R. Shanks, K. J. Gruntz, M. Cardona: Phys. Rev. B **22**, 6140 (1980)
7.58 H. Matsumara, Y. Nakagome, S. Furukawa: Appl. Phys. Lett. **36**, 439 (1980)
7.59 T. Shimada, Y. Katayama, S. Hirigome: Jpn. J. Appl. Phys. **19**, 265 (1980)
7.60 R. Tsu: Unpublished data
7.61 H. Matsumara, K. Sakai, Y. Kawakyu, S. Furukawa: J. Appl. Phys. **52**, 5537 (1981)
7.62 S. A. Solin: AIP Conf. Proc. **31**, 205 (1976)
7.63 Y. Katayama, K. Usami, T. Shimada: Phil. Mag. B **43**, 283 (1981)
7.64 J. S. Lannin: *Amorphous and Liquid Semiconductors*, ed. by J. Stuke, W. Brenig (Taylor
      and Francis, London 1974), p. 1245

7.65 J. C. Knights, R. A. Street, G. Lucovsky: J. Non-Cryst. Solids **35/36,** 279 (1980)
7.66 P. John, I. M. Odeh, M. J. K. Thomas, M. J. Tricker, J. I. B. Wilson: Phys. Stat. Sol. (b) **105,** 499 (1981)
7.67 G. H. Azarbayejani, R. Tsu, G. Lucovsky: Bull. Am. Phys. Soc. **28,** 533 (1983)
7.68 G. Lucovsky, W. B. Pollard: Physica **117** B/**118** B, 865 (1983); J. Vac. Sci. Tech. A **1,** 313 (1983)
7.69 H. Watanabe, K. Katoh, M. Yasui: Jpn. J. Appl. Phys. **21,** L 341 (1982)
7.70 H. Weider, M. Cardona, C. R. Guarniere: Phys. Stat. Sol. (b) **92,** 99 (1979)
7.71 D. K. Paul, B. von Roedern, S. Oquz, J. Blake, W. Paul: J. Phys. Soc. Jpn. Suppl. A **49,** 1261 (1980)
7.72 C. C. Tsai: Phys. Rev. B **19,** 2041 (1979)
7.73 S. C. Shen, M. Cardona: J. Phys. (Paris) C 4, **42,** 349 (1981)
7.74 S. C. Shen, M. Cardona: Phys. Rev. B **23,** 5322 (1981)

Subject Index

Applied Physics A Solids and Surfaces

Applied Physics A "Solids and Surfaces" is devoted to concise accounts of experimental and theoretical investigations that contribute new knowledge or understanding of phenomena, principles or methods of applied research.

Emphasis is placed on the following fields:

Solid-State Physics
Semiconductor Physics: **H. J. Queisser,** MPI Stuttgart
Amorphous Semiconductors: **M. H. Brodsky,** IBM Yorktown Heights
Magnetism (Materials, Phenomena): **H. P. J. Wijn,** Philips Eindhoven
Metals and Alloys, Solid-State Electron Microscopy: **S. Amelinckx,** Mol
Positron Annihilation: **P. Hautojärvi,** Espoo
Solid-State Ionics **W. Weppner,** MPI Stuttgart

Surface Science
Surface Analysis: **H. Ibach,** KFA Jülich
Surface Physics: **D. Mills,** UC Irvine
Chemisorption: **R. Gomer,** U. Chicago

Surface Engineering
Ion Implantation and Sputtering: **H. H. Andersen,** U. Aarhus
Laser Annealing: **G. Eckhardt,** Hughes Malibu
Integrated Optics, Fiber Optics, Acoustic Surface Waves: **R. Ulrich,** TU Hamburg

Coordinating Editor: **H. K. V. Lotsch,** Heidelberg

Special Features:
- Rapid publication (3–4 months)
- No page charges for concise reports
- 50 complimentary offprints
- Microform edition available

Subscription information and/or **sample copies** are available from your bookseller or directly from Springer-Verlag, Journal Promotion Dept., P.O.Box 105 280, D-6900 Heidelberg, FRG

Springer-Verlag
Berlin
Heidelberg
New York
Tokyo

Amorphous Semiconductors

Editor: **M.H.Brodsky**

1979. 181 figures, 5 tables. XVI, 337 pages
(Topics in Applied Physics, Volume 36)
ISBN 3-540-09496-2

Contents: *M.H.Brodsky:* Introduction. –
B.Kramer, D.Weaire: Theory of Electronic
States in Amorphous Semiconductors. –
E.A.Davis: States in the Gap and Defects
in Amorphous Semiconductors. –
G.A.N.Connell: Optical Properties of
Amorphous Semiconductors. – *P.Nagels:*
Electronic Transport in Amorphous Semicon-
ductors. – *R.Fischer:* Luminescence in
Amorphous Semiconductors. – *I.Solomon:*
Spin Effects in Amorphous Semiconductors. –
G.Lucovsky, T.M.Hayes: Short-Range Order
in Amorphous Semiconductors. –
P.G.LeComber, W.E.Spear: Doped
Amorphous Semiconductors. – *D.E.Carlson,
C.R.Wronski:* Amorphous Silicon Solar Cells.

Amorphous Solids

Low-Temperature Properties

Editor: **W.A.Phillips**

1981. 72 figures. X, 167 pages
(Topics in Current Physics, Volume 24)
ISBN 3-540-10330-9

Contents: *W.A.Phillips:* Introduction. –
D.L.Weaire: The Vibrational Density of States
of Amorphous Semiconductors. – *R.O.Pohl:*
Low Temperature Specific Heat of Glasses. –
W.A.Phillips: The Thermal Expansion of
Glasses. – *A.C.Anderson:* Thermal Conduc-
tivity. – *S.Hunklinger, M.v.Schickfus:* Acoustic
and Dielectric Properties of Glasses at Low
Temperatures. – *B.Golding, J.E.Graebner:*
Relaxation Times of Tunneling Systems in
Glasses. – *J.Jäckle:* Low Frequency Raman
Scattering in Glasses.

Springer-Verlag
Belin
Heidelberg
NewYork
Tokyo

Fundamental Physics of Amorphous Semiconductors

Proceedings of the Kyoto Summer Institute,
Kyoto, Japan, September 8–11, 1980
Editor: **F.Yonezawa**

1981. 91 figures. VIII, 181 pages
ISBN 3-540-10634-0

Contents: What are Non-Crystalline Semiconductors. – Defects in Covalent Amorphous
Semiconductors. – Surface Effects and Transport Properties in Thin Films of Hydrogenated Silicon. – The Past, Present and
Future of Amorphous Silicon. – Doping and
the Density of States of Amorphous Silicon. –
The Effect of Hydrogen and Other Additives
on the Electronic Properties of Amorphous
Silicon. – New Insights on Amorphous Semi-
conductors from Studies of Hydrogenated
aGe, a-Si, a-Si$_{1-x}$ Ge$_x$ and a-GaAs. – Chemical
Bonding of Alloy Atoms in Amorphous Sili-
con. – Photo-Induced Phenomena in
Amorphous Semiconductors. – Theory of
Electronic Properties of Amorphous Semicon-
ductors. – Some Problems of the Electron
Theory of Disordered Semiconductors. – The
Anderson Localisation Problem. – Summary
Talk. – Seminars Given During the KSI '80. –
Photograph of the Participants of the KSI '80.
– List of Participants.

Topological Disorder in Condensed Matter

Proceedings of the Fifth Taniguchi
International Symposium, Shimoda, Japan,
November 2–5, 1982
Editors: **F.Yonezawa, T.Ninomiya**

1983. 158 figures. XII, 253 pages
(Springer Series in Solid-State Sciences,
Volume 46)
ISBN 3-540-12663-5

Contents: Introduction. – Structural Aspects
of Topological Disorder. – Computer Simulations and Analyses. – Some Aspects of
Elementary Excitations. – Statistical Properties in Two Dimensions. – Related Problems.
– Summary of the Conference. – Index of
Contributors.